PACLITAXEL

PACLITAXEL

Sources, Chemistry, Anticancer Actions, and Current Biotechnology

Edited by

MALLAPPA KUMARA SWAMY

T. PULLAIAH

ZHE-SHENG CHEN

Academic Press is an imprint of Elsevier
125 London Wall, London EC2Y 5AS, United Kingdom
525 B Street, Suite 1650, San Diego, CA 92101, United States
50 Hampshire Street, 5th Floor, Cambridge, MA 02139, United States
The Boulevard, Langford Lane, Kidlington, Oxford OX5 1GB, United Kingdom

Copyright © 2022 Elsevier Inc. All rights reserved.

No part of this publication may be reproduced or transmitted in any form or by any means, electronic or mechanical, including photocopying, recording, or any information storage and retrieval system, without permission in writing from the publisher. Details on how to seek permission, further information about the Publisher's permissions policies and our arrangements with organizations such as the Copyright Clearance Center and the Copyright Licensing Agency, can be found at our website: www.elsevier.com/permissions.

This book and the individual contributions contained in it are protected under copyright by the Publisher (other than as may be noted herein).

Notices

Knowledge and best practice in this field are constantly changing. As new research and experience broaden our understanding, changes in research methods, professional practices, or medical treatment may become necessary.

Practitioners and researchers must always rely on their own experience and knowledge in evaluating and using any information, methods, compounds, or experiments described herein. In using such information or methods they should be mindful of their own safety and the safety of others, including parties for whom they have a professional responsibility.

To the fullest extent of the law, neither the Publisher nor the authors, contributors, or editors, assume any liability for any injury and/or damage to persons or property as a matter of products liability, negligence or otherwise, or from any use or operation of any methods, products, instructions, or ideas contained in the material herein.

Library of Congress Cataloging-in-Publication Data
A catalog record for this book is available from the Library of Congress

British Library Cataloguing-in-Publication Data
A catalogue record for this book is available from the British Library

ISBN 978-0-323-90951-8

For information on all Academic Press publications
visit our website at https://www.elsevier.com/books-and-journals

Publisher: Stacy Masucci
Acquisitions Editor: Rafael E. Teixeira
Editorial Project Manager: Timothy Bennett
Production Project Manager: Omer Mukthar
Cover Designer: Miles Hitchen

Typeset by STRAIVE, India

Working together
to grow libraries in
developing countries

www.elsevier.com • www.bookaid.org

Contents

Contributors xi

Preface xv

Chapter 1 Introduction to cancer and treatment approaches. 1
Madihalli Somashekharaiah Chandraprasad, Abhijit Dey,
and Mallappa Kumara Swamy

1.1   Introduction 1
1.2   About cancer biology: Causes and risk factors. 3
1.3   Cancer types, classification, and grading 6
1.4   Therapeutic interventions for cancer. 7
1.5   Advanced approaches for cancer treatment 19
1.6   Conclusions 22
References 23

Chapter 2 Taxol: Occurrence, chemistry, and understanding its
molecular mechanisms 29
Mallappa Kumara Swamy and Bala Murali Krishna Vasamsetti

2.1   Introduction 29
2.2   About taxol and its discovery. 31
2.3   Natural resources of taxol. 33
2.4   Chemistry of taxol 34
2.5   Mechanisms of action of taxol 35
2.6   Conclusion and future prospects 40
References 41

vi Contents

Chapter 3 Taxol: Mechanisms of action against cancer, an update with current research **47**

Pei Tee Lim, Bey Hing Goh, and Wai-Leng Lee

3.1 The discovery and evolution of Paclitaxel (Taxol) 47

3.2 Paclitaxel (Taxol) induces mitotic cell cycle arrest 48

3.3 Taxol induces gene-directed apoptosis 50

3.4 Taxol and calcium-dependent apoptosis 53

3.5 Immunomodulation effects by Taxol 57

3.6 Resistance mechanisms of Taxol 62

3.7 Conclusion 63

Acknowledgment 65

References 66

Chapter 4 Application of nanocarriers for paclitaxel delivery and chemotherapy of cancer **73**

Saloni Malla, Rabin Neupane, Sai H.S. Boddu,
Mariam Sami Abou-Dahech, Mariah Pasternak, Noor Hussein,
Charles R. Ashby, Jr, Yuan Tang, R. Jayachandra Babu,
and Amit K. Tiwari

4.1 Introduction 73

4.2 Nanoparticles 75

4.3 Liposomes 83

4.4 Dendrimers 91

4.5 Micelles 92

4.6 Nanotubes 93

4.7 Niosomes 94

4.8 Proniosomes 94

4.9 Ethosomes 95

4.10 Microparticles 95

4.11 Carbon dots 96

4.12 Clinical trials 96

4.13 Overcoming paclitaxel resistance by using Nanocarriers 105

Contents **vii**

4.14 Selected patents for paclitaxel formulations. 110

4.15 Conclusion 114

References 114

Chapter 5 Strategies for enhancing paclitaxel bioavailability for cancer treatment. **129**

Mina Salehi and Siamak Farhadi

5.1 Introduction 129

5.2 Alternative paclitaxel sources 130

5.3 Strategies of paclitaxel biosynthesis improvement in plant cell culture 134

5.4 Mathematical modeling for paclitaxel biosynthesis optimization 144

5.5 Concluding remarks and future perspectives 146

References 147

Chapter 6 Botany of paclitaxel producing plants **155**

S. Karuppusamy and T. Pullaiah

6.1 History of taxol. 155

6.2 Botany of *Taxus* 156

6.3 Enumeration of taxol producing plant species 160

6.4 Taxol from angiosperms 167

6.5 Conclusions and future direction 168

References 168

Chapter 7 Propagation of paclitaxel biosynthesizing plants **171**

T. Pullaiah, S. Karuppusamy, and Mallappa Kumara Swamy

7.1 Introduction 171

7.2 Propagation 172

7.3 Micropropagation 178

7.4 Conclusions 197

References 198

viii Contents

Chapter 8 Endophytes for the production of anticancer drug, paclitaxel **203**
Mallappa Kumara Swamy, Tuyelee Das, Samapika Nandy, Anuradha Mukherjee, Devendra Kumar Pandey, and Abhijit Dey

8.1 Introduction 203
8.2 Paclitaxel sources in nature 204
8.3 Available approaches for paclitaxel production 205
8.4 Endophytes producing paclitaxel from different host plant species 207
8.5 Anticancer properties of endophytes-derived paclitaxel 216
8.6 Conclusions 220
References 221

Chapter 9 Metabolic engineering strategies to enhance the production of anticancer drug, paclitaxel **229**
Lakkakula Satish, Yolcu Seher, Kasinathan Rakkammal, Pandiyan Muthuramalingam, Chavakula Rajya Lakshmi, Alavilli Hemasundar, Kakarla Prasanth, Sasanala Shamili, Mallappa Kumara Swamy, Malli Subramanian Dhanarajan, and Manikandan Ramesh

9.1 Introduction 229
9.2 Historical perspective of paclitaxel 231
9.3 Metabolic engineering strategies for paclitaxel production . . 234
9.4 Conclusions 246
References 246

Chapter 10 Paclitaxel and chemoresistance **251**
Zhuo-Xun Wu, Jing-Quan Wang, Qingbin Cui, Xiang-Xi Xu, and Zhe-Sheng Chen

10.1 Introduction 251
10.2 Mechanisms of chemoresistance 252
10.3 Clinical markers of paclitaxel resistance 256
10.4 Strategies to overcome paclitaxel resistance 258

Contents **ix**

10.5 Summary 260

References 260

Chapter 11 Paclitaxel and cancer treatment: Non-mitotic mechanisms of paclitaxel action in cancer therapy 269
Elizabeth R. Smith, Zhe-Sheng Chen, and Xiang-Xi Xu

11.1 Introduction 269

11.2 Microtubule stabilization and anti-mitotic mechanisms 270

11.3 Mitotic catastrophe 271

11.4 Non-mitotic mechanisms 272

11.5 Importance of micronucleation 273

11.6 Innate immunity leading to the bystander effect 274

11.7 Cellular retention of paclitaxel 275

11.8 Combination therapy. 277

11.9 Prospective: New formulation of paclitaxel and additional microtubule stabilizing drugs. 279

11.10 Conclusions 280

References 280

Chapter 12 An update on paclitaxel treatment in breast cancer 287
Tuyelee Das, Samapika Nandy, Devendra Kumar Pandey,
Abdel Rahman Al-Tawaha, Mallappa Kumara Swamy,
Vinay Kumar, Potshangbam Nongdam, and Abhijit Dey

12.1 Introduction 287

12.2 Types of breast cancer 288

12.3 Molecular mechanism of paclitaxel in breast cancer 293

12.4 Paclitaxel treatment in different types of breast cancer. 294

12.5 Adverse events and resistance due to paclitaxel treatment . . 298

12.6 Efficiency of other anti-cancer drugs over paclitaxel 300

12.7 Conclusions 300

Acknowledgment. 301

References 301

x Contents

Chapter 13 Paclitaxel conjugated magnetic carbon nanotubes induce apoptosis in breast cancer cells and breast cancer stem cells in vitro **309**
Prachi Ghoderao, Sanjay Sahare, Anjali A. Kulkarni,
and Tejashree Bhave

13.1 Introduction 309

13.2 Experimental details 313

13.3 Results and discussion 316

13.4 Conclusions 327

Acknowledgments 327

Conflict of interest statement 327

References 328

Index 333

Contributors

Mariam Sami Abou-Dahech Department of Pharmacology and Experimental Therapeutics, College of Pharmacy & Pharmaceutical Sciences, University of Toledo, Toledo, OH, United States

Abdel Rahman Al-Tawaha Department of Biological Sciences, Al-Hussein Bin Talal University, Maan, Jordon

Charles R. Ashby, Jr Department of Pharmaceutical Sciences, College of Pharmacy and Health Sciences, St. John's University, Queens, NY, United States

R. Jayachandra Babu Department of Drug Discovery & Development, Harrison School of Pharmacy, Auburn University, Auburn, AL, United States

Tejashree Bhave Department of Applied Physics, Defence Institute of Advanced Technology, Pune, India

Sai H.S. Boddu Department of Pharmaceutical Sciences, College of Pharmacy and Heath Sciences; Center of Medical and Bio-allied Health Sciences Research, Ajman University, Ajman, United Arab Emirates

Madihalli Somashekharaiah Chandraprasad Department of Biotechnology, BMS College of Engineering, Bengaluru, India

Zhe-Sheng Chen College of Pharmacy and Health Sciences, St. John's University, Queens, NY, United States

Qingbin Cui College of Pharmacy and Health Sciences, St. John's University, Queens, NY; Department of Cancer Biology, University of Toledo College of Medicine and Life Sciences, Toledo, OH, United States

Tuyelee Das Department of Life Sciences, Presidency University, Kolkata, India

Abhijit Dey Department of Life Sciences, Presidency University, Kolkata, India

Malli Subramanian Dhanarajan Department of Biochemistry and Biotechnology, Jeppiaar College of Arts and Science, Padur, TN, India

Siamak Farhadi Department of Plant Genetics and Breeding, Faculty of Agriculture, Tarbiat Modares University, Tehran, Iran

Prachi Ghoderao Department of Applied Physics, Defence Institute of Advanced Technology, Pune, India

Bey Hing Goh College of Pharmaceutical Sciences, Zhejiang University, Hangzhou, China; Biofunctional Molecule Exploratory (BMEX) Research Group, School of Pharmacy, Monash University Malaysia, Subang Jaya, Malaysia

Alavilli Hemasundar Department of Life Science, Sogang University, Seoul, South Korea

Noor Hussein Department of Pharmacology and Experimental Therapeutics, College of Pharmacy & Pharmaceutical Sciences, University of Toledo, Toledo, OH, United States

S. Karuppusamy Department of Botany, The Madura College, Madurai, India

Anjali A. Kulkarni Department of Botany, Savitribai Phule Pune University (Formerly University of Pune), Pune, India

Vinay Kumar Department of Biotechnology, Modern College (Savitribai Phule Pune University), Pune, India

Chavakula Rajya Lakshmi Department of Chemistry, Vishnu Institute of Technology, Bhimavaram, AP, India

Wai-Leng Lee School of Science, Monash University Malaysia, Subang Jaya, Malaysia

Pei Tee Lim School of Science, Monash University Malaysia, Subang Jaya, Malaysia

Saloni Malla Department of Pharmacology and Experimental Therapeutics, College of Pharmacy & Pharmaceutical Sciences, University of Toledo, Toledo, OH, United States

Anuradha Mukherjee MMHS, Joynagar, India

Pandiyan Muthuramalingam Department of Biotechnology, Alagappa University, Science Campus, Karaikudi; Department of Biotechnology, Sri Shakthi Institute of Engineering and Technology, Coimbatore, TN, India

Samapika Nandy Department of Life Sciences, Presidency University, Kolkata, India

Rabin Neupane Department of Pharmacology and Experimental Therapeutics, College of Pharmacy & Pharmaceutical Sciences, University of Toledo, Toledo, OH, United States

Potshangbam Nongdam Department of Biotechnology, Manipur University, Imphal, Manipur, India

Devendra Kumar Pandey Department of Biotechnology, Lovely Faculty of Technology and Sciences, Lovely Professional University, Phagwara, India

Mariah Pasternak Department of Pharmacology and Experimental Therapeutics, College of Pharmacy & Pharmaceutical Sciences, University of Toledo, Toledo, OH, United States

Kakarla Prasanth French Associates Institute for Agriculture and Biotechnology of Drylands, The Jacob Blaustein Institutes of Desert Research, Ben-Gurion University of the Negev, Sede-Boqer Campus, Beer Sheva, Israel

T. Pullaiah Department of Botany, Sri Krishnadevaraya University, Anantapur, India

Kasinathan Rakkammal Department of Biotechnology, Alagappa University, Science Campus, Karaikudi, TN, India

Manikandan Ramesh Department of Biotechnology, Alagappa University, Science Campus, Karaikudi, TN, India

Sanjay Sahare Department of Applied Physics, Defence Institute of Advanced Technology, Pune, India

Mina Salehi Department of Plant Genetics and Breeding, Faculty of Agriculture, Tarbiat Modares University, Tehran, Iran

Lakkakula Satish Department of Biotechnology Engineering, Ben-Gurion University of Negev; French Associates Institute for Agriculture and Biotechnology of Drylands, The Jacob Blaustein Institutes of Desert Research, Ben-Gurion University of the Negev, Sede-Boqer Campus, Beer Sheva, Israel

Yolcu Seher Faculty of Engineering and Natural Sciences, Sabanci University, Istanbul, Turkey

Sasanala Shamili French Associates Institute for Agriculture and Biotechnology of Drylands, The Jacob Blaustein Institutes of Desert Research, Ben-Gurion University of the Negev, Sede-Boqer Campus, Beer Sheva, Israel

Elizabeth R. Smith Department of Radiation Oncology, Sylvester Comprehensive Cancer Center, University of Miami Miller School of Medicine, Miami, FL, United States

Mallappa Kumara Swamy Department of Biotechnology, East West First Grade College, Bengaluru, Karnataka, India

Yuan Tang Department of Bioengineering, College of Engineering, University of Toledo, Toledo, OH, United States

Amit K. Tiwari Department of Pharmacology and Experimental Therapeutics, College of Pharmacy & Pharmaceutical Sciences, University of Toledo, Toledo, OH, United States

Bala Murali Krishna Vasamsetti Toxicity and Risk Assessment Division, Department of Agro-food Safety and Crop Protection, National Institute of Agricultural Sciences, Rural Development Administration, Wanju, Korea

Jing-Quan Wang College of Pharmacy and Health Sciences, St. John's University, Queens, NY, United States

Zhuo-Xun Wu College of Pharmacy and Health Sciences, St. John's University, Queens, NY, United States

Xiang-Xi Xu Department of Radiation Oncology, Sylvester Comprehensive Cancer Center, University of Miami Miller School of Medicine, Miami, FL, United States

Preface

Plants as sessile creatures produce numerous chemical compounds, which are together recognized as secondary metabolites. These plant-derived compounds are not crucial for a plant's growth and development; however, they are chiefly synthesized to have an ecological adaptation against both biotic and abiotic stress conditions. These compounds belong to different chemical classes, including phenolics, alkaloids, terpenes, steroids, etc. Remarkably, these compounds exhibit several pharmacological properties, and hence have become one among the better choice for preventing/treating human health issues. In particular, alkaloids containing nitrogen in their structures exhibit greater chemodiversity and pharmacological properties.

Among different classes of alkaloids, plants synthesize 20-carbon (C_{20}) polycyclic isoprenoids that are together represented as diterpenoids, and they are the signature compounds. Diterpenoids include toxoids that occur largely in *Taxus* (yew tree) species and possess a distinctive taxane (pentamethyl [9.3.1.0]3,8 tricyclopentadecane) skeleton. Nearly 400 naturally occurring taxoids have been structurally characterized and several of them are biologically very active. The anticancer drug Paclitaxel (Taxol) is one among them, and it was first obtained from the Pacific yew tree (*Taxus brevifolia*). It is very effective, and is an approved chemotherapeutic drug used widely to treat breast, ovarian, lung, bladder, prostate, melanoma, esophageal, and other types of solid tumors. It has also been used to treat Kaposi's sarcoma.

Paclitaxel is a cytoskeletal drug, which targets tubulin proteins. Paclitaxel-treated cells will have defects in chromosome segregation, mitotic spindle assembly, and cell division. Paclitaxel stabilizes the microtubule polymer and guards it from disassembly. Chromosomes are thus unable to achieve a metaphase spindle configuration. This blocks the progression of mitosis, and prolonged activation of the mitotic checkpoint triggers apoptosis or reversion to the G_0-phase of the cell cycle without cell division. At higher tonic levels, Paclitaxel suppresses microtubule detachment from centrosomes, a process usually initiated during mitosis. However, this drug also shows some common side effects, including nausea and vomiting, change in taste, loss of appetite, brittle or thinned hair, pain in the joints of arms or legs lasting 2–3 days, changes in nail color, and tingling in fingers or toes.

The slow growth of a yew tree and the occurrence of Paclitaxel only in matured tree barks make it difficult to extract Paclitaxel. As reported earlier, mature trees can be a source of only 2 kg of bark, and to extract 500 mg of Paclitaxel, nearly 12 kg of bark is required, viz., extraction cost from the plant source is very high as it occurs in low quantities, i.e., 0.01%–0.04% dry weight. The devastation of mature trees for extracting sufficient quantities of Paclitaxel may damage nature. In addition, the rising pharmaceutical demand in the present market for Paclitaxel greatly exceeds the supply, and hence looking for alternative sources for this drug molecule will be very useful and is much needed. In this regard, intensive explorations have complemented to produce Paclitaxel more effectively in alternative sources. Although few groups of investigators have successfully achieved the total chemical synthesis of Paclitaxel, the high cost of this synthetic method hinders its viable applications. The most encouraging methodologies to produce Paclitaxel in a sustainable manner include plant cell cultures at the commercial level. In addition, the metabolic engineering strategy is a powerful method to manipulate biosynthetic pathways and to control the production of plant compounds. The central biosynthetic path for Paclitaxel is through a terpenoid pathway, and using the metabolic engineering approach it has been productively transplanted into commonly producing *Escherichia coli* and yeast cells. Furthermore, endophytic fungi are found to be other alternatives for a continuous production of Paclitaxel. Thus, these biotechnological strategies have allowed the promises for large-scale production of Paclitaxel.

In this book, a comprehensive description of Paclitaxel is provided. This book consists of topics, such as chemistry, chemical synthesis, biosynthesis, anticancer activities, bioavailability, mechanism of action, paclitaxel resistance, currently undergoing experimental phases, and biotechnological methods, including cell cultures as well as metabolic engineering in heterologous microbial and plant systems to enhance its production. Overall, this book is a worthy material for students, teachers, and healthcare experts involved in cancer biology, biomedicine, natural products research, and pharmacological investigations. We are thankful to all contributors of this book volume for sharing their understanding on Paclitaxel. We also thank the Elsevier group for their constant support at all stages of book publication.

Mallappa Kumara Swamy, Editor
T. Pullaiah, Editor
Zhe-Sheng (Jason) Chen, Editor

Introduction to cancer and treatment approaches

Madihalli Somashekharaiah Chandraprasad[a], Abhijit Dey[b], and Mallappa Kumara Swamy[c]

[a]Department of Biotechnology, BMS College of Engineering, Bengaluru, Karnataka, India, [b]Department of Life Sciences, Presidency University, Kolkata, India, [c]Department of Biotechnology, East West First Grade College, Bengaluru, India

1.1 Introduction

Cancer is the uninhibited growth and development of abnormal cells in the body, and is one of the foremost reasons of deaths throughout the world (Paul and Jindal, 2017). These abnormal cells are commonly designated as cancerous cells, tumorous cells, or malignant cells. In 2018, cancer accounted for an estimated 9.6 million deaths. The predominant cancers in men are related to lung, colorectal, prostate, liver, and stomach, while in women the breast, lung, colorectal, thyroid, and cervical cancers are predominant. Cancer is mostly associated with a group of over 100 distinctive disorders. In addition to genetic causes, several other causes for cancers in individuals include their lifestyle or habits (for instance, alcohol and tobacco consumption), exposure to carcinogens (both chemicals and radiations), exposure to infective agents like *Helicobacter pylori*, diet, ethnicity, etc. (Paul and Jindal, 2017; Santosh et al., 2017). The carcinogens induce certain damages at the cellular or genetic level. Nevertheless, the precise basis of few cancers is yet to be known. As per the statistics of World Health Organization, cancer was solely responsible for about 88 lakhs deaths in 2015. Also, it has been predicted that cancer related deaths around the world will reach to 132 lakhs by 2030, and new cases of cancers are projected to intensify to over 203 lakhs by 2030 (Ferlay et al., 2010; Bray et al., 2012; Paul and Jindal, 2017). As per the survey, developing nations are mostly at greater risk of cancers, and about 63% of cancer-associated deaths were recorded mainly from the developing nations.

Cancer, a multifactorial malady involves multifarious alterations in the genome due to interactions with the individual's environment.

Paclitaxel. https://doi.org/10.1016/B978-0-323-90951-8.00010-2
Copyright © 2022 Elsevier Inc. All rights reserved.

Cancer hallmarks include uninhibited replication, inability to respond to growth signals, arrest of the cell division, continuous angiogenesis, evasion of apoptosis, and lastly the ability to infiltrate other tissues, i.e., metastasis (Hanahan and Weinberg, 2011). While, mutations at the molecular level may induce uncontrolled division of normal cells through modifying the cell cycle, and lead to growth of mass of abnormal cells, which is recognized as tumor, the benign tumor remains confined in its original site of manifestation, and fails to spread to nearby tissues of the body (Lodish et al., 2000; Santosh et al., 2017; Abbas and Rehman, 2018). However, dysregulation of different regulatory proteins in the cellular environment of cells, manifested by benign tumors plays a major role in the manifestation and advancement of cancers (Santosh et al., 2017; Abbas and Rehman, 2018).

The classification of cancers is based on the type of cells, tissues, or organs that are affected. For example, sarcomas are the connective tissue cancers, affecting muscle, bone, and cartilage. Likewise, the malignant type of the epithelial cells represents carcinoma. Leukemia and lymphoma are the tumors of blood-forming tissues and lymphatic system, respectively. Further, some cancers are categorized on the basis of their tissue of origin, for instance, lung and the breast cancer (Lodish et al., 2000; Lahat et al., 2008; Santosh et al., 2017).

In earlier times, surgery was regarded as the most ideal choice of cancer treatment. Later, radiation treatment was introduced for controlling cancers. Although chemo-drugs efficiently impede tumor cells (chemotherapy), they persuade toxicity and additional adverse health complications in patients. However, these individual treatment options were found to be not effective, and hence considered their use in combination for controlling cancers (Shewach and Kuchta, 2009; Santosh et al., 2017; Kroschinsky et al., 2017). Currently, the, cancers can be cured by means of both conventional tonic approaches (i.e., surgery, radiation therapy, and chemotherapy) and nonconventional or complementary therapeutic methods, including hormone therapy, immunotherapy, nanotherapy, etc. Nevertheless, these existing therapeutic interventions cause adverse side effects, and largely distress the normal cells, tissues, and organs. The side effects could vary from one patient to another, and even amongst those individuals who receive the similar treatment. In addition, a therapy may show less or more side effects in different patients. Selecting the best cancer therapy approach depends on various factors, such as the type of malignancies, growth stages, age, management frequencies, dosage of medicines, and healthiness of patients (Santosh et al., 2017).

The selection of tumor treatment options depends on tumor type, its origin, and stages of the disorder. In cancer management, surgery is done after diagnosis by biopsy to prevent cancer progression, and in limited situations as a treatment choice (Faries and Morton, 2007).

Side effects of post-surgery include pain, microbial infections, blood loss and clotting, injury to neighboring tissues or organs to name a few. Radiation therapy involves the usage of ionizing radiations to destroy the tumorous cells. The highly energetic radiations terminate the divisions of cells and prevent their proliferative property by destroying the genetic material. Before the surgery, generally the radiation treatment is given to patients for condensing cancer lump. While, radiotherapy is administered to destroy the remaining cancerous cells and decrease the tumor relapse, after the surgery (Delaney et al., 2005; Abbas and Rehman, 2018). Radiation therapy in combination with chemotherapy is employed as a better choice for treating cancers. Chemotherapy is known to be the best treatment option and widely used in several cancer types, however exhibits severe side effects (Hausheer et al., 2006; Aslam et al., 2014; Kampan et al., 2015). Chemo-drugs target the cancerous cells and induce the production of reactive oxygen species, leading to the death of cancerous cells via their genotoxic activity (DeVita and Chu, 2008; Santosh et al., 2017).

Variations in the secretion of hormone may also cause few types of cancers, for instance, breast, prostate, and uterine cancers. Thus, hormone therapy could be useful in preventing the secretion of such sex hormones and restrict their use by cells for their further development (Ellis and Perou, 2013). Similarly, the gene therapy, involving in situ transfer of exogenous genes into the cancerous cells might offer as a potent curative method to treat benign cancers. The application of stem cell therapy, i.e., introducing modified in vitro stem cells with specific genes having the potential anticancer activities can be another promising tool for cancer treatment. The advanced cancer therapy options include immunotherapy, which is used to manipulate the immune system so as to eliminate tumorous cells (Palucka and Banchereau, 2012; Jiang et al., 2012; Abbas and Rehman, 2018). In this chapter, cancer prevalence, types, available cancer therapeutic approaches, and their adverse side effects are discussed in detail.

1.2 About cancer biology: Causes and risk factors

Cancer, a multifactorial disorder is caused by the building up of epigenetic and genetic modifications within cells that lead to abnormality in cellular appearance and transformed cell growth and development. Cells transform into cancerous because of the buildup of cellular defects, and/or mutations in their genetic material. Cancer is an unusual physiological event, wherein cells propagate rapidly in an uninhibited way by disregarding the cell cycle and cell division control mechanisms. Various cell signals that control the normal divisions of cells, and signals that regulate routine programmed cell death, i.e., apoptosis fails, carcinogenesis occurs, leading to uncontrolled growth and proliferation of cancer cells (Hanahan and Weinberg, 2011). The uncontrolled cancerous cell proliferation may lead to mortality. Indeed, over 90% of cancer-related deaths are because of the wide spreading or penetration of cancerous cells into other tissue parts, which is commonly known as metastasis.

The normal cells' growth and division occurs interdependently during mitosis, which generally depend on the influence of several external signals or growth factors (as reviewed by Witsch et al., 2010). When these regulating signals are limited or restricted, cells fail to reproduce. While, cancer cells grow and develop independently, i.e., without the influence of any signals growth control factors (Lum et al., 2005). Several years of investigations have proved the fact that normal cells growing in a two-dimensional culture plate establishes the cell-to-cell contacts and activate to inhibit further multiplying of cells, resulting to form cell monolayers. Notably, such contact inhibition is absent in numerous types of cancerous cells. This suggests that contact inhibition is a mechanism to safeguard usual tissue homeostasis, and the failure of which may lead to carcinogenesis. Till date, the mechanisms involved in this type of growth regulation remain unclear with only few investigational understanding (Hanahan and Weinberg, 2011).

A normal cells' lifespan is limited, and is well-programmed, i.e., after a definite cycle of cell divisions, cells' death occurs via apoptosis mechanism, and new cells will be replaced. This apoptosis mechanism serves as a regular barrier for the development of cancers (Lowe et al., 2004; Adams and Cory, 2007; Hanahan and Weinberg, 2011). Various physiological stresses or signals, such as antiapoptotic regulators (e.g., Bcl-2 and Bcl-xL), proapoptotic factors (e.g., Bim, Bax, Puma), survival signal, i.e., Igf1/2 trigger apoptosis in normal cells. However, imbalance in these apoptosis-inducing signals causes tumorigenesis. Further, this programmed cell demise is in accordance with DNA replication efficacy. In a normal cell, the telomeric DNA length controls the succeeding cell generations that its progeny can pass through before telomere is being largely damaged, and thus lost its shielding roles, prompting cell senescence. Further, a repetitive DNA replication in normal cells causes shortening of telomeric DNA sequences. In contrast, cancerous cells exhibit higher telomerase enzyme activity, which constantly replaces the disappeared, worn-out ends of telomere, permitting unrestricted multiplying of cells (Hanahan and Weinberg, 2011; Pavlova and Thompson, 2016; Abbas and Rehman, 2018). Therefore, cell death is a protective barrier to tumorigenic growth that could be elicited by different proliferation-linked irregularities, together with elevated levels of oncogenic signals and shortening of telomeric DNA sequences.

Cancer cells are capable of proliferating independently without the influence of growth regulators or signaling molecules, however they need oxygen and nutrients for their development. Normally, cells are supplied with an adequate amount of nutrients and oxygen via capillary networks. With the progression of pathogenesis, fresh blood vessels are formed by cancer cells via angiogenesis, a physiological process to facilitate reaching of nutrients to the cells situated at the center of the cancer lump, having accessibility to usual blood vessels (Baeriswyl and Christofori, 2009; Carmeliet and Jain, 2011).

Cancer is caused by a definite change in genes that regulate cell functions, specifically how they grow, develop, and undergo division. Genes possess the messages to produce proteins for cell's functioning. However, certain gene changes in cells may result in evading of regular growth control mechanisms and grow into cancer. Mutations in tumor suppressor genes and oncogenes lead to cancer development. For instance, the normal functions, such as DNA synthesis, gene transcription, apoptosis, and DNA repair mechanisms in cells are controlled by p53 tumor suppressor gene, and any modifications and mutations in p53 can induce carcinogenesis. Further, the normal function of p53 gene involves a multifarious biochemical pathway, which is executed by a complex protein structure. In some case, the viral oncoproteins alter these active molecules, and hinder the binding and other interactions with p53 and other proteins in cells, leading to an abnormality in cellular regulatory functions. Likewise, certain types of mutations in genes controlling the cell division may cause duplication or deletion of chromosomal segments, and converts normal cells toward abnormality (Greenblatt, 1994; Ralph et al., 2010; Rivlin et al., 2011; Burrell et al., 2013; Abbas and Rehman, 2018).

Cancer is caused by anything that induces normal body cells to grow abnormally. Some cancer sources remain unidentified, whereas other cancers develop from more than one identified reason, including environmental and lifestyle triggers (Ames et al., 1995). In some cases, initiation of cancer is influenced by an individual's genetic makeup. However, development of cancer in some persons is because of a combination of these reasons. It is often tough or difficult to conclude the instigating events that may trigger the development of cancer in an individual. However, investigational studies have exposed several internal and external factors that may act in initiating cancer. Some of the external factors include exposure to ionizing radiations (ultraviolet rays from sunlight, radiation from α, β, γ, and X-ray-radiating sources), chemical (benzene, vinyl chloride, nickel, asbestos, cadmium, tobacco, or cigarette smoke, etc.) exposure, and pathogens (human papillomavirus, Epstein-Barr virus, Kaposi's sarcoma-associated herpes virus, hepatitis viruses B and C, *Schistosoma* spp., and *Helicobacter pylori*, and many others) attack. Within the cell, internal factors, such as hormones, mutations, immune conditions, and aging are also responsible for the origin and advancement of the tumors (https://www.medicinenet.com/cancer/article.htm). Some of the specific cancer types, such as breast, colorectal, ovarian, prostate, and melanoma have been associated with human genes (https://www.cancer.gov/about-cancer/causes-prevention/genetics).

1.3 Cancer types, classification, and grading

A series of stages can be observed during the autonomous proliferation of cancer cells. Firstly, an outsized mass of cells, identified as hyperplasia is formed due to uninhibited cell divisions. Later, cell growth is added with irregularities, and this condition is known as dysplasia. Several other alterations arise in the subsequent phase, recognized as anaplasia, where these abnormal cells begin spreading to nearby tissues, and fail to do their original roles. This stage is considered as benign, as the carcinogenesis is not invasive. In the later advanced stage, the cancerous cells attain the capabilities to rapidly spread to nearly regions of the body as well as those body parts located far away by the way of bloodstream. The process of this rapid invasion is known as metastasis, and this advanced phase is identified as malignancy, which is rarely can be treated. If a cancer is diagnosed in the beginning stage only, the progress of cancers to malignant stage can be prevented (DeBerardinis et al., 2008; Abbas and Rehman, 2018).

Cancers are categorized on the basis of cell types from where they are originated. For example, modifications in the epithelial cells results in carcinomas, and is the most common type of cancers. The cancerous growth in bones, muscles, and connective tissues are represented as sarcomas. Leukemia is the abnormalities found in the white blood cells. The malignant form of the lymphatic cells or system, which is originated from the bone marrow is called as lymphoma. Likewise, uncharacteristic plasma cells developed in the bone marrow reproduce themselves very rapidly, and this condition is illustrated as myeloma or multiple myeloma. In general, grades, from 1 to 4 are given to represent the increase in cancer severity with respect to their neighboring regular tissues. Lower grade tumors include well-differentiated cells that are very similar to normal cells, and high-grade tumors include inappropriately differentiated cells that are extremely abnormal with respect to their surrounding tissues. In grade 1 tumors, well-differentiated cells will have minor abnormalities. Grade 2 tumorous cells show higher abnormalities, and are discreetly differentiated. In grade 3, tumor cells are inappropriately differentiated and exhibit increased abnormalities, including DNA damages and mutations. In this stage, abnormal cells tend to secrete several detrimental chemicals that disturb proximate cells and may possibly pass into the bloodstream. In contrast, cells and tissues of grade 4 stage are undifferentiated with extreme abnormalities. This is the highest grade, where tumorous cells generally grow and spread more rapidly as compared to lower grade tumors.

1.4 Therapeutic interventions for cancer

Early therapeutic intervention is the need of the hour to impede cancer, a fatal health complication. The treatment techniques should maximize the efficacy and reduce their possible side effects. The type, location and the progression of a cancer determines the method of therapeutic intervention to be employed. At present, various curative quality treatment trials are being investigated with the main focus of preventing cancer (Abbas and Rehman, 2018). The pathogenesis of benign tumor can be prevented, if diagnosed at its early stage as compared to the malignant tumor. The major difference between the two being localized or restricted to a tissue or organ, and able to spread to the nearby tissues and organ makes them to respond differently to treatment options (Kassi et al., 2019). Various treatment techniques exist with diverse efficacies, and can be effective in removing or hindering the cancerous cells. Surgical removal of the targeted cancer tissue, being the old traditional method provides an immediate remedy from the cancer. The advanced techniques hitherto attempted include chemotherapy and the radiotherapy. Drug delivery vehicles targeting the cancerous tissue include nanocarriers, liposomes as extracellular vesicles (EVs), anti-oxidants and phytochemicals. Even the bio-based surfactants of microbial origin, such as mannosylerythritol lipids, trehaloses, and cellobiose lipids are also reported to have anticancerous properties (Madihalli and Doble, 2019). More recently, various molecular-based approaches are being increasingly researched, including gene therapy, targeted silencing by siRNAs, expression of genes triggering apoptosis and wild tumor suppressors (Pucci et al., 2019). An overview of different therapeutic interventions of the past and the present to successfully neutralise the tumors are discussed henceforth.

1.4.1 Surgical excision: A traditional and the effective one

Surgery remains as the most traditional and preliminary approach, and involves tumor tissue excision to permanently get rid of the complications or to prolong the progression of the cancer to metastasize. It causes the minimum effect to the surrounding tissues, when compared to radiotherapy and chemotherapy. The surgical intervention to treat cancer depends on various factors, including type, mass, site, stage, and grade of tumors. In addition, the overall health aspects, such as physical fitness, age, and other illness of the patient determine the possibility of surgery (Isogai et al., 2017). The kind of the surgery, whether open or minimally invasive depends on the factors, such as reason for surgery, part of the body that should undergo surgery, amount of cancerous tissue to be removed and of course the decision by the patient. The surgery also depends on stages of the cancer, and involve the removal of the entire tumor from a part, debulking the cancerous mass to prevent the further effect to the body part or minimal invasive intervention to ease the pain or intense pressure on the body part (Abbas and Rehman, 2018).

To overcome the complications associated with surgical removal of tumors, recently new approaches are being introduced, i.e., thermal ablation and magnetic hyperthermia. Both these techniques work by targeting to a very narrow and precise areas, and could be an efficient alternative to the traditional surgical methods (Brace, 2011; Hervault and Thanh, 2014).

Though the surgical removal of cancerous tissue is the immediate solution in many instances, the underlying side effects are of important concern. The most prevailing post-surgery side effects include blood clotting, loss of blood, pain, impairment of tissues and infections. While the mentioned side-effects are tentative in many cases, the long-lasting and permanent deformities caused by the surgical treatment are reported elsewhere. For instance, male patients had lost the control of urine flow after undergoing prostatectomy, the removal of bony tumor lead to loss of limb function, resection of acoustic tumor resulted in loss of ear function and loss of vision was reported after surgery for orbital tumors. The general and short-term side effects are usually managed with drug interventions using pain killers and antibiotics. Surgical treatment is the more effective in prevention and diagnosis of various tumors, and is efficiency can be further enhanced in combination with chemotherapeutic drugs, immunotherapy and radiation therapy (Santosh et al., 2017).

1.4.1.1 Robotic-assisted surgery

The United States Food and Drug Administration (FDA) has approved robotic-assisted surgery, which has the potential to specifically excise out the tumor part, thus providing more improved patient care. The advanced surgical robots have greatly contributed to better optical imagining and enhanced surgical maneuvering for retraction, exposure, and resection of tissue. The technique has found much potential in resection of most tumors, but in case of spine surgery it has demerit taking longer operative times. The technique has recently found potential applications in surgical treatment of endometrial, cervical, cardiothoracic and otolaryngo tumors (Sayari et al., 2019).

1.4.2 Radiotherapy

Radiotherapy (RT) has been potentially a successful approach in managing the cancer treatment, when used alone or as a part of combinatorial treatment with chemotherapy, surgery or immunotherapy (Glatstein et al., 2008). The major types of radiations used in the therapy include electron beams, X-rays, and Gamma rays, having high energy and penetration potential to reach the targets. The high energy ionizing radiations target the tumor regions and ionize the cell's DNA, thus completely damaging the cells. The energy and the amount of the radiation dose (in Gy) for the therapy will be balanced, considering the amount of tissue to be damaged. The radiotherapy specifically targets the tumor area, and ensures very minimal damage to the surrounding healthy tissues. Targeting specifically the tumor tissues and directing the radiation dose is very crucial for maintaining the intact body structures (Terasawa et al., 2009).

There exist two major RT techniques, the external-beam RT, and the internal-beam RT, of which the former one is most commonly used. In case of the external-beam RT, the radiations will be applied outside the body to cover most of the body regions. Further the external-beam RT is sub-classified into five types, namely three-dimensional conformal RT (3D-CRT), intensity-modulated RT (IMRT), proton beam therapy, image-guided RT (IMRT) and stereotactic RT (SRT). Permanent implants and temporary internal RT are the two main techniques under internal-beam RT (Sadeghi et al., 2010).

Many cancers are treated with this technique, and the radiotherapy utilization rate is been used to estimate the percentage of the population undergoing this treatment. The radiotherapy utilization rate is defined as the percentage of the population that receives at least one radiation dose in the entire cancer treatment regime. The RT utilization rate varies from time to time, and is also based on types and stages of cancer. In one of the studies in US, the RT utilization rates during the initial course of cancer (breast, central nervous system, gastrointestinal, genitourinary, gynecologic, musculoskeletal, skin, and thoracic cancer) management was determined for the period from 2004 to 2014. In all these cases, the utilization rate of radiotherapy was significantly declined from 33.9% to 31.2%. While, the systemic therapeutic approach and surgical treatment practice rate increased from 37.3% to 44.1% and 67.7% to 67.5%, correspondingly. According to their study, a decreased percent of cancer patients receiving radiotherapy in their initial phase of cancer management was noticed. While, the use of combinatorial treatment of cancer using surgery or systemic therapy was prominently increased (Royce et al., 2018). In another study, in case of treatment of lung cancer the optimal RT utilization rate was ranged from 61%–82% as estimated during 2009-2019 (Liu et al.,

2019). In one of the multimodal approaches, the radiotherapy was combined with gene therapy (recombinant adenovirus carrying p53) at a clinical stage to induce complete disease regression in head and neck squamous cell cancer, and it was found to be successful (Raty et al., 2008). Researchers have reported that the survival rates of patients with RT has significantly improved from 30% to 80% in the case of head and neck cancers, when compared to two decades ago (Chen and Kuo, 2017).

1.4.2.1 Side effects of radiotherapy

The targeted treatment of cancer tissues with a great precision by RT is essential to minimize the damages caused to the surrounding normal tissues. Though the RT has advanced with more accuracy of targeting the tumors, the most worrying aspect of radiation treatment is insensitivity and intrinsic resistance of some tumor cells. These negative aspects of the tumor cells are responsible for most of the recurrence of tumor after the treatment with radiations. It has been well studied in the case of breast cancer, wherein about 50% of the patients showed recurrence of the tumor at the same site after irradiation and conserving surgery. Circulating tumor cells (CTCs) that escape from the primary and metastatic sites are stimulated by the cytokine, granulocyte-macrophage colony stimulating factor in re-colonizing the original sites for recurrence of tumor (Chen and Kuo, 2017). The growth of secondary tumors can be possibly caused by RT in cancer patients, for instance, TR given to young women with Hodgkin lymphoma was found to develop secondary breast cancers (Horning, 2008; Santosh et al., 2017).

Another important drawback of the RT is the radiation-induced DNA damage of normal tissues. Though it is the principal mechanism of killing the tumor cells, the same damage caused to the nearby normal cells leads to many biological consequences. The major reparations include single and clustered damage sites and double stranded breaks, which are very difficult to be repaired. The effects caused by the radiation are very similar to the ones induced by the reactive oxygen and nitrogen species. As reported, the ionizing radiations induce in mammalian cells around 1000 single stranded breaks, 450 purine lesions, 20–40 double stranded breaks and 850 pyrimidine lesions (Lomax et al., 2013). Likewise, RT-induced side-effect in cancerous patients can lead to ovarian failure and infertility (Santosh et al., 2017). The ovaries are usually exposed to irradiation during treatment for cervical and rectal cancer and CNS malignancies as well. The radiation dosage, which could damage 50% of the follicles LD_{50}) is estimated to be 4 Gy. Further studies by Lushbaugh and Casarett (1976) revealed that women below the age of 40 years are more resistant to radiation-induced follicle damage, when compared to older women.

In another study, involving the treatment of cancer in women of childhood and adolescence, the total body exposure to a radiation dose of 14.4 Gy caused ovarian failure in 75% of the women (Meirow and Nugent, 2001). In many circumstances, lymphedema can be noticed after the treatment with radiations (Santosh et al., 2017).

1.4.3 Immunotherapy

The human immune system is a complex cellular structure, and involves complex molecular interactions. It is an excellent defense mechanism, fighting against the most invading pathogens. Cellular-mediated and humoral-mediated immunities are the two main principal defense mechanisms, which function by producing cytotoxic T-cells and the antibodies, respectively. The immune system is also known to provide defense against tumor formation, either by preventing the formation or by destroying the cancerous cells.

1.4.3.1 Immune checkpoint inhibitors

These are the class of monoclonal antibodies, which arrest the inhibitory action of T-cell molecules allowing the T-cells to identify and kill the tumor cells. The two main T-cell molecules that block the T-cell functioning against the cancerous cells are cytotoxic T-lymphocyte associated protein-4 (CTLA-4) and programmed death receptor-1 (PD-1). The very first attempt in immunotherapy is the use of anti-CTLA antibody against patients, who received cancer vaccines for advanced melanoma and ovarian cancer (Hodi et al., 2003). This study had initiated larger clinical trials, resulting in higher efficiency of treatment and ultimately paving the way for approval in 2011 for the usage of such monoclonal antibodies against metastatic melanoma (Hodi et al., 2003). It has been reported that the inhibition of the interaction of PD-1 with its ligand on T-cell by specific antibodies could increase the antitumor response. These observations later led to the development of two specific antibodies, namely pembrolizumab (Keytruda) and nivolumab (Opdivo) for the treatment of cancer. After their approval in 2014, these antibodies were successfully tried against Hodgkin lymphoma, small cell cancer, non-small cell lung cancer, squamous cell carcinoma of head and neck and many more advanced melanomas. In some of the cases, patients initially sensitive to treatment were observed to develop resistance to either of the checkpoint inhibitors by immune or genetic mechanisms. The recent research outcome involving a combination of both immune checkpoint inhibitors has increased the response rates. Owing to their non-overlapping mechanisms of action, the co-blockade of CTLA-4 and PD-1 resulted in increase of survival rate by 3 years, when compared to treatment with anti-PD-1 antibodies alone (Zaidi and Jaffee, 2019).

1.4.3.2 Immunotherapy boosted by metronomic chemotherapy

Metronomic chemotherapy involves continuous administration of anti-cancerous agents without a period of rest to overcome the frequent resistance to anticancer drugs. When combined with immunotherapy, it can stimulate the actions of the immune system against the immunosuppressive reactions elicited by the tumor cells. The studies indicated that when immunotherapy is provided in combination with chemotherapeutic agents, it actively stimulates the production of stable and effective cytotoxic T-lymphocytes to destroy tumor cells. The preliminary data obtained from various studies has hinted the feasibility of applying this multimodal approach at the clinical level for treatment of cancer in the near future (Chen et al., 2017).

1.4.4 Chemotherapy

Chemotherapy (CT) is a type of systemic therapy that uses drug molecules, having the potential of destroying fast-growing cancerous cells. Currently, the term CT is usually denoted for cancer treatment, since the cancerous cells are the rapidly proliferating cells among others. CT has been well adopted as an effective therapeutic intervention in the treatment of both benign and malignant tumors (Santosh et al., 2017).

1.4.4.1 History of chemotherapy

In the early 1900s, Paul Ehrlich, a German chemist was the first person to coin the term chemotherapy, and it was defined as the treatment of diseases using chemical compounds. In addition, he was the first person to establish the use of various animal models to screen the effectiveness of a wide array of chemicals against the diseases. His initial attempts to cure cancer with chemicals went unsuccessful for various reasons. In the 1960s, cancer treatment was dominated by surgery and radiotherapy with a successful rate of only around 33%, since these procedures left behind heretofore-unappreciated micrometastases, which could potentially recur into tumors. The initial CT processes were hampered, as it was difficult to develop the ideal animal models to screen the vast majority of chemicals to narrow down to a few and very effective ones. The other hurdle was a limited access to the clinical facilities for testing the efficacy of selected chemo-drugs (DeVita and Chu, 2008).

It was Murray Shear, who in 1935 had set up the first National Cancer Institute for the very purpose of screening anticancer drugs. With this program, he could screen around 3000 chemicals using murine S37 as the model system. The result of this attempt was only two drugs could enter into the clinical trials, and even those were dropped as they showed significant amounts of unacceptable cytotoxicity. The most successful attempt during the period was hormonal therapy by Charles Huggins for treatment of prostate cancer in men. Though this exciting work could fetch Charles Huggins Nobel prize, the hormones were treated differently, and hence the question of chemicals acting as anti-cancerous agents remained unanswered (Huggins and Hodges, 2002).

It was not until 1943, when Alfred Gilman and Louis Goodman conducted active research with mustard compounds (nitrogen mustard) on mice with transplanted lymphoid tumor and observed marked regression in the cancerous growth. This was followed by testing on lymphoma patients, and the results had set off a platform to synthesize and test several classes of alkylating compounds, of which chlorambucil and cyclophosphamide were the predominant ones (DeVita and Chu, 2008).

1.4.4.2 The principal mechanisms of chemotherapy

Most chemical compounds exert their therapeutic effect by disrupting the cell cycle by various mechanisms (Fig. 1.1). The uncontrolled cell division in tumors is a result of mutations in proto-oncogenes and tumor suppressor genes, and the overall failure of the control of normal cell cycle. Anti-cancerous chemical agents cause the cell death by apoptosis, which is mediated by direct action on DNA, or by directly

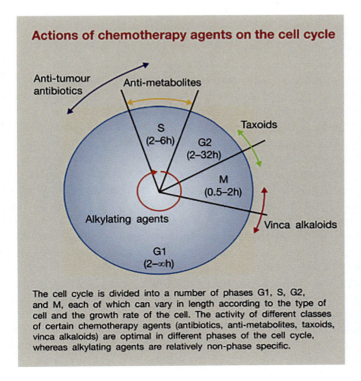

Fig. 1.1 Mechanisms of action of chemotherapy drugs. Adapted from Dickens, E., Ahmed, S., 2018. Principles of cancer treatment by chemotherapy. Surgery 36(3), 134–138.

targeting the key proteins needed for cell division (Dickens and Ahmed, 2018). The most targeted processes include enzymatic actions, hormone interactions and angiogenesis. Based on the mode of action, chemotherapeutic agents are classified in two ways; one being the effect on cell cycle and the other based on biochemical properties. Classifying chemocompounds based on the cell cycle specificity is useful in programming the schedule and combining the drugs for better efficacy. For instance, 5-fluorouracil (5-FU) a phase-specific agent acts by inhibiting thymidylate synthase or can get mis-incorporated into DNA in place of thymidine during synthesis of DNA. This action of 5-FU require the tumor cells to be exposed to this drug during their S-phase, thus making the drug administration programmed for days and weeks (Santosh et al., 2017).

1.4.4.3 Classes of chemotherapeutic agents and their uses

There exists a broad classification for anti-cancerous agents under chemotherapy, hormonal therapy, and immunotherapy. In particular, the chemicals for CT are categorized based on their mode of action and structural moieties. The major groups include antibiotics, alkylating agents, antimetabolites, mitosis inhibitors, topoisomerase I and II inhibitors, platinum compounds (for example, cisplatin and carboplatin) and others.

Alkylating agents

Alkylating agents are one of the primitive and most widely used anti-cancerous drugs in CT. The usage of these drugs dates back to the 1940s with nitrogen mustard being the first compound to be screened at the experimental and clinical levels. They act directly on the DNA by crosslinking, which in turn leads to DNA strand breaks. This affects the cell by wrong base pairing and eventually inhibiting division of cells. Alkylating agents are basically classified into six groups namely (i) ethyleneimines and methylmelamines (hexamethylmelamine), (ii) triazenes (dacarbazine), (iii) alkyl sulfonates (busulfan), (iv) nitrogen mustards (cyclophosphamide), (v) nitrosoureas (carmustine) and (vi) piperazines (Ralhan and Kaur, 2007). Many of these chemical compounds have been used in combination for effective treatment of the cancer. Synergistic combination of Cyclophosphamide with 5,6-dimethylxanthenone-4-acetic acid an antivascular agent, significantly reduced the tumor blood flow without any side-effects. An attempt was made in combining alkylating agent (name not revealed) with cyclooxygenase-2 inhibitor (Celecoxib) to treat colon cancer with reduced incidences of nausea and vomiting that was experienced with treatment by alkylating agent alone (Ralhan and Kaur, 2007). The major cancers treated mostly with alkylating agents are lymphoma, leukemia, multiple myeloma, Hodgkin's disease, and sarcomas.

Though alkylating agents are promising remedy for cancer treatment, they exert certain severe side effects in the patients. It has been shown that the long-term use of alkylating agents leads to permanent infertility in both males (decrease in sperm count) and females (cessation of menstruation). Years of treatment with alkylating agents also lead to the development of secondary cancers like myeloid leukemia. In addition, many of the alkylating agents demonstrated severe toxicity affecting kidney, GI tract, cardiac system, lungs and blood circulatory systems (Iqbal et al., 2017; Ralhan and Kaur, 2007).

Antimetabolites

The antimetabolites, usually act as structural analogues of DNA or RNA components. These molecules get inserted in the place of nucleotides during replication, thus inhibiting the process, and ultimately lead to cell death. Currently, most of the antimetabolites under use are purine and pyrimidine analogues. As of 2009, there exists around 14 FDA approved antimetabolites used in cancer treatment, which amounts to 20% of overall cancer treatment using CT (Parker, 2009). Some of the clinically important antimetabolites include decitabine, nelarabine, clofarabine and capecitabine. Though many of these drugs have almost similar structural features and share the common metabolic pathway, they exhibit diverse clinical activities. For instance, clofarabine that differs from its near relative cladribine in just having one fluorine atom demonstrated higher therapeutic efficacy against relapsed and refractory pediatric acute lymphoblastic leukemia.

Anthracyclines (Anti-tumor antibiotics)

Anthracyclines are also well-established drugs, since the 1960s in the treatment of cancer. Between 1950s and 1960s, actinomycin D was proved to be very efficient drug in the treatment of pediatric tumors (Pinkel, 1959). These are the secondary metabolites of the fungi, mainly produced by the genus *Streptomyces*. In addition to anthracyclines, other anti-tumor antibiotics include bleomycin and mitoxantrone. Most anti-tumor antibiotics act by intercalating DNA at the specific sites, leading to double or single strand breaks. Further, anthracyclines, preferably act by inhibiting the function of topoisomerases I and II required for uncoiling of the DNA during replication. Another mode of action of anthracyclines is the production of hydroxyl free radicals during their enzymatic reduction (Hortobagyi, 1997). Since their inception into clinical practice, these have been the promising anticancer agents in the treatment of breast cancer. Apart from antibiotics, some microbial metabolites such as glycolipid biosurfactants have showed to possess pharmaceutical property, the anticancerous being one among them (Madihalli et al., 2020).

Although anthracyclines are proved to be one of the most efficient anti-cancerous agents, the cardiotoxic effects caused has limited their usage. The free radicals produced by enzymatic processing of anthracyclines are responsible for side effects, which outweighs the efficiency of cancer treatment. Currently, efforts are under progress to reduce the risk of cardiotoxicity by using cardioprotective (Dexrazoxane—the only FDA approved drug) strategies (Raber and Asnani, 2019).

Plant-derived anticancerous agents

Since ancient times, many plants and their products are used for the medicinal purpose to cure health complications. Also, the medicinal plants and their phytochemicals proved to be successful in treating various cancers (Ravichandra et al., 2018; Tan and Norhaizan, 2018; Swamy and Akhtar, 2019). Phytochemicals and other derived products are present in various parts of the plants and occur in different chemical forms of either primary or secondary metabolites, such as terpenes, alkaloids, saponis, flavonoids, lignans, gums, glycosides, or many more (Akhtar and Swamy, 2018a; Lee et al., 2018; Swamy, 2020). Even today, plants serve as the major natural resource to obtain anticancer compounds (Akhtar and Swamy, 2018a; Akhtar and Swamy, 2018b). Few of the very important and widely used plant products in treatment of cancer are outlined below.

Vinca Alkaloids (VA) Vinca alkaloids (Vas) are the phytochemicals that are naturally found in *Catharanthus roseus* and the well-known anticancer compounds include vinorelbine, vindesine, vincristine, and vinblastine. Currently around 64 varieties of *C. roseus* are screened for phytochemicals. The modes of action of VAs involve binding at the specific sites on microtubulin heterodimers in turn arresting the actively dividing cell at metaphase (Singh et al., 2013). Owing to their tedious and expensive extraction procedure, currently many semi-synthetic derivatives of VAs are synthesized, such as vinorelbine, vindesine, vinfosiltine, and vinovelbine, which are under clinical usage. Apart from plants, few endophytic fungi from *C. roseus* are known to produce VAs. *Alternaria* sp. is reported to produce vinblastine, *Fusarium oxysporum* is known to produce vincristine, while *F. solani* is known to produce mixture of Vas (Uzma et al., 2018). Recently, researchers have successfully enhanced the production of VAs from the cell suspension culture of *C. roseus*. Addition of cell extracts of endophytic fungi namely *Fusarium solani* RN1 and *Chaetomium funicola* RN3 to the cell suspension culture has further increased the yield of Vas (Linh et al., 2021). Currently VAs are used in the treatment of the cancers associated with blood, breast, liver and lung.

Taxanes Taxanes are the class of chemical compounds obtained from the inner bark of the pacific yew tree (*Taxus brevifolia*), and chemically they are polycyclic diterpenoids. Initial studies in 1962 involved the use of the crude extract to determine the cytotoxic activity. Later it took many years to purify the Pre-Approval Inspection (PAI) form of the extract, i.e., paclitaxel, which is sold in the trade name of taxol, and is the first anticancerous diterpenoid to be extracted in the pure form, and currently it is a promising drug to treat breast, ovary and Kaposi's sarcoma (Lee et al., 2018; Uzma et al., 2018). It is also regarded as the first or the second line treatment for various types of cancers. Docetaxel is another taxane isolated and purified from the sp. of *Taxus* plant.

The taxanes usually act by interfering with formation of microtubules and arresting the cell division. Microtubules, the cytoskeleton components constitute repeating sub-units of α- and β-tubulin proteins play crucial cellular activities, including the cell shape regulation, vesicular transport, trafficking of transcription factors, segregation of chromosomes during mitosis and mitochondrial metabolism. Mode of action of paclitaxel is well studied, wherein it binds to β-tubulin in the lumen of microtubules and significantly reduces microtubule dynamics. This eventually leads to arrest of the cell cycle at metaphase stage (Iqbal et al., 2017). In particular, paclitaxel attaches to the N-terminal end of β-tubulin sub-unit, and reduces the threshold level of tubulin sub-units required for microtubule assembly, and thereby creating imbalance in tubulin subunit levels (Fig. 1.2). Further, paclitaxel interacts with microtubules directly and stabilize them against cold- and calcium-induced depolymerization. Consequently, taxane-treated cancerous cells exhibit mitotic arrest, due to defects in bipolar spindle formation. In this course of treatment, the spindle assembly checkpoint is stimulated to prevent the cell cycle advancement, particularly, the chromosome segregation is affected due to the non-availability of active microtubules that can attach to kinetochore. Paclitaxel-treated cell fates may be different. In one scenario, cell death may occur as a result of the persistent mitotic arrest. Alternatively, cancerous cells exit from paclitaxel-prompted the mitotic arrest, also called as mitotic slippage to escape consequent cell death leading to the formation of a tetraploid G1 cell. However, mechanisms that determine the effect require further exploration (Weaver, 2014; Mikuła-Pietrasik et al., 2019).

From the time of its inception into clinical uses, paclitaxel is the best-selling drug with a market size of $ 1.6 billion in 2005 and till date it is the most sold drug with few competitors such as Abraxane. As reported, to treat one cancer patient six yew trees of 100 years old (0.01-0.03% dry weight of paclitaxel) were required. The constraints, such as availability and relatively low concentration led to the development of alternate drugs for paclitaxel, 10-deacetyl-baccatin III being one such compound isolated from *T. baccata*. 10-deacetyl-baccatin III is presently used as precursor in the synthesis of paclitaxel by semi-synthetic

18 Chapter 1 Introduction to cancer and treatment approaches

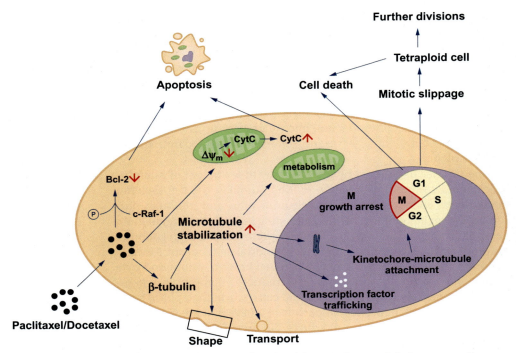

Fig. 1.2 Mechanisms of paclitaxel and docetaxel cytotoxicity in cancer cells. Adapted from Mikuła-Pietrasik, J., Witucka, A., Pakuła, M., Uruski, P., Begier-Krasińska, B., Niklas, A., Tykarski, A., Książek, K., 2019. Comprehensive review on how platinum-and taxane-based chemotherapy of ovarian cancer affects biology of normal cells. Cell. Mol. Life Sci. 76(4), 681–697.

method. Further scientific investigations had explored many endophytic fungi as rich sources of paclitaxel. The major taxa of the fungi which were researched to produce paclitaxel include *Alternaria, Aspergillus, Botryodiplodia, Botrytis, Cladosporium, Ectostroma, Fusarium, Metarhizium, Monochaetia, Mucor, Ozonium* and many more. Among many fungi studied, *Metarhizium anisopliae* isolated from *T. chinensis* was able to produce 846.1 μg/L of paclitaxel when cultured in liquid media (Uzma et al., 2018).

Though paclitaxel is a gold standard chemotherapeutic agent for cancer treatment, few toxic effects, such as myelotoxicity, neurotoxicity and tumor resistance is a major concern in the current treatment regime. To overcome these complications, recently several novel approaches are in clinical practice for controlled release of the drug and minimize the toxicity. The major approaches include soluble polymers, paclitaxel-soluble pro-drugs, micellar drug carriers, folic acid conjugated paclitaxel, and polymeric nano-capsules (Dhanasekaran, 2019).

Camptothecin (CPT) and its derivatives Camptothecin (CPT) is one the alkaloid class of chemical compounds extracted from the bark portion of the Chinese happy tree, *Camptotheca acuminata*. CPT was the first drug to be isolated in 1966, while screening various plants for the isolation of novel steroids. This compound was shown to exhibit antitumor activity in both in vitro systems and mouse leukemia models. From the ancient times, the Chinese traditional medicine system uses this compound (in extract form) as natural remedy against cancers (Iqbal et al., 2017; Swamy et al., 2021). In the later days, the presence of this drug was traced in many other unrelated genera, such as *Tabernaemontana heyneana, Nothapodytes nimmoniana, Ophiorrhiza pumila, O. mungos,* etc (Swamy et al., 2021). Further, several entophytic fungi (*Fusarium solani, Entrophospora infrequens,* etc.) associated with these plants were identified to produce CPT in small quantity (Kaushik et al., 2015; Prakash et al., 2016; Alurappa et al., 2018; Pu et al., 2019). The four other derivatives of CPT, which are currently in use are topotecan, irinotecan, belotecan and trastuzumab deruxtecan. The mode of action of CPT involves inhibition of topoisomerase 1 (TOP 1), which is known to be present in higher amounts in cancer cells when compared to normal tissues. The formation of TOP I-CPT complex leads to an irreversible strand breaks in the DNA eventually causing the death of cells (Martino et al., 2017; Swamy et al., 2021). The initial usage of CPT and its derivatives was targeted to gastrointestinal tumors, and further slowly explored in the treatment of breast, ovarian, colon, lung, and stomach cancers. Its low solubility and resistance by the cancer cells has reduced its usage.

1.5 Advanced approaches for cancer treatment

From the ancient time, various approaches are in place starting from surgical intervention to hormonal therapy to treat cancer. Many of these approaches are very potential and successful in eliminating or killing the cancerous cells of benign and malignant tumors. Though, the present techniques are very efficient and provide a long-term remedy in treating cancers, they pose one or the other side effects, mainly toxicity to the healthy tissues and organs. Apart from these, other drawbacks of treatment techniques include low specificity, low bioavailability, and solubility issues. Hence, recently a very innovative and molecular-based methods are being researched, and some of them are already in clinical practices. The recent approaches include nano medicines, encapsulation by extracellular vesicles, gene therapy, thermal ablation, and magnetic hyperthermia. These methods are opening new opportunities for the precision medicine, and help in overcoming the drawbacks faced with the previous methods. The other two in-silico tools, which assist in the fastening of the scientific research on cancer include the radiomics and pathomics (Pucci et al., 2019). A brief overview to some of these well-known approaches are explained below.

1.5.1 Nanomedicines

Nanoparticles are the smallest particles measuring in nanometer (nm), and exhibit versatile physico-chemical properties, which may be attributed to their small size and high surface-to-volume-ratio as well (Rudramurthy and Swamy, 2018). The overall benefits of encapsulating the anti-cancerous drugs in nanoparticles are increased solubility, bioavailability, stability, controlled release, and precise delivery of drugs to the targeted tissues (Patra et al., 2018; Akhtar and Swamy, 2018c). Inorganic compounds like quantum dots are encapsulated as nanoparticles, and used in the cancer diagnosis. Their electronic and optical properties make them highly fluorescent in turn increasing the sensitivity of detection and imaging of the tumors (Matea et al., 2017). Nanoparticles consisting of superparamagnetic iron oxide show remarkable interactions under magnetic field and are recently exploited for cancer treatment (glioblastoma) using magnetic hyperthermia method (Boyer et al., 2010). Similarly, photodynamic therapy has explored the optical, electrical, and low toxic properties of gold nanoparticles in the treatment of breast cancer. The nanoshell formulation consisted of silica core and gold shell with polyethylene glycol (PEG) encapsulation (Pucci et al., 2019). Lipids are also shown to be excellent carriers of drugs due to their amphiphilic nature. Biosurfactants like mannosylerythritol lipids form nano-sized micelles above their CMC (critical micellar concentration), and are proven to be ideal carriers in drug delivery of water-insoluble drugs (Madihalli et al., 2020). Many lipid micelles are usually used to encapsulate hydrophobic drugs. For instance, doxil, doxorubicin-loaded PEGylated liposomes were the first nanoparticles approved by the FDA in 1995 in treatment of HIV-associated Kapsosi's sarcoma. From there on many more molecules, including biocompatible natural polymeric compounds were approved as encapsulating agents in nanoformulations of anticancer agents (Nasir et al., 2015; Patra et al., 2018; Akhtar and Swamy, 2018c).

1.5.2 Extracellular vesicles as drug delivery systems and diagnostic tool

Exosomes, a type of membrane bound extracellular vesicles (EVs) act as the prospective tool in diagnosis, and assist in the treatment of cancer, wherein they function as drug carriers. These are formed by the fusion of multivesicular bodies (MVBs), and mostly found in extracellular bodily fluids are responsible for the molecular trafficking between cells. EVs are shown to be consisting of informative molecules, such as dsDNS, mRNA, miRNA, proteins and lipids, and thus serve as one of the potential tool in diagnosis of cancer (Pucci et al., 2019). In cancer therapy, the exosomes are exploited as nanocarriers for drug delivery. Being natural, biocompatible, and low immunogenicity, they will further increase their applications in treating cancers. In a recent in vitro and in vivo study, the exosomes incubated with doxorubicin were used to target the liver cancer cells, and was successful in inhibiting the cell growth. The researchers have isolated exosomes derived from macrophages expressing aminoethyl anisamide-PEG, and used them to target sigma receptor in lung cancer cells. These exosomes loaded with paclitaxel drug could effectively target liver cancer cells and inhibit the metastatic growth (Qi et al., 2016; Kim et al., 2018).

1.5.3 Genome-based methods for cancer treatment

Gene therapy involves the replacement of a defective gene with a normal gene, thus restoring the function of the targeted system. The first clinical application of gene therapy dates back to 1990 that involved retroviral-mediated gene transfer, targeting T-cells to treat severe combined immunodeficiency (SCID) in patients. From thereon, gene therapy has demonstrated to be a promising remedy for many rare and chronic disorders of humans, including cancer treatment. As reported in a recent survey, around 66.6% of gene therapy experimentation is related to cancer treatment. Various strategies designed in gene therapy are based on the expression or sensitization of anti-cancer genes or natural cell death mechanism (Pucci et al., 2019). Some of the successful gene therapy interventions include expression of thymidine kinase through stimulation by the prodrug ganciclovir to treat prostate cancer and glioma, vectors carrying tumor suppressor p53 gene targeted against head and neck squamous cell cancer (Freeman et al., 1993; Raty et al., 2008). Though gene therapy has potential to treat the cancer alone or in multimodal approach, the technique poses problems such as accurate expression or suppression, ideal gene delivery systems, genome integration and suppression by immune system.

The small interfering RNAs (siRNAs) are known to target functional mRNAs, and thus are used to produce a targeted gene silencing. The strategies of siRNA-based treatment involve silencing the production of anti-apoptotic proteins, transcription factors or silencing of cancer mutated genes (Mansoori et al., 2014). The current clinical trials are based on local administration of siRNA oligonucleotides to targeted tissues and organs. The merits of the siRNA strategy in cancer treatment are safety, high efficacy, target specificity, few side-effects and cost-effective. Currently cationic liposomes are exploited in encapsulation of siRNAs through electrostatic interactions thus increasing their stability (Rautela et al., 2021).

1.5.4 Thermal ablation and magnetic hyperthermia

To overcome the complications associated with surgical removal of tumors, recently new approaches are introduced such as thermal ablation and magnetic hyperthermia. Both the techniques works by targeting to a very narrow and precise areas and could be an efficient alternative to traditional surgical methods (Brace, 2011; Hervault and Thanh, 2014). Thermal ablation employs heat or cold to destroy neoplastic tissues. Further, it is shown that the cancerous cells are highly sensitive to extreme temperatures when compared to normal cells. Thermal ablation by high temperature uses radiofrequency, microwave and laser radiations (Pucci et al., 2019).

Another new way to heat the cancer tissues is through magnetic hyperthermia. The method involves generation of heat by superparamagnetic nanoparticles by subjecting into strong and alternating magnetic field. As discussed earlier the Superparamagnetic Iron Oxide Nanoparticles (SPIONs) are deflected to the most extent under magnetic field which is made use in disease diagnosis through Magnetic resonance imaging (MRI) (Hervault and Thanh, 2014). These particles could be encapsulated in biocompatible polymer/lipids and bound with specific ligands for target identification. Till date, only a formulation of iron oxide nanoparticles (15 nm) coated with aminosilane (Nanotherm) has the approval to be used for treating glioblastoma (Sanchez et al., 2011; Lorkowski et al., 2021).

1.6 Conclusions

Cancer is one of the most rapidly emerging diseases, and affecting over 80% of the global people. It is a multifactorial malady, involving multifarious alterations in the genome, due to interactions with the individual's environment. Surgery, radiotherapy, and chemotherapy are regarded as the ideal choice for controlling cancers. Though, chemo-drugs efficiently impede tumor cells, they persuade toxicity and additional adverse health complications in patients. However, these individual treatment options were found to be not effective when used individually, and hence considered their use in combination for controlling cancers. Nonetheless, these curative practices induce adverse side effects that could vary from one patient to another, and even amongst those individuals, who receive the similar treatment. Selecting the best cancer therapy approach depends on various factors, such as the type of malignancies, growth stages, age, management frequencies, quantity of medicines, and healthiness of patients. To overcome these adverse side effects of conventional treatment modalities to the normal body cells, some improvements in cancer management approaches, such as immunotherapy, hormone therapy, gene therapy, stem cell therapy can be an option for cancer treatment. If traditional treatments combined with these advanced techniques are used appropriately, one can expect a higher chance cancer cure, and the possibilities of relapse in cancer patients can be prevented.

References

Abbas, Z., Rehman, S., 2018. An overview of cancer treatment modalities. Neoplasm 1, 139–157.

Adams, J.M., Cory, S., 2007. The Bcl-2 apoptotic switch in cancer development and therapy. Oncogene 26 (9), 1324–1337.

Akhtar, M.S., Swamy, M.K., 2018a. Anticancer Plants: Mechanisms and Molecular Interactions. vol. 4 Springer, Singapore, https://doi.org/10.1007/978-981-10-8417-1.

Akhtar, M.S., Swamy, M.K., 2018b. Anticancer Plants: Properties and Application. Springer, Singapore, https://doi.org/10.1007/978-981-10-8548-2.

Akhtar, M.S., Swamy, M.K., 2018c. Anticancer Plants: Clinical Trials and Nanotechnology. Springer, https://doi.org/10.1007/978-981-10-8216-0.

Alurappa, R., Chowdappa, S., Narayanaswamy, R., Sinniah, U.R., Mohanty, S.K., Swamy, M.K., 2018. Endophytic fungi and bioactive metabolites production: an update. In: Patra, J.K., Das, G., Shin, H.S. (Eds.), Microbial Biotechnology. Springer, Singapore, pp. 455–482, https://doi.org/10.1007/978-981-10-7140-9_21.

Ames, B.N., Gold, L.S., Willett, W.C., 1995. The causes and prevention of cancer. Proc. Natl. Acad. Sci. 92 (12), 5258–5265.

Aslam, M.S., Naveed, S., Ahmed, A., Abbas, Z., Gull, I., Athar, M.A., 2014. Side effects of chemotherapy in cancer patients and evaluation of patients opinion about starvation based differential chemotherapy. J. Cancer Ther. 5 (8), 1–6. https://doi.org/10.4236/jct.2014.58089.

Baeriswyl, V., Christofori, G., 2009. The angiogenic switch in carcinogenesis. Semin. Cancer Biol. 19 (5), 329–337. https://doi.org/10.1016/j.semcancer.2009.05.003.

Boyer, C., Whittaker, M.R., Bulmus, V., Liu, J., Davis, T.P., 2010. The design and utility of polymer-stabilized iron-oxide nanoparticles for nanomedicine applications. NPG Asia Mat. 2 (1), 23–30.

Brace, C., 2011. Thermal tumor ablation in clinical use. IEEE Pulse 2, 28–38.

Bray, F., Jemal, A., Grey, N., Ferlay, J., Forman, D., 2012. Global cancer transitions according to the human development index (2008-2030), a population-based study. Lancet Oncol. 13, 790–801.

Burrell, R.A., McGranahan, N., Bartek, J., Swanton, C., 2013. The causes and consequences of genetic heterogeneity in cancer evolution. Nature 501 (7467), 338–345.

Carmeliet, P., Jain, R.K., 2011. Molecular mechanisms and clinical applications of angiogenesis. Nature 473 (7347), 298–307.

Chen, Y.L., Chang, M.C., Cheng, W.F., 2017. Metronomic chemotherapy and immunotherapy in cancer treatment. Cancer Lett. 400, 282–292.

Chen, H.H.W., Kuo, M.T., 2017. Improving radiotherapy in cancer treatment: promises and challenges. Oncotarget 8, 62742–62758.

DeBerardinis, R.J., Lum, J.J., Hatzivassiliou, G., Thompson, C.B., 2008. The biology of cancer: metabolic reprogramming fuels cell growth and proliferation. Cell Metab. 7 (1), 11–20.

Delaney, G., Jacob, S., Featherstone, C., Barton, M., 2005. The role of radiotherapy in cancer treatment. Cancer 104 (6), 1129–1137.

DeVita, V.T., Chu, E., 2008. A history of cancer chemotherapy. Cancer Res. 68, 8643–8653. https://doi.org/10.1158/0008-5472.can-07-6611.

Dhanasekaran, S., 2019. Augmented cytotoxic effects of paclitaxel by curcumin induced overexpression of folate receptor-α for enhanced targeted drug delivery in HeLa cells. Phytomedicine 56, 279–285.

Dickens, E., Ahmed, S., 2018. Principles of cancer treatment by chemotherapy. Surgery 36 (3), 134–138.

Ellis, M.J., Perou, C.M., 2013. The genomic landscape of breast cancer as a therapeutic roadmap. Cancer Discov. 3, 27–34.

Faries, M.B., Morton, D.L., 2007. Surgery and sentinel lymph node biopsy. Semin. Oncol. 34, 498–508.

Ferlay, J., Shin, H.R., Bray, F., Forman, D., Mathers, C., Parkin, D.M., 2010. Estimates of worldwide bur-den of cancer in 2008: GLOBOCAN 2008. Int. J. Cancer 127, 2893–2917.

Freeman, S.M., Abboud, C.N., Whartenby, K.A., Packman, C.H., Koeplin, D.S., Moolten, F.L., Abraham, G.N., 1993. The bystander effect: tumor regression when a fraction of the tumor mass is genetically modified. Cancer Res. 53, 5274–5283.

Glatstein, E., Glick, J., Kaiser, L., Hahn, S.M., 2008. Should randomized clinical trials be required for proton radiotherapy? An alternative view. J. Clin. Oncol. 26, 2438–2439.

Greenblatt, M.S., 1994. Mutations in the p53 tumor suppressor gene: clues to cancer etiology and molecular pathogenesis. Cancer Res. 54, 4855–4878.

Hanahan, D., Weinberg, R.A., 2011. Hallmarks of cancer: the next generation. Cell 144 (5), 646–674.

Hausheer, F.H., Schilsky, R.L., Bain, S., Berghorn, E.J., Lieberman, F., 2006. Diagnosis, management and evaluation of chemotherapy induced peripheral neuropathy. Semin. Oncol. 33, 15–49.

Hervault, A., Thanh, N.T.K., 2014. Magnetic nanoparticle-based therapeutic agents for thermo-chemotherapy treatment of cancer. Nanoscale 6, 11553–11573.

Hodi, F.S., Mihm, M.C., Soiffer, R.J., Haluska, F.G., Butler, M., Seiden, M.V., Davis, T., Henry-Spires, R., MacRae, S., Willman, A., Padera, R., 2003. Biologic activity of cytotoxic T lymphocyte-associated antigen 4 antibody blockade in previously vaccinated metastatic melanoma and ovarian carcinoma patients Proc. Nat. Acad. Sci. 100, 4712–4717.

Horning, S., 2008. Hodgkin's lymphoma. In: Abeloff, M.D., Armitage, J.O., Niederhuber, J.E., Kastan, M.B., MCKenna, W.G. (Eds.), Abeloff's Clinical Oncology, fourth ed. Churchill Livingstone, Philadelphia, pp. 2253–2268.

Hortobagyi, G.N., 1997. Anthracyclines in the treatment of cancer. Drugs 54, 1–7.

Huggins, C., Hodges, C.V., 2002. Studies on prostatic cancer: I. The effect of castration, of estrogen and of androgen injection on serum phosphatases in metastatic carcinoma of the prostate. J. Urol. 167, 948–951.

Iqbal, J., Abbasi, B.A., Mahmood, T., Kanwal, S., Ali, B., Shah, S.A., Khalil, A.T., 2017. Plant-derived anticancer agents: a green anticancer approach. Asian Pac. J. Trop. Biomed. 7, 1129–1150.

Isogai, T., Yasunaga, H., Matsui, H., Tanaka, H., Hisagi, M., Fushimi, K., 2017. Factors affecting in-hospital mortality and likelihood of undergoing surgical resection in patients with primary cardiac tumors. J. Cardiol. 69, 287–292.

Jiang, W., Peng, J., Zhang, Y., Cho, W., Jin, K., 2012. The implications of cancer stem cells for cancer therapy. Int. J. Mol. Sci. 13 (12), 16636–16657.

Kampan, N.C., Madondo, M.T., McNally, O.M., Quinn, M., Plebanski, M., 2015. Paclitaxel and its evolving role in the management of ovarian cancer. Biomed. Res. Int. 2015. https://doi.org/10.1155/2015/413076, 413076.

Kassi, M., Polsani, V., Schutt, R.C., Wong, S., Nabi, F., Reardon, M.J., Shah, D.J., 2019. Differentiating benign from malignant cardiac tumors with cardiac magnetic resonance imaging. J. Thorac. Cardiovasc. Surg. 157, 1912–1922.e1912.

Kaushik, P.S., Swamy, M.K., Balasubramanya, S., Anuradha, M., 2015. Rapid plant regeneration, analysis of genetic fidelity and camptothecin content of micropropagated plants of *Ophiorrhiza mungos* Linn.—a potent anticancer plant. J. Crop. Sci. Biotechnol. 18 (1), 1–8.

Kim, M.S., Haney, M.J., Zhao, Y., Yuan, D., Deygen, I., Klyachko, N.L., Kabanov, A.V., Batrakova, E.V., 2018. Engineering macrophage-derived exosomes for targeted paclitaxel delivery to pulmonary metastases: in vitro and in vivo evaluations. Nanomedicine 14 (1), 195–204.

Kroschinsky, F., Stölzel, F., von Bonin, S., Beutel, G., Kochanek, M., Kiehl, M., Schellongowski, P., 2017. New drugs, new toxicities: severe side effects of modern targeted and immunotherapy of cancer and their management. Crit. Care 21, 89. https://doi.org/10.1186/s13054-017-1678-1.

Lahat, G., Lazar, A., Lev, D., 2008. Sarcoma epidemiology and etiology: potential environmental and genetic factors. Surg. Clin. North Am. 88, 451–481.

Lee, K.W., Ching, S.M., Hoo, F.K., Ramachandran, V., Swamy, M.K., 2018. Traditional medicinal plants and their therapeutic potential against major cancer types. In: Akhtar, M.S., Swamy, M.K. (Eds.), Anticancer Plants: Natural Products and Biotechnological Implements. Springer, Singapore, pp. 383–410.

Linh, T.M., Mai, N.C., Hoe, P.T., Ngoc, N.T., Thao, P.T.H., Ban, N.K., Van, N.T., 2021. Development of a cell suspension culture system for promoting alkaloid and vinca alkaloid biosynthesis using endophytic fungi isolated from local *Catharanthus roseus*. Plants 10, 672. https://doi.org/10.3390/plants10040672.

Liu, W., Liu, A., Chan, J., Boldt, R.G., Munoz-Schuffenegger, P., Louie, A.V., 2019. What is the optimal radiotherapy utilization rate for lung cancer?—a systematic review. Transl. Lung Cancer Res. 8, S163–S171. https://doi.org/10.21037/tlcr.2019.08.12.

Lodish, H., Berk, A., Zipursky, S.L., Matsudaira, P., Baltimore, D., Darnell, J., 2000. Molecular Cell Biology, fourth ed. W. H. Freeman, New York. https://www.ncbi.nlm.nih.gov/books/NBK21590/.

Lomax, M.E., Folkes, L.K., O'Neill, P., 2013. Biological consequences of radiation-induced DNA damage: relevance to radiotherapy. Clin. Oncol. 25, 578–585.

Lorkowski, M.E., Atukorale, P.U., Ghaghada, K.B., Karathanasis, E., 2021. Stimuli-responsive iron oxide nanotheranostics: a versatile and powerful approach for cancer therapy. Adv. Healthc. Mater. 10 (5), 2001044.

Lowe, M., Lane, J.D., Woodman, P.G., Allan, V.J., 2004. Caspase-mediated cleavage of syntaxin 5 and giantin accompanies inhibition of secretory traffic during apoptosis. J. Cell Sci. 117 (7), 1139–1150.

Lum, J.J., Bauer, D.E., Kong, M., Harris, M.H., Li, C., Lindsten, T., Thompson, C.B., 2005. Growth factor regulation of autophagy and cell survival in the absence of apoptosis. Cell 120 (2), 237–248.

Lushbaugh, C.C., Casarett, G.W., 1976. The effects of gonadal irradiation in clinical radiation therapy: a review. Cancer 37, 1111–1120.

Madihalli, C., Doble, M., 2019. Microbial production and applications of mannosylerythritol, cellobiose and trehalose lipids. In: Banat, I.M., Thavasi, R. (Eds.), Microbial Biosurfactants and Their Environmental and Industrial Applications. CRC Press, USA, p. 81. eBook.

Madihalli, C., Sudhakar, H., Doble, M., 2020. Production and investigation of the physico-chemical properties of MEL-A from glycerol and coconut water. World J. Microbiol. Biotechnol. 36, 1–11.

Mansoori, B., Shotorbani, S.S., Baradaran, B., 2014. RNA interference and its role in cancer therapy. Adv. Pharma Bull. 4 (4), 313. https://doi.org/10.5681/apb.2014.046.

Martino, E., Della Volpe, S., Terribile, E., Benetti, E., Sakaj, M., Centamore, A., Sala, A., Collina, S., 2017. The long story of camptothecin: from traditional medicine to drugs. Bioorg. Med. Chem. Lett. 27, 701–707.

Matea, C.T., Mocan, T., Tabaran, F., Pop, T., Mosteanu, O., Puia, C., Iancu, C., Mocan, L., 2017. Quantum dots in imaging, drug delivery and sensor applications. Int. J. Nanomedicine 12, 5421–5431.

Meirow, D., Nugent, D., 2001. The effects of radiotherapy and chemotherapy on female reproduction. Hum. Reprod. Update 7, 535–543.

Mikuła-Pietrasik, J., Witucka, A., Pakuła, M., Uruski, P., Begier-Krasińska, B., Niklas, A., Tykarski, A., Książek, K., 2019. Comprehensive review on how platinum-and taxane-based chemotherapy of ovarian cancer affects biology of normal cells. Cell. Mol. Life Sci. 76 (4), 681–697.

Nasir, A., Kausar, A., Younus, A., 2015. A review on preparation, properties and applications of polymeric nanoparticle-based materials. Polym.—Plast. Technol. Eng. 54, 325–341.

Palucka, K., Banchereau, J., 2012. Cancer immunotherapy via dendritic cells. Nat. Rev. Cancer 12 (4), 265–277.

Parker, W.B., 2009. Enzymology of purine and pyrimidine antimetabolites used in the treatment of cancer. Chem. Rev. 109, 2880–2893. https://doi.org/10.1021/cr900028p.

Patra, J.K., Das, G., Fraceto, L.F., Campos, E.V.R., del Pilar Rodriguez-Torres, M., Acosta-Torres, L.S., Diaz-Torres, L.A., Grillo, R., Swamy, M.K., Sharma, S., Habtemariam, S., 2018. Nano based drug delivery systems: recent developments and future prospects. J. Nanobiotechnol. 16 (1), 1–33.

Paul, A.T., Jindal, A., 2017. Nano-natural products as anticancer agents. In: Akhtar, M.S., Swamy, M. (Eds.), Anticancer Plants: Clinical Trials and Nanotechnology. Springer, Singapore, https://doi.org/10.1007/978-981-10-8216-0_2.

Pavlova, N.N., Thompson, C.B., 2016. The emerging hallmarks of cancer metabolism. Cell Metab. 23 (1), 27–47.

Pinkel, D., 1959. Actinomycin D in childhood cancer: a preliminary report. Pediatrics 23, 342–347.

Prakash, L., Middha, S.K., Mohanty, S.K., Swamy, M.K., 2016. Micropropagation and validation of genetic and biochemical fidelity among regenerants of *Nothapodytes nimmoniana* (Graham) Mabb. employing ISSR markers and HPLC. 3Biotech 6 (2), 1–9. https://doi.org/10.1007/s13205-016-0490-y.

Pu, X., Zhang, C.R., Zhu, L., Li, Q.L., Huang, Q.M., Zhang, L., Luo, Y.G., 2019. Possible clues for camptothecin biosynthesis from the metabolites in camptothecin-producing plants. Fitoterapia 134, 113–128.

Pucci, C., Martinelli, C., Ciofani, G., 2019. Innovative approaches for cancer treatment: current perspectives and new challenges. Ecancer Med. Sci. 13, 961. https://doi.org/10.3332/ecancer.2019.961.

Qi, H., Liu, C., Long, L., Ren, Y., Zhang, S., Chang, X., Qian, X., Jia, H., Zhao, J., Sun, J., Hou, X., 2016. Blood exosomes endowed with magnetic and targeting properties for cancer therapy. ACS Nano 10 (3), 3323–3333.

Raber, I., Asnani, A., 2019. Cardioprotection in cancer therapy: novel insights with anthracyclines. Cardiovasc. Res. 115, 915–921.

Ralhan, R., Kaur, J., 2007. Alkylating agents and cancer therapy. Expert Opin. Ther. Patents 17, 1061–1075. https://doi.org/10.1517/13543776.17.9.1061.

Ralph, S.J., Rodríguez-Enríquez, S., Neuzil, J., Saavedra, E., Moreno-Sánchez, R., 2010. The causes of cancer revisited: "mitochondrial malignancy" and ROS-induced oncogenic transformation—why mitochondria are targets for cancer therapy. Mol. Asp. Med. 31 (2), 145–170.

Raty, J.K., Pikkarainen, J.T., Wirth, T., Yla-Herttuala, S., 2008. Gene therapy: the first approved gene-based medicines, molecular mechanisms and clinical indications. Curr. Mol. Pharmacol. 1, 13–23.

Rautela, I., Sharma, A., Dheer, P., Thapliyal, P., Sahni, S., Sinha, V.B., Sharma, M.D., 2021. Extension in the approaches to treat cancer through siRNA system: a beacon of hope in cancer therapy. Drug Deliv. Transl. Res., 1–15.

Ravichandra, V.D., Ramesh, C., Swamy, M.K., Purushotham, B., Rudramurthy, G.R., 2018. Anticancer plants: chemistry, pharmacology, and potential applications. In: Anticancer Plants: Properties and Application. Springer, Singapore, pp. 485–515.

Rivlin, N., Brosh, R., Oren, M., Rotter, V., 2011. Mutations in the p53 tumor suppressor gene: important milestones at the various steps of tumorigenesis. Genes Cancer 2 (4), 466–474.

Royce, T.J., Qureshi, M.M., Truong, M.T., 2018. Radiotherapy utilization and fractionation patterns during the first course of cancer treatment in the United States from 2004 to 2014. J. Am. Coll. Radiol. 15, 1558–1564.

Rudramurthy, G.R., Swamy, M.K., 2018. Potential applications of engineered nanoparticles in medicine and biology: an update. J. Biol. Inorg. Chem. 23 (8), 1185–1204.

Sadeghi, M., Enferadi, M., Shirazi, A., 2010. External and internal radiation therapy: past and future directions. J. Cancer Res. Ther. 6, 239–248.

Sanchez, C.M., Belleville, P., Popall, M., Nicole, L., 2011. Applications of advanced hybrid organic-inorganic nanomaterials: from laboratory to market. Chem. Soc. Rev. 40, 696–753.

Santosh, S., Rajagopalan, M.D., Pallavi, B.A., Rudramurthy, G.R., Rajashekar, V., Sridhar, K.A., Swamy, M.K., 2017. Cancer therapies: current scenario, management, and safety aspects. In: Akhtar, M.S., Swamy, M. (Eds.), Anticancer Plants: Clinical Trials and Nanotechnology. Springer, Singapore, pp. 1–25, https://doi.org/10.1007/978-981-10-8216-0_1.

Sayari, A.J., Pardo, C., Basques, B.A., Colman, M.W., 2019. Review of robotic-assisted surgery: what the future looks like through a spine oncology lens. Ann. Transl. Med. 7 (10), 224. https://doi.org/10.21037/atm.2019.04.69.

Shewach, D.S., Kuchta, R.D., 2009. Introduction to cancer chemotherapeutics. Chem. Rev. 109, 2859–2861.

Singh, S., Jarial, R., Kanwar, S.S., 2013. Therapeutic effect of herbal medicines on obesity: herbal pancreatic lipase inhibitors. Wudpecker J. Med. Plants 2, 53–65.

Swamy, M.K., 2020. Plant-Derived Bioactives. Springer, Singapore, https://doi.org/10.1007/978-981-15-2361-8.

Swamy, M.K., Akhtar, M.S., 2019. Natural Bio-Active Compounds: Volume 2: Chemistry, Pharmacology and Health Care Practices. Springer Nature, https://doi.org/10.1007/978-981-13-7205-6.

Swamy, M.K., Purushotham, B., Sinniah, U.R., 2021. Camptothecin: occurrence, chemistry and mode of action. In: Pal, D., Nayak, A.K. (Eds.), Bioactive Natural Products for Pharmaceutical Applications. Advanced Structured Materials, vol. 140. Springer, Cham, https://doi.org/10.1007/978-3-030-54027-2_9.

Tan, B.L., Norhaizan, M.E., 2018. Plant-derived compounds in cancer therapy: traditions of past and drugs of future. In: Akhtar, M.S., Swamy, M.K. (Eds.), Anticancer Plants: Properties and Application. Springer, Singapore, pp. 91–127.

Terasawa, T., Dvorak, T., Ip, S., Raman, G., Lau, J., Trikalinos, T.A., 2009. Systematic review: charged-particle radiation therapy for cancer. Ann. Intern. Med. 151, 556–565.

Uzma, F., Mohan, C.D., Hashem, A., Konappa, N.M., Rangappa, S., Kamath, P.V., Singh, B.P., Mudili, V., Gupta, V.K., Siddaiah, C.N., Chowdappa, S., 2018. Endophytic fungi—alternative sources of cytotoxic compounds: a review. Front. Pharmacol. 9. https://doi.org/10.3389/fphar.2018.00309.

Weaver, B.A., 2014. How taxol/paclitaxel kills cancer cells. Mol. Biol. Cell 25 (18), 2677–2681. https://doi.org/10.1091/mbc.E14-04-0916.

Witsch, E., Sela, M., Yarden, Y., 2010. Roles for growth factors in cancer progression. Physiology 25 (2), 85–101. https://doi.org/10.1152/physiol.00045.2009.

Zaidi, N., Jaffee, E.M., 2019. Immunotherapy transforms cancer treatment. J. Clin. Invest. 129, 46–47.

2

Taxol: Occurrence, chemistry, and understanding its molecular mechanisms

Mallappa Kumara Swamy[a] and Bala Murali Krishna Vasamsetti[b]

[a]Department of Biotechnology, East West First Grade College, Bengaluru, Karnataka, India, [b]Toxicity and Risk Assessment Division, Department of Agrofood Safety and Crop Protection, National Institute of Agricultural Sciences, Rural Development Administration, Wanju, Korea

2.1 Introduction

The invention of a key antitumor a diterpene plant alkaloid, taxol has prospered the interest among plant scientists, phytochemists, biologists, medical doctors, drug discovery institutions, and pharmaceutical industries (Slichenmyer and Von Hoff, 1991; Isah, 2016; Weaver, 2014). It is one of the most recognized plant-derived anticancer drug molecule (Fridlender et al., 2015; Akhtar and Swamy, 2018b). The generic name of taxol is paclitaxel, and it is mainly obtained from the Pacific Yew (*Taxus brevifolia*) tree bark. This is one of the highly valued natural compounds, approved by the FDA (Food and Drug Administration), USA for treating several types of cancers, including breast, ovarian, lung cancer, and Kaposi's sarcoma (Akhtar and Swamy, 2018b; Isah, 2016). The mode of action of taxol in cancerous cells involves stabilizing cellular microtubules by attaching β-tubulin subunits, leading to interruption with their regular breakdown during cell divisions with subsequent safeguarding of polymers through protecting them from dis-assembly (Löwe et al., 2001; Isah, 2016; Jordan and Wilson, 2004). Taxol drug arrests cells during mitotic stage due to the manifestation of un-attached kinetochores (Waters et al., 1998).

Taxol was discovered and characterized for the first time from the bark extracts of *T. brevifolia* at Research Triangle Institute in North Carolina, while screening several natural compounds against cancer cells as part of the collaborative program initiated by the National Cancer Institute (NCI) and the US Department of Agriculture (USDA), USA (Wani, 1972; Kampan et al., 2015; Isah, 2016). Taxol is sequestered

Paclitaxel. https://doi.org/10.1016/B978-0-323-90951-8.00009-6
Copyright © 2022 Elsevier Inc. All rights reserved.

at low quantities from bark, seeds, and needles of *T. brevifolia* and *T. baccata* (European yew) trees. Further, some angiosperms, gymnosperms and many novel endophytic microbes are being reported in recent times to produce taxol (Lichota and Gwozdzinski, 2018). However, the yield of taxol is very low in all these natural resources. The obtainable amount of taxol from these sources differs with tissue types, genotypes, seasons, environmental conditions, and the use of extraction methods (Vidensek et al., 1990; Wheeler et al., 1992; Wianowska et al., 2009; Isah, 2016).

Though, this natural compound is effective in exhibiting cytotoxicity, numerous critical issues are presented by taxol, such as less water-solubility, stability in addition to unpredictable adverse effects in patients (Isah, 2016). Further, inadequate supply has hindered its expansion into an expedient chemotherapy drug molecule. To overcome these problems, several attempts have been made to develop novel chemical derivatives, which are readily soluble through semi-synthetic approaches involving the basic structural modifications. The first anticancer drug to have the status of "blockbuster" compound is the taxol with sales over $1 billion in 1997. Later, its sales have increased drastically over the years. As assessed, the global taxol market was valued at US$ 78.77 million in 2017, and predicted to reach US$ 161.66 by the end of 2025 (https://www.marketwatch.com/press-release/global-paclitaxel-market-will-reach-16166-million-usd-by-the-end-of-2025-growing-at-a-cagr-of-94-2019-11-14). It is one of the top-selling drugs that are being ranked on the basis of overall global sales/revenue as reported for 2018 by bio/pharma companies (https://www.genengnews.com/a-lists/top-10-best-selling-cancer-drugs-of-2018/). Unfortunately, tons of trees are required to obtain a small quantity (milligram) of taxol. Hence, the chemical exploration of taxol has resulted in a total synthesis of this molecule (Holton et al., 1995). Further, another anticancer compound, 10-deacetylbaccatin III derived from needles of *Taxus* species in higher quantity has supplemented the demand for taxol (Holton et al., 1995). An effective chemosynthesis and semisynthesis approaches have been explored to generate congeners (Isah, 2016). However, obtaining these synthesized congeners was not cost effective. Hence, these approaches are regarded incompatible to meet the demand. Research efforts are being extensively carried out to discover the possibilities to overcome the side effects of taxol in cancerous individuals by varying the drug administration, improving its solubility, safe and effective delivery into the patient's body. In this direction, new taxol formulations are being created, for instance, using albumin nanoparticles or pro-drugs. Currently, taxol production is achieved by semisynthesis from a precursor, microbial production using fermentation technology, and the use plant cell culture approaches (Isah, 2016; Akhtar and Swamy, 2018a).

This chapter summarizes the occurrence of taxol in natural resources, chemistry, its mode of action and pharmacological significance. Overall, this comprehensive review encourages research interest among chemists and biologists to explore more on this novel compound.

2.2 About taxol and its discovery

Dr. Mansukh Wani and Dr. Monroe Wall paved a way to the discovery of novel anticancer compounds as a means for chemotherapy from the research lab of Research Triangle Institute at North Carolina, USA by discovering taxol from Pacific yew tree. Taxol discovery is not by chance, but it is the outcome of a collective work of USDA (the US Department of Agriculture) and NCI (the National Cancer Institute), USA during the period from 1960 to 1981 to screen plant-derived compounds against different cancers. In this collaborative plant screening investigations over the period between 1960 and 1981, plants of 15,000 plant species were collected and evaluated nearly 115,000 different plant extracts to recognize phytocompounds possessing anticancer properties (Kampan et al., 2015; Kundu et al., 2017). In 1962, a botanist, Arthur Barclay from USDA collected Pacific yew tree (*T. brevifolia*) twigs, needles, bark, and fruits. Later, Mansukh Wani and Monroe Wall started to explore *T. brevifolia* crude extracts along with other plant species in their lab for their cytotoxicity effects. In the year 1967, they could able to observe the cytotoxicity effects of *T. brevifolia* bark extract. Considering this exploration, they isolated/extracted bioactive compound, named as taxol, on the basis of its origin from plant species and the occurrence of free hydroxyl groups (Perdue, 1969; Wall and Wani, 1995; Weaver, 2014). Finally, in the year 1971, the chemical structure of taxol was proposed, and it was used to develop drugs by NCI (Weaver, 2014; Kampan et al., 2015; Kundu et al., 2017). The screening program for anticancer compounds was concluded in 1981, and only taxol (paclitaxel) entered into clinical trial stages by showing effectiveness against P388 leukemia and mouse tumor models (Kampan et al., 2015; Renneberg, 2007; Weaver, 2014; Tuma, 2003). However, taxol presented mixed results during pre-clinical trials. Hence, it was not equivalently measured as the most encouraging plant ingredient against cancers. Further, water insolubility imposed taxol's preparation with polyethoxylated castor oil (cremophor EL), and caused diminished enthusiasm as an anticancer drug as it may result in causing increased anaphylactic reactions (Weaver, 2014).

Noteworthy to mention that though, several clinical trials were conducted, and few investigations were hampered due to taxol's scarcity in its supply. The slowly growing *T. brevifolia* tree was the only resource of obtaining taxol at that time. Interestingly, some clinical studies conducted on ovarian cancer patients were shown to be positive and nearly 30% of ovarian cancerous individuals responded well to taxol treatments (McGuire et al., 1989; Weaver, 2014). Thus, it became one the vital phytoconstituents in the cancer therapy and related investigations. This valued diterpene alkaloid basically occurs in the bark of several *Taxus* species. Unfortunately, in plants, taxol occurs in very small amount, i.e., 0.01% to 0.04% dry weight (dw), and hence the cost of its extraction is extremely very high. As reported earlier, a mature *T. brevifolia* tree may yield only about 2 kg of bark. Moreover, to obtain 500 mg of taxol, about 12 kg of bark is required (Wani, 1972; Goodman and Walsh, 2001; Kundu et al., 2017). As there was a high demand for taxol, matured *Taxus* species were severely cut down, and resulted in species loss in nature. To safeguard its natural source, *T. brevifolia* tree was included in the list of endangered species in 1990 (Goodman and Walsh, 2001; Weaver, 2014; Kampan et al., 2015).

In relation to growing long-term prospects for taxol, and to overcome the challenges of producing taxol from nature, i.e., *T. brevifolia* and environmental issues concerned, the total chemical synthesis of taxol was initiated by more than 30 research laboratories globally in 1988. Nevertheless, owing to its complex and unique chemostructure, the chemosynthesis of this molecule was not successful up to 1994 (Holton et al., 1994; Nicolaou et al., 1994; Weaver, 2014).

In due course, a number of approaches for total chemosynthesis were developed, which almost involved approximately 40 steps of chemical reactions. Later, a practical approach, i.e., semi-synthetic procedure was established, which became the best way for its synthesis. Later, NCI decided to transfer taxol chemosynthesis to a pharmaceutical company, Bristol-Myers Squibb (BMS) to commercialize its production in 1992. This company trademarked the name, "Taxol" and generated new name as "Paclitaxel." This happened only after the long term of over 20 years, and during which taxol name was cited in more than 600 research manuscripts, suggesting that the trademark could have been not given, and BMS company should surrender its rights on it (Walsh and Goodman, 1999, 2002a, b; Weaver, 2014).

In 1991 and 1993, congressional hearings raised the issues related to the monopoly of converting taxol to Paclitaxel by BMS on a natural resource. In these meetings, the higher prices for taxol drug were questioned as it was recognized and developed using public money rather than private funds. The sub-committee team resolved the agreements stuck between NCI and BMS. In total, taxol is one of the highest profited chemotherapy drugs in history. It is the only chemodrug, which is in clinical utility at present acknowledged by the plant-screening program (Walsh and Goodman, 2002a, b; Weaver, 2014). In 1992, Paclitaxel was registered as the chemotherapeutic drug to treat ovarian cancer. Scientists furthermore verified the efficiency of Paclitaxel in treating advanced breast cancer. Succeeding clinical investigations have confirmed its effectiveness against cancers. In 1994, the Food and Drug Authority (FDA), USA approved the drug, Taxol to treat breast cancer. Presently, it is used either singly or together with other drugs like carboplatin or cisplatin to treat ovarian, breast and non-small-cell lung cancers (Piccart and Cardoso, 2003; Bonomi et al., 1997; Kampan et al., 2015).

2.3 Natural resources of taxol

For the first time in 1963, the occurrence of taxol was found in the bark extract of *T. brevifolia* tree at NCI, USA. *T. brevifolia* is an unusual low-growing perennial tree, located in the forests North-Western Pacific, a geographic region in the west part of the North America covered by the Pacific Ocean to the west side and the Rocky Mountains on the eastern part. Later, various pre-clinical investigations proved cytotoxic activity to many cancer types. The increased demand for taxol drug was not met by its main source, *T. brevifolia* tree due to various reasons, including low yield, scarcity in nature, difficulty in isolating, higher cost for extracting, etc. Thus, *T. brevifolia*-mediated taxol extraction became impractical in addition to creating financial burden (Walsh and Goodman, 1999; Holton et al., 1994). An effective semi-synthetic method to synthesize taxol was proposed. Later, this semi-synthetically formulated taxol was approved by the FDA in 1992, and even today also this approach is extensively used for its bulk production (Holton et al., 1995; Walsh and Goodman, 2002b).

There are few *Taxus* species, including *T. brevifolia, T. baccata, T. wallichiana, T. floridana, T. cuspidata, T. globosa, T. sumatrana, T. fauna, T. chinensis, T. canadensis* to name a few in nature that are mostly irregularly distributed in northern temperate zones. However, Southern yew tree (*Austrotaxus spicata*) is located in the regions of the southern hemisphere. However, taxol occurrence in these species varies from one species to another and different tissues, and it ranges between 0% and 0.069% (Guo et al., 2006). Noteworthy to mention here that taxoids biosynthesis pathway is similar in all these different species and their tissues of the same species (Isah, 2015). As indicated above, *Taxus* species have become endangered and taxol is found only in the mature tree, and yields low content of taxol. Hence, its isolation from these tree species is inadequate. Currently, seedling cultures and improving forestation are regularly considered as the best practical approaches to produce taxol and precursors required for its chemical synthesis. Simultaneously, genetically improved *Taxus* species producing high content of *Taxus* in needles provides good choices for large scale production (Guo et al., 2006; Isah, 2015).

In 1993, a novel endophytic fungal species, *Taxomyces andreanae* was isolated from *T. brevifolia* bark, and was screened to produce taxol compound. They developed a method to mass-produce taxol from *T. andreanae* cultures (Stierle et al., 1993). Though, the yield was in small amount (24–50 ng/mL), this endophytic fungal discovery generated increased curiosity in scientists. So far, many endophytic fungi have been reported to yield taxol (Guo et al., 2006; Zhou et al., 2010; Naik, 2019; Isah, 2015). Microbial fermentation approaches are very promising in recent times to produce taxol at commercial scale. Endophytic fungi have the capability of producing therapeutic products and can self-sufficiently biosynthesize bioactive metabolites that are similar to those produced by their host plants. Numerous research groups have investigated various factors that improve the fungal-mediated taxol production. However, fungal-mediated taxol synthesis at the commercial scale is yet to be made successful (Zhou et al., 2010; Naik, 2019). The major concern with the fungal-mediated production is the low quantity of taxol yield and its unpredictability. Generally, the yields range between 24 ng and 70 µg/l (Nikolic et al., 2011). In this direction, different strategies, including the optimization of fungal fermentation cultures by supplementing different carbon and nitrogen sources, precursor molecules, inducers, and inhibitors. Different biotechnological approaches like gene cloning, genetic transformation and gene mutations are other few promising technologies to enhance the production of taxol from endophytic fungi in large scale, and more details are reviewed by earlier reports (Naik, 2019; Isah, 2015; Nikolic et al., 2011).

2.4 Chemistry of taxol

Taxol is one of the leading plant-derived alkaloids having antineoplastic properties. The empirical formula of taxol is $C_{47}H_{51}NO_{14}$. Its molecular weight is 853.9 g/mol. The IUPAC name is (2α,4α,5β,7β,10β,13α)-4,10-bis(acetyloxy)-13-{[(2R,3S)-3-(benzoylamino)-2-hydroxy-3 phenylpropanoyl]oxy}-1,7-dihydroxy-9-oxo-5,20-epoxytax-11-en-2-yl benzoate. The molecular structure of taxol is shown in Fig. 2.1. Taxol has two molecules; (a) a taxane ring containing four-membered oxetane side ring positioned at C4 and C5, and (b) a homochiral ester side chain positioned at C13, which is the active site of the compound, and attaches to microtubules, stabilizes the tubulin proteins, and activates tubulin depolymerization in GTP (guanosine triphosphate)-independent manner (Kampan et al., 2015). Consequently, cell multiplication is repressed due to inhibition of metaphase/anaphase stages of the cell cycle and by inducing the destabilization of the microtubules. Further, detailed studies have established that both taxane ring and ester side chain plays a crucial role

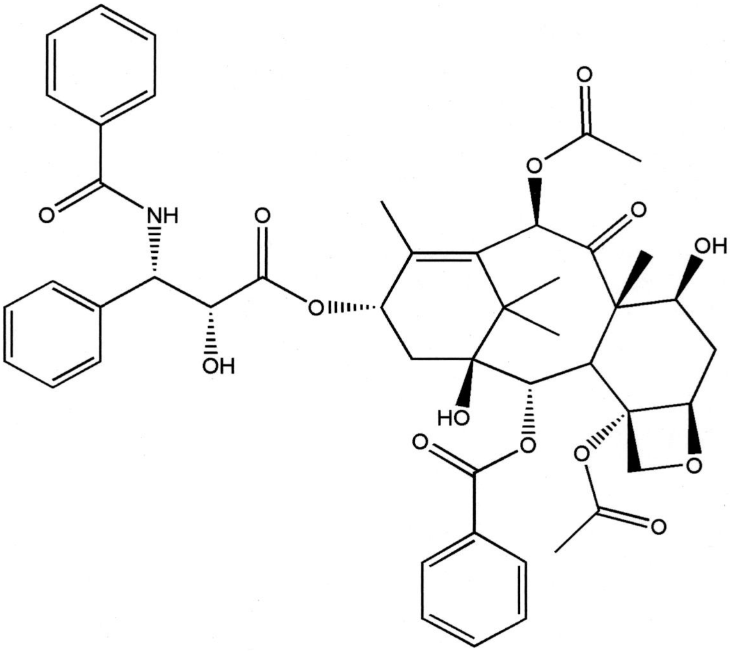

Fig. 2.1 Chemical structure of taxol.

of cytotoxicity activity (Guenard et al., 1993; Kingston, 1994; Nikolic et al., 2011; Kampan et al., 2015).

Taxol is a white crystalline powder, soluble in DMSO (dimethyl sulfoxide) (50 mg/ml). Also, it can be dissolved in methanol (50 mg/mL). After 14 days at room temperature, it undergoes transesterification and hydrolysis. It can be rapidly damaged in strong acidic solutions like concentrated hydrochloric acid or methanolic solutions. Acetic acid (0.1%) added to methanol for dissolving taxol can prevent degradation up to one week at room temperature, and further, it can be stored intact up to 3 months at 4°C. Ethanol and acetonitrile can also be used to dissolve taxol (Nikolic et al., 2011). Taxol possesses lower solubility in water, and quickly gets destroyed in a weak alkaline aqueous solution. Its melting point is at about 217°C. At pH ranging between 3 and 5 is ideal for taxol solution. Taxol preparations (1 mg/mL) in dextrose (5%) injection remain active up to 3 days 22°C. Likewise, taxol solutions (1 mg/mL) in sodium chloride (0.9%) injection remains active for 3 days at 32°C (Xu et al., 1994; Nikolic et al., 2011).

2.5 Mechanisms of action of taxol

Taxol is the antineoplastic drug, which targets microtubules, i.e., the major components of cytoskeleton structure. Microtubules are the polymers that consist of repeating sub-units of α- and β-tubulin (~ 25 to 30 nm in diameter) heterodimers. Microtubules play a significant role in forming the mitotic spindle fibers during cellular divisions. Also, they offer stability to cell organization and cytoplasmic movement inside the cell (Kampan et al., 2015). Taxol increases the tubulin subunit levels, but suppresses the assembling of purified tubulin subunits that are required for polymerization to form microtubules. Besides, polymerized microtubules attached with taxol are protected from both cold and calcium treatments that normally induces disassembly (Schiff et al., 1979; Weaver, 2014). Taxol-exposed HeLa cells and fibroblasts challenged with taxol at 0.25 and 10 μmol/mL, respectively, blocked the G2 and M-phase of the cell cycle, and therefore disrupting cell division to cause apoptosis (Schiff and Horwitz, 1980; Nikolic et al., 2011; Abu Samaan et al., 2019). Remarkably, these properties were unambiguously different to early acknowledged microtubule a drug (colchicine and vinca alkaloids) that prevents microtubule polymerization of the monomers, and does not require the occurrence of guanosine triphosphate (GTP) or microtubule-associated proteins (MAPs). Numerous studies have indicated that taxol-blocked cells are found in M-phase, and consist of near-normal, bipolar spindle fibers. Microscopic images of taxol treated cells will have chromosomes that are aligned at the cell equator, with few instances of chromosomal misalignment (Jordan and Wilson, 1998; Waters et al., 1998; Weaver, 2014). Several studies have reported the induction of multipolar spindles by taxol treatment. From a majority of study reports, it can be concluded that the anticancer effects of taxol is due to arresting of cells during bipolar spindles formation at M-phase (Chen and Horwitz, 2002; Hornick et al., 2008; Weaver, 2014). Moreover, taxol attaches to microtubules at a specific binding site. It attaches precisely in a revocable mode to the N-terminal 31 amino acids of β-tubulin subunits in the microtubule instead of dimer structure (De Brabander et al., 1981; Weaver, 2014; Kampan et al., 2015; Nikolic et al., 2011; Vyas and Kadow, 1995; Zhang et al., 2014; Rao et al., 1995; Abu Samaan et al., 2019). Taxol-treatment arrests cells in mitosis because of the existence of a small number of unattached kinetochores (Waters et al., 1998). Overall, taxol hampers microtubule polymerization and microtubules dynamic stability. This interrupts the progress of mitosis via prompting the failure of chromosomal segregation, leading to arresting of mitosis and induction of apoptosis (Jordan et al., 1993; Jordan and Wilson, 1998; Long and Fairchild, 1994; Giannakakou et al., 2001).

In vitro studies have shown the dose dependent cytotoxicity activity of taxol. As documented earlier, the concentration of taxol beyond 12 nM causes G2/M-phase arrest in A549 (lung carcinoma) and MCF-7 (breast cancer) cell lines. Remarkably, lesser quantity (3 to 6 nM) of taxol was shown to suppress the spread of cancerous cells by encouraging apoptosis (Giannakakou et al., 2001; Abu Samaan et al., 2019). Taxol at low concentration (10 nM) exerted anti-cancer properties in breast cancer cell line (MDA-MB-231) by expressing voltage-reliant sodium channels. This effect could be attributed to the regulation of intracellular signaling pathways comprising sodium channels (Tran et al., 2009). Recently, a therapeutic strategy was examined by treating a mixture consisting of taxol at low concentration and XAV939, a Wnt signaling inhibitor against different breast cancer cells, including TNBC cells (MDA-MB-468, BT549, MDA-MB-231) and ER + ve cells (T-47D and MCF-7). The study results have shown that taxol-amalgamated XAV939 regimen effectively induced programmed cell death and inhibited Wnt signaling pathway, and resulted in suppressing of the expression of epithelial-mesenchymal transition and angiogenesis (Shetti et al., 2019).

Research studies have reported that taxol activates Raf-1 kinase enzymes, which plays an essential role in Bcl-2 (B-cell lymphoma 2) phosphorylation and programmed cell death (Blagosklonny et al., 1997). Likewise, others also have stated that both Raf-1 and Bcl-2 phosphorylation are closely interconnected to mitotic arrest (Torres and Horwitz, 1998). Taxol-prompted M-phase cell cycle arrest and cell death are associated with c-Mos gene expression stimulation in ovarian cancer cells (Ling et al., 1998). Some investigators come to an understanding that the cell toxicity effect of taxol relies in its capability of causing Bcl-2 hyperphosphorylation. While, other researchers explain about the Bcl-2 dephosphorylation along with apoptosis (Abu Samaan et al., 2019). A distinctive caspase-3 and caspase-9 independent pathways are stimulated by the treatment of taxol to encourage programmed cell death in breast cancer MCF7 and human ovarian cancer SKOV3 cell lines (Ofir et al., 2002).

Taxol-induced apoptosis is directly linked to the influence on the mitochondria. It is stated that the side effects of taxol are correlated to their role in the calcium signal cascade. As reported earlier, taxol bind to Bcl-2, leading to influence the mitochondrial permeability transition pore (PTP) (Shimizu et al., 1999; Shimizu et al., 2001). The reduction of calcium ions from the mitochondrial reserve via mitochondrial permeability transition pore (PTP) is encouraged by the treatment of taxol. This condition prompts PTP to discharge the apoptogenic factor, Cytochrome C (Cyto C) into the cytosol to initiate cell apoptosis (Blajeski et al., 2001; Varbiro et al., 2001). Taxol-treated cells rapidly releases calcium ions from the mitochondria and cause reduced mitochondrial membrane potential (Kidd et al., 2002). Further, high doses of taxol damages mitochondria to release Cyto C, and initiates apoptosis without the efflux of calcium ions (Kidd et al., 2002). At lower levels of taxol exhibit apoptotic activity, which is independent of the extracellular levels of calcium ion concentration (Pan et al., 2014).

A model of human skin cancer metastasis was used to investigate the mode of actions of taxol in inhibiting metastasis (Wang et al., 2003). According to them, IP injection of taxol (5 mg/kg per day) for 21 days effectively inhibits metastases development in the lungs of mice. Taxol was shown to induce melanogenesis and programmed cell death in cancerous skin cells. In melanoma tissue lesions, angiogenesis was inhibited. Further, it abridged the expression of vascular endothelial growth factor (VEGF). Equally, there was an increase in the expression of the metastases suppressor gene, nm23 mRNA and E-cadherin expression levels (Wang et al., 2003). A fluorescence time-lapse microscopic examination was utilized to record microtubule dynamic instability behavior in paclitaxel-dependent mutant cell lines, Tax 11-6 (mutation in α-tubulin) and Tax 18 (mutation in β-tubulin). The results have concluded that taxol alters plus-end microtubule dynamics and rescues mutant cell division by preventing the detachment of microtubule minus ends from centrosomes (Ganguly et al., 2010). Likewise, it is also being reported that a kinesin-related protein, MCAK (mitotic centromere-associated kinesin) destabilizes microtubules, and plays a vital role in microtubules detachment (Ganguly et al., 2011).

MicroRNAs (miRNAs), which are a small non-coding nuclear RNAs play vital regulatory roles in the expression of genes. Moreover, miRNAs play a role in the process of cancer growth and metastasis. miRNAs may be modulated by different antitumor drug molecules, including taxol. According to several investigations, the application of taxol exhibits altered miRNA expression profiles. For instance, in taxol-treated breast cancer cell lines (SKBR-3, MCF-7, MDA-MB-231, and BT-474), there was an altered expression levels of both miR-205 and let-7a miRNAs with tumor suppressor potential targeting KRAS protein connected with the RAS/MAPK signaling pathway (Asghari et al., 2018). Low-dose metronomic (LDM) chemotherapy demonstrated a good anti-angiogenic property on breast cancers with minimum side effects. In a study, miRNAs were reported to reduce the level of let-7f, encouraging the overexpression of anti-angiogenic factor, thrombospondin-1 (TSP-1) in taxol-LDM therapy of breast cancers (Tao et al., 2015).

In both HeLa and Chinese hamster ovary cell lines, the inhibition of extracellular signal-regulated kinase (ERK) activity potentiated taxol-induced poly(ADP-ribose) polymerase (PARP) cleavage and phosphatidylserine externalization. These results suggested that ERK activity corresponded with the cytotoxicity effects of taxol (McDaid and Horwitz, 2001). It has been established that the mechanisms of taxol-mediated cell inhibition are dependent on the concentrations. As stated above, taxol alters the activation of Raf-1 kinase, involved in intracellular signal transduction events that are required for drug-prompted apoptosis. A study results suggest that at lower concentrations of taxol (<9 nM), the cell death occurs after an aberrant mitosis by a Raf-1 independent pathway, while at higher concentrations (≥9 nM), the cell death is connected with the terminal mitotic arrest, due to a Raf-1-dependent pathway, leading to apoptosis (Torres and Horwitz, 1998). Further, the induction of apoptosis is also under the influence of tumor suppressor proteins, p21 and p53 (Kampan et al., 2015).

Studies also have reported the immunomodulatory effect of taxol, involving both stimulation and suppression of the cells of immune system linked with tumor growth. The destruction of immune cells might show undesirable immune responses in the host against tumor growth (Javeed et al., 2009; Abu Samaan et al., 2019). As suggested by earlier researchers, taxol stimulates macrophages and participates of in regulating host immunity. Taxol stimulates the secretion of cytokines, namely tumor necrosis factor alpha (TNF-α) and interleukin 12 (IL-12) that mediates inflammatory responses. Further, it induces the activation of NK (natural killer) cells, dendric cells, and cytotoxic T-lymphocytes to encourage the eradication of cancer cells (Wanderley et al., 2018; Larionova et al., 2019). A concentration-dependent administration of taxol induced higher levels of MHC (major histocompatibility complex) class II molecules that usually occur only on specialized antigen-presenting cells like dendritic cells. Likewise, taxol was shown to promote the maturation of antigen-presenting cells by binding to Toll-like receptors (TLR-1) present on the surface of cells (Emens and Jaffee, 2005; John et al., 2010; Pfannenstiel et al., 2010). Studies also have suggested the impact of taxol on immunomodulatory effect in NK cells. The cytotoxicity activity of NK cells improved by the addition of taxol in a dose-dependent manner, which was observed by increased levels of perforin, a crucial effector protein (Kubo et al., 2005; Abu Samaan et al., 2019). Taxol induces ROS (reactive oxygen species) generation and upsurge the production of peroxols through enhancement of the NADPH (nicotinamide adenine dinucleotide phosphate) oxidase activity, contributing to oxidative stresses. This can increase the potency of taxol's antineoplastic activity (Alexandre et al., 2007; Hadzic et al., 2010). Taxol exerts its mechanism of action by activating multiple signal-transduction pathways, such as c-Jun N-terminal kinase (JNK), MAPK, nuclear factor kappa B (NF-κB), Janus kinase (JAK)-signal transducer and signal transducer and activator of transcription. All these pathways mediate either anti-apoptotic or pro-apoptotic signals (Wang et al., 2006; Szakács et al., 2006; Pfannenstiel et al., 2010; Abu Samaan et al., 2019). However, detailed research efforts are needed to understand the modulation of the immune response in cancer cells treated with taxol. The mechanisms of action of taxol in anticancer events that are described above are summarized in Fig. 2.2.

40 Chapter 2 Taxol

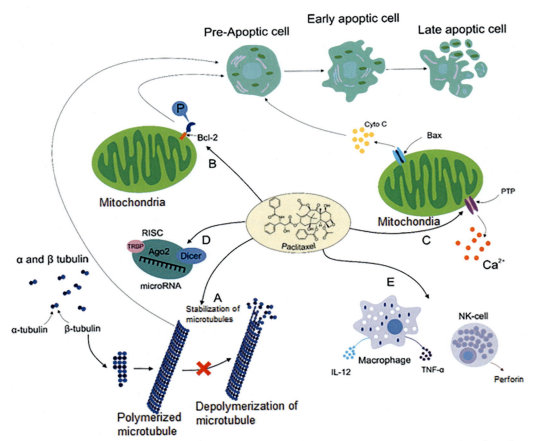

Fig. 2.2 Mechanism of action of taxol. Anti-tumor mechanism of action of taxol leading to stabilization of microtubule, cell arrest, and subsequent apoptosis (A). Taxol also causes activation of the immune response contributing to tumor eradication (B). The ability of taxol to inactivate Bcl-2 via phosphorylation of the anti-apoptotic protein resulting in apoptosis (C). Participation of taxol in the regulation of certain miRNAs associated with the modulation of tumor progression (D). Regulation of calcium signaling by taxol results in taxol-induced release of cytochrome C from the mitochondria and programmed cell death. Adapted from Abu Samaan, T.M., Samec, M., Liskova, A., Kubatka, P., Büsselberg, D., 2019. Paclitaxel's mechanistic and clinical effects on breast cancer. Biomolecules 9 (12), 789. https://doi.org/10.3390/biom9120789.

2.6 Conclusion and future prospects

Taxol is approved by the FDA to treat different types of cancers, including breast, prostate, ovarian and lung cancers. It was widely isolated from plant sources, especially *T. brevifolia* and other related species. However, its yield from plant sources has certain limitations, including low yield, loss of natural resources and other environmental issues. Presently, it is also widely extracted from alternative sources, such as microbial endophytes and plant cell culture approaches. In addition to the microbiological route, synthetic procedures are adopted to meet the global demand. Further, applications of plant metabolic engineering strategies together with simple and cost-effective methods for extracting and purifying taxol from microbial cells, plant cells/organs/tissues should be of more priority in the present time, and also more importance has to be given in future too. Furthermore, bioprospecting of microbial species and plant species should be encouraged to identify and produce taxol to fulfill the supply crisis. Likewise, research efforts should focus on improving or designing novel technological approaches to produce taxol in large quantities to meet to ever increasing worldwide demand for taxol. Likewise, various kinds of carrier molecules, including nanoparticles should be used appropriately to increase the effectiveness of taxol to cure cancers. Also, improving solubility and stability of taxol, safety aspects and different formulations of taxol are other areas of research interest to be considered in future. Taxol affects cancer cells by various mechanisms of action, such as Bcl-2 phosphorylation, modulating microtubules polymerization, mitochondrial calcium ion concentrations influx or efflux, regulation of miRNAs expression cascades, etc. Further, studies have suggested the possibilities of taxol directly impacting on the immune systems during carcinogenesis. However, more investigations are needed to understand various molecular mechanisms of action to improve the present approach of taxol chemotherapy against different cancer types.

References

Abu Samaan, T.M., Samec, M., Liskova, A., Kubatka, P., Büsselberg, D., 2019. Paclitaxel's mechanistic and clinical effects on breast cancer. Biomolecules 9 (12), 789. https://doi.org/10.3390/biom9120789.

Akhtar, M.S., Swamy, M.K., 2018a. Anticancer Plants: Natural Products and Biotechnological Implements. vol. 2 Springer, https://doi.org/10.1007/978-981-10-8064-7.

Akhtar, M.S., Swamy, M.K., 2018b. Anticancer Plants: Properties and Application. vol. 1 Springer, https://doi.org/10.1007/978-981-10-8548-2.

Alexandre, J., Hu, Y., Lu, W., Pelicano, H., Huang, P., 2007. Novel action of paclitaxel against cancer cells: bystander effect mediated by reactive oxygen species. Cancer Res. 67 (8), 3512–3517.

Asghari, F., Haghnavaz, N., Shanehbandi, D., Khaze, V., Baradaran, B., Kazemi, T., 2018. Differential altered expression of let-7a and miR-205 tumor-suppressor miRNAs in different subtypes of breast cancer under treatment with Taxol. Adv. Clin. Exp. Med. 27 (7), 941–945.

Blagosklonny, M.V., Giannakakou, P., El-Deiry, W.S., Kingston, D.G., Higgs, P.I., Neckers, L., Fojo, T., 1997. Raf-1/bcl-2 phosphorylation: a step from microtubule damage to cell death. Cancer Res. 57 (1), 130–135.

Blajeski, A.L., Kottke, T.J., Kaufmann, S.H., 2001. A multistep model for paclitaxel-induced apoptosis in human breast cancer cell lines. Exp. Cell Res. 270 (2), 277–288.

Bonomi, P., Kim, K., Kugler, J., 1997. Comparison of survival for stage IIIB versus stage IV non-small cell lung cancer (NSCLC) patients with etoposide-cisplatin versus taxol-cisplatin: an Eastern Cooperative Oncology Group (ECOG) trial. In: Proceedings of the American Society of Clinical Oncology, p. 454.

Chen, J.-G., Horwitz, S.B., 2002. Differential mitotic responses to microtubule-stabilizing and-destabilizing drugs. Cancer Res. 62 (7), 1935–1938.

De Brabander, M., Geuens, G., Nuydens, R., Willebrords, R., De Mey, J., 1981. Taxol induces the assembly of free microtubules in living cells and blocks the organizing capacity of the centrosomes and kinetochores. Proc. Natl. Acad. Sci. 78 (9), 5608–5612.

Emens, L.A., Jaffee, E.M., 2005. Leveraging the activity of tumor vaccines with cytotoxic chemotherapy. Cancer Res. 65 (18), 8059–8064.

Fridlender, M., Kapulnik, Y., Koltai, H., 2015. Plant derived substances with anti-cancer activity: from folklore to practice. Front. Plant Sci. 6, 799. https://doi.org/10.3389/fpls.2015.00799.

Ganguly, A., Yang, H., Cabral, F., 2010. Paclitaxel-dependent cell lines reveal a novel drug activity. Mol. Cancer Ther. 9 (11), 2914–2923.

Ganguly, A., Yang, H., Pedroza, M., Bhattacharya, R., Cabral, F., 2011. Mitotic centromere-associated kinesin (MCAK) mediates paclitaxel resistance. J. Biol. Chem. 286 (42), 36378–36384.

Giannakakou, P., Robey, R., Fojo, T., Blagosklonny, M.V., 2001. Low concentrations of paclitaxel induce cell type-dependent p53, p21 and G1/G2 arrest instead of mitotic arrest: molecular determinants of paclitaxel-induced cytotoxicity. Oncogene 20 (29), 3806–3813.

Goodman, J., Walsh, V., 2001. The Story of Taxol: Nature and Politics in the Pursuit of an Anti-Cancer Drug. Cambridge University Press.

Guenard, D., Gueritte-Voegelein, F., Dubois, J., Potier, P., 1993. Structure-activity relationships of Taxol and Taxotere analogues. J. Natl. Cancer Inst. Monogr. 15, 79–82.

Guo, B., Kai, G., Jin, H., Tang, K., 2006. Taxol synthesis. African J. Biotech. 5 (1), 15–20.

Hadzic, T., Aykin-Burns, N., Zhu, Y., Coleman, M.C., Leick, K., Jacobson, G.M., Spitz, D.R., 2010. Paclitaxel combined with inhibitors of glucose and hydroperoxide metabolism enhances breast cancer cell killing via H_2O_2-mediated oxidative stress. Free Radic. Biol. Med. 48 (8), 1024–1033.

Holton, R.A., Biediger, R.J., Boatman, P.D., 1995. Semisynthesis of Taxol and Taxotere. CRC Press, Boca Raton, FL.

Holton, R.A., Kim, H.B., Somoza, C., Liang, F., Biediger, R.J., Boatman, P.D., Shindo, M., Smith, C.C., Kim, S., 1994. First total synthesis of taxol. 2. Completion of the C and D rings. J. Am. Chem. Soc. 116 (4), 1599–1600.

Hornick, J.E., Bader, J.R., Tribble, E.K., Trimble, K., Breunig, J.S., Halpin, E.S., Vaughan, K.T., Hinchcliffe, E.H., 2008. Live-cell analysis of mitotic spindle formation in taxol-treated cells. Cell Motil. Cytoskeleton 65 (8), 595–613.

Isah, T., 2015. Natural sources of taxol. Br. J. Pharm. Res. 6 (4), 214–227.

Isah, T., 2016. Anticancer alkaloids from trees: development into drugs. Pharmacogn. Rev. 10 (20), 90–99.

Javeed, A., Ashraf, M., Riaz, A., Ghafoor, A., Afzal, S., Mukhtar, M.M., 2009. Paclitaxel and immune system. Eur. J. Pharm. Sci. 38 (4), 283–290.

John, J., Ismail, M., Riley, C., Askham, J., Morgan, R., Melcher, A., Pandha, H., 2010. Differential effects of Paclitaxel on dendritic cell function. BMC Immunol. 11 (1), 14. https://doi.org/10.1186/1471-2172-11-14.

Jordan, M.A., Toso, R.J., Thrower, D., Wilson, L., 1993. Mechanism of mitotic block and inhibition of cell proliferation by taxol at low concentrations. Proc. Natl. Acad. Sci. 90 (20), 9552–9556.

Jordan, M.A., Wilson, L., 1998. Microtubules and actin filaments: dynamic targets for cancer chemotherapy. Curr. Opin. Cell Biol. 10 (1), 123–130.

Jordan, M.A., Wilson, L., 2004. Microtubules as a target for anticancer drugs. Nat. Rev. Cancer 4 (4), 253–265.

Kampan, N.C., Madondo, M.T., McNally, O.M., Quinn, M., Plebanski, M., 2015. Paclitaxel and its evolving role in the management of ovarian cancer. BioMed Res. Intern. 2015. https://doi.org/10.1155/2015/413076.

Kidd, J.F., Pilkington, M.F., Schell, M.J., Fogarty, K.E., Skepper, J.N., Taylor, C.W., Thorn, P., 2002. Paclitaxel affects cytosolic calcium signals by opening the mitochondrial permeability transition pore. J. Biol. Chem. 277 (8), 6504–6510.

Kingston, D.G., 1994. Taxol: the chemistry and structure-activity relationships of a novel anticancer agent. Trends Biotechnol. 12 (6), 222–227.

Kubo, M., Morisaki, T., Matsumoto, K., Tasaki, A., Yamanaka, N., Nakashima, H., Kuroki, H., Nakamura, K., Nakamura, M., Katano, M., 2005. Paclitaxel probably enhances cytotoxicity of natural killer cells against breast carcinoma cells by increasing perforin production. Cancer Immunol. Immunother. 54 (5), 468–476.

Kundu, S., Jha, S., Ghosh, B., 2017. Metabolic engineering for improving production of taxol. In: Jha, S. (Ed.), Transgenesis and Secondary Metabolism. Springer, Cham, pp. 463–484, https://doi.org/10.1007/978-3-319-28669-3_29.

Larionova, I., Cherdyntseva, N., Liu, T., Patysheva, M., Rakina, M., Kzhyshkowska, J., 2019. Interaction of tumor-associated macrophages and cancer chemotherapy. Oncoimmunology 8 (7), e1596004.

Lichota, A., Gwozdzinski, K., 2018. Anticancer activity of natural compounds from plant and marine environment. Int. J. Mol. Sci. 19 (11), 3533.

Ling, Y.-H., Yang, Y., Tornos, C., Singh, B., Perez-Soler, R., 1998. Paclitaxel-induced apoptosis is associated with expression and activation of c-Mos gene product in human ovarian carcinoma SKOV3 cells. Cancer Res. 58 (16), 3633–3640.

Long, B.H., Fairchild, C.R., 1994. Paclitaxel inhibits progression of mitotic cells to G1 phase by interference with spindle formation without affecting other microtubule functions during anaphase and telephase. Cancer Res. 54 (16), 4355–4361.

Löwe, J., Li, H., Downing, K., Nogales, E., 2001. Refined structure of $\alpha\beta$-tubulin at 3.5 Å resolution. J. Mol. Biol. 313 (5), 1045–1057.

McDaid, H.M., Horwitz, S.B., 2001. Selective potentiation of paclitaxel (taxol)-induced cell death by mitogen-activated protein kinase kinase inhibition in human cancer cell lines. Mol. Pharmacol. 60 (2), 290–301.

McGuire, W.P., Rowinsky, E.K., Rosenshein, N.B., Grumbine, F.C., Ettinger, D.S., Armstrong, D.K., Donehower, R.C., 1989. Taxol: a unique antineoplastic agent with significant activity in advanced ovarian epithelial neoplasms. Ann. Intern. Med. 111 (4), 273–279.

Naik, B.S., 2019. Developments in taxol production through endophytic fungal biotechnology: a review. Orient Pharm Exp Med 19 (1), 1–13. https://doi.org/10.1007/s13596-018-0352-8.

Nicolaou, K., Yang, Z., Liu, J.J., Ueno, H., Nantermet, P., Guy, R., Claiborne, C., Renaud, J., Couladouros, E., Paulvannan, K., 1994. Total synthesis of taxol. Nature 367 (6464), 630–634.

Nikolic, V.D., Savic, I.M., Savic, I.M., Nikolic, L.B., Stankovic, M.Z., Marinkovic, V.D., 2011. Paclitaxel as an anticancer agent: isolation, activity, synthesis and stability. Central Eur. J. Med. 6 (5), 527–536.

Ofir, R., Seidman, R., Rabinski, T., Krup, M., Yavelsky, V., Weinstein, Y., Wolfson, M., 2002. Taxol-induced apoptosis in human SKOV3 ovarian and MCF7 breast carcinoma cells is caspase-3 and caspase-9 independent. Cell Death Differ. 9 (6), 636–642.

Pan, Z., Avila, A., Gollahon, L., 2014. Paclitaxel induces apoptosis in breast cancer cells through different calcium—regulating mechanisms depending on external calcium conditions. Int. J. Mol. Sci. 15 (2), 2672–2694.

Perdue, R.E., 1969. Search for plant sources of anticancer drugs. Morris Arboretum Bull. 20, 35–58.

Pfannenstiel, L.W., Lam, S.S., Emens, L.A., Jaffee, E.M., Armstrong, T.D., 2010. Paclitaxel enhances early dendritic cell maturation and function through TLR4 signaling in mice. Cell. Immunol. 263 (1), 79–87.

Piccart, M., Cardoso, F., 2003. Progress in systemic therapy for breast cancer: an overview and perspectives. Eur. J. Cancer Suppl. 1 (2), 56–69.

Rao, S., Orr, G.A., Chaudhary, A.G., Kingston, D.G., Horwitz, S.B., 1995. Characterization of the taxol binding site on the microtubule 2-(m-azidobenzoyl) taxol photolabels a peptide (amino acids 217-231) of β-tubulin. J. Biol. Chem. 270 (35), 20235–20238.

Renneberg, R., 2007. Biotech history: yew trees, paclitaxel synthesis and fungi. Biotech. J. Healthc. Nutr. Tech. 2 (10), 1207–1209.

Schiff, P.B., Fant, J., Horwitz, S.B., 1979. Promotion of microtubule assembly in vitro by taxol. Nature 277 (5698), 665–667.

Schiff, P.B., Horwitz, S.B., 1980. Taxol stabilizes microtubules in mouse fibroblast cells. Proc. Natl. Acad. Sci. 77 (3), 1561–1565.

Shetti, D., Zhang, B., Fan, C., Mo, C., Lee, B.H., Wei, K., 2019. Low dose of paclitaxel combined with XAV939 attenuates metastasis, angiogenesis and growth in breast cancer by suppressing Wnt signaling. Cells 8 (8), 892. https://doi.org/10.3390/cells8080892.

Shimizu, S., Matsuoka, Y., Shinohara, Y., Yoneda, Y., Tsujimoto, Y., 2001. Essential role of voltage-dependent anion channel in various forms of apoptosis in mammalian cells. J. Cell Biol. 152 (2), 237–250.

Shimizu, S., Narita, M., Tsujimoto, Y., 1999. Bcl-2 family proteins regulate the release of apoptogenic cytochrome c by the mitochondrial channel VDAC. Nature 399 (6735), 483–487.

Slichenmyer, W.J., Von Hoff, D.D., 1991. Taxol: a new and effective anti-cancer drug. Anticancer Drugs 2 (6), 519–530.

Stierle, A., Strobel, G., Stierle, D., 1993. Taxol and taxane production by *Taxomyces andreanae*, an endophytic fungus of Pacific yew. Science 260 (5105), 214–216.

Szakács, G., Paterson, J.K., Ludwig, J.A., Booth-Genthe, C., Gottesman, M.M., 2006. Targeting multidrug resistance in cancer. Nat. Rev. Drug Discov. 5 (3), 219–234.

Tao, W.-Y., Liang, X.-S., Liu, Y., Wang, C.-Y., Pang, D., 2015. Decrease of let-7f in low-dose metronomic paclitaxel chemotherapy contributed to upregulation of thrombospondin-1 in breast cancer. Int. J. Biol. Sci. 11 (1), 48–58.

Torres, K., Horwitz, S.B., 1998. Mechanisms of Taxol-induced cell death are concentration dependent. Cancer Res. 58 (16), 3620–3626.

Tran, T.-A., Gillet, L., Roger, S., Besson, P., White, E., Le Guennec, J.-Y., 2009. Non-antimitotic concentrations of taxol reduce breast cancer cell invasiveness. Biochem. Biophys. Res. Commun. 379 (2), 304–308.

Tuma, R.S., 2003. Taxol's journey from discovery to use: lessons & updates. Oncol. Times 25 (18), 52–57.

Varbiro, G., Veres, B., Gallyas Jr., F., Sumegi, B., 2001. Direct effect of Taxol on free radical formation and mitochondrial permeability transition. Free Radic. Biol. Med. 31 (4), 548–558.

Vidensek, N., Lim, P., Campbell, A., Carlson, C., 1990. Taxol content in bark, wood, root, leaf, twig, and seedling from several Taxus species. J. Nat. Prod. 53 (6), 1609–1610.

Vyas, D.M., Kadow, J.F., 1995. 6. Paclitaxel: a unique tubulin interacting anticancer agent. In: Progress in Medicinal Chemistry. vol. 32. Elsevier, pp. 289–337.

Wall, M.E., Wani, M.C., 1995. Camptothecin and taxol: discovery to clinic—thirteenth Bruce F. Cain Memorial Award Lecture. Cancer Res. 55 (4), 753–760.

Walsh, V., Goodman, J., 1999. Cancer chemotherapy, biodiversity, public and private property: the case of the anti-cancer drug Taxol. Soc. Sci. Med. 49 (9), 1215–1225.

Walsh, V., Goodman, J., 2002a. The billion dollar molecule: taxol in historical and theoretical perspective. In: Biographies of Remedies. Brill Rodopi, pp. 245–267.

Walsh, V., Goodman, J., 2002b. From Taxol to Taxol®: the changing identities and ownership of an anti-cancer drug. Med. Anthropol. 21 (3-4), 307–336.

Wanderley, C.W., Colon, D.F., Luiz, J.P.M., Oliveira, F.F., Viacava, P.R., Leite, C.A., Pereira, J.A., Silva, C.M., Silva, C.R., Silva, R.L., 2018. Paclitaxel reduces tumor growth by reprogramming tumor-associated macrophages to an M1 profile in a TLR4-dependent manner. Cancer Res. 78 (20), 5891–5900.

Wang, F., Cao, Y., Zhao, W., Liu, H., Fu, Z., Han, R., 2003. Taxol inhibits melanoma metastases through apoptosis induction, angiogenesis inhibition, and restoration of E-cadherin and nm23 expression. J. Pharmacol. Sci. 93 (2), 197–203.

Wang, T., Chan, Y., Chen, C., Kung, W., Lee, Y., Wang, S., Chang, T., Wang, H., 2006. Paclitaxel (Taxol) upregulates expression of functional interleukin-6 in human ovarian cancer cells through multiple signaling pathways. Oncogene 25 (35), 4857–4866.

Wani, M., 1972. Plant antitumor agents. VI. The isolation and structure of taxol, a novel antileukemic and antitumor agent from *Taxus brevifolia*. J. Am. Chem. Soc. 19, 2325–2326.

Waters, J.C., Chen, R.-H., Murray, A.W., Salmon, E., 1998. Localization of Mad2 to kinetochores depends on microtubule attachment, not tension. J. Cell Biol. 141 (5), 1181–1191.

Weaver, B.A., 2014. How Taxol/paclitaxel kills cancer cells. Mol. Biol. Cell 25 (18), 2677–2681. https://doi.org/10.1091/mbc.e14-04-0916.

Wheeler, N.C., Jech, K., Masters, S., Brobst, S.W., Alvarado, A.B., Hoover, A.J., Snader, K.M., 1992. Effects of genetic, epigenetic, and environmental factors on taxol content in *Taxus brevifolia* and related species. J. Nat. Prod. 55 (4), 432–440.

Wianowska, D., Hajnos, M.Ł., Dawidowicz, A.L., Oniszczuk, A., Waksmundzka-Hajnos, M., Głowniak, K., 2009. Extraction methods of 10-deacetylbaccatin III, paclitaxel, and cephalomannine from *Taxus baccata* L. twigs: a comparison. J. Liq. Chromatogr. Relat. Technol. 32 (4), 589–601.

Xu, Q., Trissel, L.A., Martinez, J.F., 1994. Stability of paclitaxel in 5% dextrose injection or 0.9% sodium chloride injection at 4, 22, or 32 C. Amer. J. Health-System Pharm. 51 (24), 3058–3060.

Zhang, D., Yang, R., Wang, S., Dong, Z., 2014. Paclitaxel: new uses for an old drug. Drug Design Dev. Ther. 8, 279–284.

Zhou, X., Zhu, H., Liu, L., Lin, J., Tang, K., 2010. A review: recent advances and future prospects of taxol-producing endophytic fungi. Appl. Microbiol. Biotechnol. 86 (6), 1707–1717.

3

Taxol: Mechanisms of action against cancer, an update with current research

Pei Tee Lim[a], Bey Hing Goh[b,c], and Wai-Leng Lee[a]

[a]School of Science, Monash University Malaysia, Subang Jaya, Malaysia,
[b]College of Pharmaceutical Sciences, Zhejiang University, Hangzhou, China,
[c]Biofunctional Molecule Exploratory (BMEX) Research Group, School of Pharmacy, Monash University Malaysia, Subang Jaya, Malaysia

3.1 The discovery and evolution of Paclitaxel (Taxol)

Paclitaxel or better known as Taxol is anti-mitotic drug classified under taxane drugs that originates from Pacific yew tree, *Taxus brevifolia*, in the 1960s (Weaver, 2014). The drug was discovered from a plant-screening program for new anti-cancer agents and it was the only plant that was enlisted into clinical trial afterward (Kampan et al., 2015). In 1978, Taxol was found to exert tumor regression in mammalian tumor xenograft which subsequently prompted the discovery of the drug's unique mechanism of action which targets the microtubules assembly by Dr Susan Horwitz at Albert Einstein Medical College in 1979 (Martin, 1993). Finally, in 1992, Taxol was approved and registered by The Food and Drug Administration (FDA) for treating ovarian cancer in 1992 and for breast cancer in 1994 (Kampan et al., 2015). It is now used either as a single chemotherapeutic agent or combined with other chemo-drugs for treating ovarian cancer, breast cancer and non-small-cell lung cancer (Kampan et al., 2015; Abu Samaan et al., 2019). Taxol comprises a homochiral ester side chain at C13 that acts as the active portion for microtubules binding (Kampan et al., 2015).

Paclitaxel. https://doi.org/10.1016/B978-0-323-90951-8.00007-2
Copyright © 2022 Elsevier Inc. All rights reserved.

3.2 Paclitaxel (Taxol) induces mitotic cell cycle arrest

Taxol targets microtubules and causes its polymerization and stabilization to disrupt mitotic cell in living cancer cells (Weaver, 2014). Microtubules are built up by tubulin into long and filamentous polymers in a dynamic equilibrium consisting alpha and beta protein subunits (Lee et al., 2012). Microtubules are important elements of the mitotic spindle apparatus in many critical cellular functions like cell mobility, attachment, maintaining cell shape, and cell transport (Martin, 1993). In addition, it is important in the partitioning of replicated chromosomes during mitosis as the microtubule dynamicity regulates the precise attachment of kinetochores on chromosomes to the spindle at pro-metaphase, and the proper movements and alignment of chromosomes to metaphase plate, and finally the synchronized separation in anaphase followed by telophase checkpoint (Mukhtar et al., 2014).

3.2.1 Taxol induces microtubules stabilization

Taxol is the drug which exhibits this mechanism by disrupting the microtubules dynamic and induces cell apoptosis. It has a binding pocket in β-tubulin, which binds to GTP molecule, the hydrolysis of GTP allows depolymerization of microtubules (Yusuf et al., 2003). Taxol promotes the conformational changes in M-loop of β-tubulin, which results in stable lateral interactions between proto-filaments, leading to prevention in the microtubule depolymerization or also known as microtubules stabilization (Lee et al., 2012; Mukhtar et al., 2014). The failure of chromosomes to attach with microtubules halts the cell from proceeding to the next phase, disrupts the mitotic spindle assembly which induces spindle assemble checkpoint (SAC), causing mitotic arrest which eventually will end up in apoptosis (Huang et al., 2009; Mukhtar et al., 2014). It is assumed that the consequence of microtubule dysfunctionality is the G2/M phase arrest, which is needed for cell death in cancer (Mukhtar et al., 2014). A summary of the cell cycle arrest caused by Taxol is shown in Fig. 3.1.

3.2.2 Mitotic slippage

However, cancer cells tend to resist the apoptotic effect of Taxol by escaping the mitotic arrest and induce a premature exit from the mitosis process before the cell apoptosis happens (Cheng and Crasta, 2017). This escape plan of the cancer cells is known as mitotic slippage which the cells exits mitosis and "slip" into tetraploid G1 phase as there is no proper chromosome segregation and cytokinesis (Huang et al., 2009;

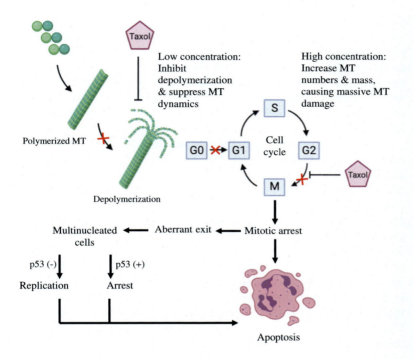

Fig. 3.1 Mechanism of action of Taxol in microtubules. At low concentration, Taxol which targets the β-tubulin inhibits the depolymerization of microtubules and suppress its dynamics, leading to cell cycle arrest at G0/G1 or G2/M phase. At higher concentration, Taxol causes massive microtubules damage which leads to cell cycle arrest, and eventually cell apoptosis. In some circumstances when the mitotic spindle checkpoint could not sustain the arrest, aberrant exit or mitotic slippage happens, causing production of multinucleated cells which eventually will undergo cell apoptosis. MT: microtubules; G0: resting phase, G1; cells enlarge and make new protein, S phase: DNA replication, G2: preparation for division, M phase: cell division/mitosis.

Cheng and Crasta, 2017). The subsequent event of mitotic slippage is either the cells got arrested in G1 phase, post-slippage cell death, or continue the cell cycle but in a genomically unstable mode (Cheng and Crasta, 2017). This highlights that the entry of cells into mitosis is a prerequisite event for Taxol killing effect but the apoptosis is not limited to be occurring from G2/M phase arrest only (Mukhtar et al., 2014).

3.2.3 Paclitaxel's effect is dose-dependent

Taxol had been claimed to exert different mitotic effects in low concentrations, it produces aneuploidy cells in the absence of mitotic block (Orr et al., 2003). In vitro studies suggested that Taxol's mechanism is dose-dependent, and the Taxol-induced cell death could be independent from mitotic arrest (Fan, 1999; Giannakakou et al., 2001). For example, Giannakakou et al. (2001) found that in low concentration, Taxol utilized other pathways to inhibit cell proliferation without arresting mitosis. They showed that in A549 cells, at very low concentration of Taxol (3–6 nM) it is insufficient to inhibit mitotic cycle but it induced p53 and p21 proteins to cause the G1 and G2 arrest instead (Giannakakou et al., 2001). But at higher concentrations (100 nM) Taxol will predominantly activate G2/M phase arrest. Additionally, another study found that Taxol-mediated cell death was triggered without mitotic arrest as long as the cells were exposed to Taxol for 1 h, the apoptosis was not necessarily a result from mitotic arrest and two events could occur independently (Fan, 1999). They justified that the apoptotic events require longer time to induce as compared to mitotic arrest hence it always seem to be occurring immediately following mitotic arrest (Fan, 1999). This had shown that Taxol exerts different effects depending on the dose, time as well as the cell types.

However, the mechanism of action adopted by Taxol at low or high concentrations remains confusing as there was report saying that low concentration of Taxol (20 nM) induced microtubule dynamics suppression and G2/M phase arrest (Wang et al., 2000). This did not compliment with the previous findings which indicated that low Taxol concentration activates cell apoptosis without causing mitotic arrest (Fan, 1999; Giannakakou et al., 2001). A suggested pathway for the Taxol mechanism at lower concentration is by arresting G2/M phase of cell cycle, causing cell apoptosis, while in some circumstances Taxol could inhibit the mitotic spindle checkpoint and lead to abnormal early exit from the mitotic arrest and this will cause formation of multinucleated cells (Long and Fairchild, 1994; Wang et al., 2000). This phenomenon is similar with the 'mitotic slippage' as discussed earlier. Then the multinucleated cells either undergo multiple round of DNA replication and eventually die if the cells contain mutant p53 or they would undergo cell arrest and apoptosis directly if they are p53-positive cells, as shown in Fig. 3.1 (Sorger et al., 1997).

On the other hand, high dose of Taxol had been shown to modulate the microtubule dynamics differently and cause massive microtubule damage (Wang et al., 2000). At higher dosage, instead of inhibiting the microtubules depolymerization, Taxol increases microtubules mass and number to cause its stabilization (Kampan et al., 2015). Additionally, the high Taxol dosage was also observed to regulate certain gene expressions or signaling pathways like Bcl-2 proteins (BAK, BAX), p34, JNK/SAPK pathway and NFκB pathway which will be discussed in the following sections.

3.3 Taxol induces gene-directed apoptosis

Several apoptosis-related or survival signaling genes were shown to be activated by Taxol for instance JNK, p34, NFκB, tumor necrosis factor-α (TNFα) and Bcl-2 proteins (Fan, 1999; Giannakakou et al., 2001; Kampan et al., 2015; Whitaker and Placzek, 2019). One of the gene-directed pathways that was extensively discussed for Taxol's mechanism of action is the Bcl-2 family of proteins which are apoptotic regulators that control cell survival (Srivastava et al., 1999).

3.3.1 Roles of Bcl-2 family in Taxol-induced apoptosis

It was reported that phosphorylation by Taxol on serine residues of Bcl-2 induces its inactivation and leads to inhibition of its anti-apoptotic effect (Haldar et al., 1995; Haldar et al., 1996; Kroning and Lichtenstein, 1998). Proteins in the Bcl-2 family is divided into pro-apoptotic and anti-apoptotic, while pro-apoptotic Bcl-2 members can be further divided into BH-domain effector proteins and BH3-only proteins (Campbell and Tait, 2018). Examples of anti-apoptotic Bcl-2 proteins are Bcl-X, MCL-1, BFL1 and Bcl-2 proteins which are differentially expressed in different cell types or tissues (Campbell and Tait, 2018). On the other hand, BIM, BAD, BID are examples of BH3-only proteins whereas BAX, BAK are examples of BH-domain effector proteins which have similar three-dimensional conformation with pro-survival Bcl-2 protein (Campbell and Tait, 2018). The Bcl-2 family resides predominantly in mitochondria membranes but was also seen on membranes for endoplasmic reticulum and nucleus (Scatena et al., 1998). The crosstalk between the pro-apoptotic and anti-apoptotic Bcl-2 proteins has been implicated in the biochemical pathway for cell apoptosis induced by Taxol (Whitaker and Placzek, 2019).

In the presence of microtubule inhibiting agent like Taxol, Bcl-2 will be phosphorylated at serine residues (Scatena et al., 1998). The phosphorylation of Bcl-2 causes its inactivation and this allows BAX or BAK protein to activate mitochondrial outer membrane permeabilization and opening of pores (Scatena et al., 1998; Miller et al., 2013; Campbell and Tait, 2018). The formation of pores or channels on mitochondrial membrane or better known as mitochondrial permeability transition pore (mPTP) allows the release of cytochrome C which in turn activates caspases to initiate apoptosis (Scatena et al., 1998; Campbell and Tait, 2018). The dimerization of Bcl-2 with BAX will prevent BAX-mediated apoptosis thus the phosphorylation of Bcl-2 by Taxol causes its dissociation from BAX, allowing BAX to induce apoptosis (Scatena et al., 1998). It was found that Taxol increased cell apoptosis and phosphorylation of Bcl-2, but the dimerization of Bcl-2/BAX complex was diminished in Bcl-2 positive cells (Haldar et al., 1996). This highlighted that Bcl-2 is a negative regulator for BAX protein and its inhibition is a potential target for sensitizing the Taxol cytotoxicity in cell apoptosis. The Taxol-induced Bcl-2 phosphorylation hence was said to be critical for Taxol-induced cell death.

However, more recent studies demonstrated that the subgroup of Bcl-2 family, BH3-only proteins which are another group of pro-apoptotic proteins, play more significant role in the taxane-induced apoptosis pathway (Sunters et al., 2003; Janssen et al., 2007; Kutuk and Letai, 2010). The crosstalk between pro-survival Bcl-2 proteins (Bcl-2,

Bcl-xL, MCL-1) with pro-apoptotic Bcl-2 members (BAX, BAK) as well as activator BH3-only proteins (BIM, BID, Puma) is inter-related among each other in the regulation of Taxol-induced apoptotic mechanism (Janssen et al., 2007; Kutuk and Letai, 2010). This was shown by the delay or attenuation of Taxol-induced apoptosis when BIM was downregulated by siRNA in cell based models (Miller et al., 2013). Additionally, the knockdown of Bcl-2 in breast cancer and NSCLC cell lines did not show significant difference in Taxol-induced killing as compared to the effect of BIM knockdown by siRNA (Li et al., 2005). In contrast to Bcl-2, BIM is a tumor suppressor, and it activates and acts upstream of BAX and BAK protein (Kutuk and Letai, 2010). BIM could either antagonize the pro-survival Bcl-2 or directly activate the dimerization of BAX and BAK which leads to the induction of apoptotic cascade via mitochondrial outer membrane permeabilization (Tan et al., 2005; Czernick et al., 2009; Gogada et al., 2013).

The inactivation of Bcl-2 in triggering cell apoptosis has been contradicting as some suggested it is a critical event but there are studies who suggested that it could merely be a bystander event (Yamamoto et al., 1999; Li et al., 2005). Besides the increase of pro-apoptotic or decrease of anti-apoptotic Bcl-2 proteins, another novel mechanism involving the serial displacement on BH3-only proteins was found to be responsible for the cell death by Taxol (Kutuk and Letai, 2010). Another group of BH3-only proteins called sensitizers comprises Bad, Noxa, Bik, Puma and Bmf does not have the function to directly activate BAX or BAK but could indirectly activate them instead. In a breast cancer study, a displacement mechanism was proposed to be a novel mechanism of action for Taxol, Bmf which is a BH3-only protein was seen to displace BIM from the anti-apoptotic proteins (Bcl-2 or MCL-1) to promote cell death (Kutuk and Letai, 2010). They found that the level of BIM in breast cancer cells is not the key factor to triggering cell apoptosis but the displacement of BIM is rather a significant step instead (Kutuk and Letai, 2010).

3.3.2 Taxol activates JNK/SAPK to promote cell apoptosis

Other than its role in Bcl-2 family, Taxol is known to activate c-Jun N-terminal kinase (JNK) which serves as an important step for cell death (Akimov and Belkin, 2001; Jo et al., 2016). Bcl-2 family was correlated to JNK-dependent apoptosis as it is one of the JNK substrates (Jo et al., 2016). It was suggested that the activation of JNK/stress-activated protein kinase (SAPK) pathway by Taxol mediates the cell death in a dose- and time-dependent manner, through its regulation on Bcl-2 phosphorylation (Lee et al., 1998; Wang et al., 1998; Yamamoto et al., 1999; Wang et al., 2000). However it had been deduced that there is none common or typical apoptotic pathway as it still depends on cell type, time or dosage (Kolomeichuk et al., 2008). There is contradictory finding regarding the significance and necessity of activated JNK/SAPK in Bcl-2 phosphorylation.

In ovarian and leukemia cancer cell lines, the Bcl-2 phosphorylation was found to be independent from the activation of JNK pathway (Attalla et al., 1998; Wang et al., 1998). However, multiple evidence indicated that Taxol activated JNK signaling pathway is necessary to induce Bcl-2 inactivation and subsequent cell apoptosis (Lee et al., 1998; Yamamoto et al., 1999; Kolomeichuk et al., 2008; Jo et al., 2016). Commonly, Taxol will activate JNK which leads to phosphorylation of Bcl-2, followed by its dissociation from BAX and translocation of BAX to mitochondria to induce mPTP and cell apoptosis as shown in Fig. 3.2 (Jo et al., 2016). But the signaling proteins located downstream of JNK could differ among different cell types, for example JNK-dependent cell apoptosis could be activator protein-1 (AP-1)-dependent or independent (Lee et al., 1998; Kolomeichuk et al., 2008). Otherwise, it could be mediated by apoptosis signaling regulating kinase 1 (ASK1) which leads to caspase-independent but apoptotic inducing factor (AIF)-dependent apoptosis in human ovarian cancer (SKOV3) cells (Jo et al., 2016).

In overall, the JNK activation and Bcl-2 phosphorylation still acts as a common feature in Taxol-induced cell death although the interrelation between both is an issue of great complexity (Lee et al., 1998; Wang et al., 2000). Conclusively, the role of members of the Bcl-2 family is less updated so far and the former studies did not provide a precise mechanism on how these apoptosis-related proteins affect Taxol's efficacy. Notably, different categories of Bcl-2 proteins possess distinguished roles depending on the cell lines thus it is challenging to conclude one finalized mechanism, hence further investigation into this is much needed.

3.4 Taxol and calcium-dependent apoptosis

It is well established that internal calcium homeostasis is significant for normal cell survival and regulation of cell physiology (Baggott et al., 2012; Pan et al., 2014). It was suggested that Ca^{2+} acts as a double-sided sword in the regulation of cell proliferation as it could trigger mitotic division in certain cell types but also triggers cell death (Pinton et al., 2008). Impaired calcium homeostasis causes toxicity to cells and leads to cell death (Pan et al., 2014). Most of the chemotherapeutic agents induce cell death mainly by inducing cell apoptosis, which could be activated by a multi-pathway process (Pan and Gollahon, 2013). There are two major signaling pathways involved in the apoptotic process.

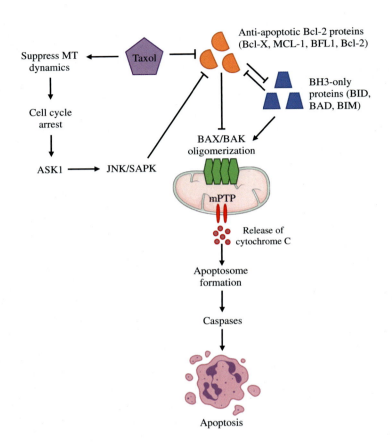

Fig. 3.2 Interaction of Bcl-2 family proteins regulates mitochondria outer membrane permeabilization (MOMP) to induce apoptosis. Bcl-2 like proteins are anti-apoptotic and they inhibit BAX/BAK oligomerization to prevent cell apoptosis whereas BH3-only proteins are pro-apoptotic which promote BAX/BAK oligomerization. The activation of BAX/BAK allows MOMP and opening of mitochondria permeability transition pore (mPTP) which allows release of cytochrome C and apoptosome formation. This will attract caspases which activate cell apoptosis. ASK1/JNK pathway could also be activated upon cell cycle arrest which causes inactivation of Bcl-2 proteins and apoptotic cell death. MT: Microtubules; ASK1: apoptosis signaling regulating kinase 1; JNK/SAPK: c-Jun N-terminal kinase/Stress-activated protein kinase; mPTP: mitochondrial permeability transition pore.

Firstly, the intrinsic pathway which is marked by increase in mitochondrial Ca^{2+} and release of cytochrome c, leading to activation of caspases to initiate cell death (Varghese et al., 2019). Whereas the extrinsic apoptosis pathway involves the activation of death receptors or transmembrane receptors for example trimeric Fas ligand (FasL) and TNF that binds to Fas receptor or TNF receptor (Rathore et al., 2017). In fact, Bcl-2 proteins regulates the intrinsic cell apoptosis pathway through the modulation of Ca^{2+} homeostasis in the endoplasmic reticulum (ER), plasma membrane and mitochondria (Varghese et al., 2019). There is an important crosstalk between the Ca^{2+} store, mitochondria, and Bcl-2 proteins in the regulation of cell apoptosis.

3.4.1 ER-calcium dependent apoptosis

The mechanism on how Taxol influences the Ca^{2+} homeostasis in the cells remain controversial and there are a few mechanisms being suggested. Firstly, Taxol treatment was shown to produce ceramide or reactive oxygen species (ROS) from the mitochondrial matrix which could then activate the release of Ca^{2+} from the ER showed that Taxol is indirectly involved in the calcium-dependent apoptosis (Rao et al., 2004; Colina et al., 2005; Pan and Gollahon, 2011). Besides, it was suggested that Taxol can directly attack the internal ER calcium store to release apoptosis-activating calcium signals (Pan and Gollahon, 2011, 2013). Taxol could act on ER release channels like inositol 1,4,5-trisphosphate receptor (IP3R) which was reported to bind directly with Taxol to stimulate calcium release in neuronal cells (Boehmerle et al., 2006). The rapid calcium change then serves as a signal to activate specific caspase, sensitize mitochondria, bypassing the mitotic arrest and leading to an end effect of apoptosis (Pan and Gollahon, 2011). Besides, some calcium-dependent enzymes that regulate the apoptosis-related genes or proteins could indirectly affect the cell apoptosis too (Pan and Gollahon, 2011).

Another study indicated a close interaction between ER, mitochondria and microtubules which leads to interruption in the Ca^{2+} dynamics and causes apoptosis (Mironov et al., 2005). They suggested that the microtubules stabilized by Taxol would promote the opening of mPTP on mitochondria, and due to the close apposition between mitochondria and ER, the Ca^{2+} released via mPTP would trigger the spontaneous release of Ca^{2+} from ER (Mironov et al., 2005). The mechanism is known as Ca^{2+} induced Ca^{2+} release (CICR) (Mironov et al., 2005). However, according to the study, CICR was observed in neuronal cells where the mitochondria and ER are located in close proximity and its role in cancer cells remains undetermined (Mironov et al., 2005).

Also, low dose Taxol suppresses the microtubules dynamic to cause mitotic arrest but high dose Taxol cause massive microtubules damage and mitotic-arrest independent apoptosis (Wang et al., 2000). Therefore, this highlights that the cell apoptosis mechanism induced by Taxol is highly dependent on its dosage and Taxol-induced $ERCa^{2+}$ release could be a signal to bypass the mitotic arrest during high dosage treatment (Pan and Gollahon, 2011). Additional to that, ER-calcium dependent apoptosis could be the pathway that triggers the death of cells in the presence of high dose Taxol. A study found that the calcium changes in Taxol-induced apoptosis were only inducible by high dose treatment and long term exposure in breast cancer cells (Pan and Gollahon, 2013). The study found that high dosage of Taxol at 2.5 µM induced a rapid and significant increase in Ca^{2+} release from ER, followed by a cytoplasmic Ca^{2+} level increase, whereas low dosage at 0.2 µM did not cause a significant change (Pan and Gollahon, 2013). Moreover, they observed that rapid ER Ca^{2+} release induced by high dose Taxol promotes apoptosis while low dose Taxol did not (Pan and Gollahon, 2013).

3.4.2 Crosstalk between Bcl-2 and calcium homeostasis

Additionally, there is a controversy on the relationship between the action mechanism of Taxol in regards Bcl-2 proteins and calcium homeostasis. Bcl-2 could regulate the calcium homeostasis either toward anti-apoptotic or pro-apoptotic directions. There are a few models being hypothesized for the inhibition of Bcl-2 on calcium release from ER: (1) Bcl-2 expression increases sacro/endoplasmic reticulum Ca^{2+} ATPase (SERCA) (2) Bcl-2 inhibits IP3R on the ER and thus attenuating the initial calcium signal (3) Bcl-2 reduces the ER calcium content by increasing the calcium leaking from ER (Bonneau et al., 2013). There is evidence indicating that Bcl-2 could inhibit ER calcium release by targeting the IP3R on the ER and thus attenuating the initial calcium signal for the induction of apoptosis (Ferlini et al., 2009; Rong et al., 2009). This indicated that the presence of anti-apoptotic Bcl-2 proteins at the ER could attenuate Taxol-induced cell death (Pan et al., 2014).

On the contrary, pro-apoptotic Bcl-2 proteins are also able to localize at the ER and regulate the ER calcium changes (Bonneau et al., 2013). BAX and BAK were shown to mediate ER calcium release, which then sensitizes the mitochondria toward calcium-mediated cytochrome c release and an end result of apoptosis (Nutt et al., 2002a; Nutt et al., 2002b). But another study observed that the absence of BAX and BAK proteins in the cells enhanced the action of Bcl-2 on IP3R, promoting calcium leakage from ER while the presence of BAX or BAK can inhibit the anti-apoptotic Bcl-2 and in turn changing to a pro-apoptotic effect (Oakes et al., 2005; Bonneau et al., 2013). Conclusively, the Bcl-2 family plays a conflicting role in the ER Ca^{2+} homeostasis and Taxol-induced cell apoptosis.

Pan and Gollahon (2011) also demonstrated that Bcl-2 positive breast cancer cells showed higher resistance toward Taxol treatment and have higher releasable ER calcium as compared to Bcl-2 negative cells, indicating that Bcl-2 expression does have an inhibitory effect on apoptosis. However, upon Taxol treatment, ER calcium release was observed to be higher in Bcl-2 positive cells, suggesting that Taxol attacks ER store directly independent of Bcl-2 status (Pan and Gollahon, 2013). Based on the studies, it can be concluded that the dosage of Taxol plays a significant role in regulating its effect on the ER apoptotic pathway, while the expression of Bcl-2 suppresses and inhibits the ER calcium release to prevent cell apoptosis (Pan and Gollahon, 2011, 2013; Pan et al., 2014). However high dosage Taxol is able to act directly on ER calcium store and overcome the Bcl-2's inhibitory effect and thus counteracting the Bcl-2 resistance for Taxol (Pan and Gollahon, 2011, 2013; Pan et al., 2014). Despite the findings, it is hard to conclude a common mechanism of action for Taxol and co-regulation by Bcl-2 at the ER as the effects might not be applicable to all cell types, but it remains clear that ER calcium store is a common target for Taxol and Bcl-2 and both are influenced by the drug dosage (Pan and Gollahon, 2013).

3.5 Immunomodulation effects by Taxol

Immune system has been associated with cancer progression for a long time as immunity protects the human body against "nonself" antigens from unrecognized pathogens, infections or malignancies and it is supposed to destroy any of these invaders to protect the host (Pandya et al., 2016). The immune system comprising innate (macrophages, natural killers cells, dendritic cells) and adaptive mechanisms (mainly T-lymphocytes) contributes to the anti-tumor immunity in the cancer environment (Kampan et al., 2015). However, when the balance between immune surveillance and cancer progression is disrupted, the cancer cells escape immune response and progress to further cancer growth and metastasis (Pandya et al., 2016). As immune cells possess the ability to prevent tumor growth, recent scientific advances focus on the enhancement of immunological responses in cancer patients and cancer immunotherapy came into the light (Javeed et al., 2009).

Interestingly, Taxol which is initially being described as a mitotic inhibitor and microtubule stabilizer, was found to exert effect on host's immune system as well (Javeed et al., 2009; Kampan et al., 2015). Various studies had shown that Taxol has regulating effect on immune cells such as effector T cells (Teff) , regulatory T cells (Treg), macrophages, dendritic cells and others (Tsavaris et al., 2002; Javeed et al., 2009; Kampan et al., 2015; Zhu and Chen, 2019). Taxol treatment has both stimulatory and suppressive effects on the immune system, standard dose of Taxol is immunosuppressive and inhibits a group of immune cells involved in tumor elimination (Kampan et al., 2015; Demeckova et al., 2017). But lower dose of Taxol exerts an opposite effect and promotes anti-tumor immunity which stimulated its potential role in immunogenic effects (Javeed et al., 2009; Kampan et al., 2015). Thus, the understanding of the role of Taxol in immunomodulation could potentially provide an improved therapeutic regimen for cancer treatment.

3.5.1 Taxol and regulatory T cells (Tregs)

Regulatory T cells (Tregs) are a subpopulation of CD4[+] T cells that maintains the immune tolerance and defence against auto-immune disease (Smyth et al., 2006; Zhang et al., 2008). They suppress both innate and adaptive immune responses at the oncogenic environment thus inhibiting the anti-tumor effect (Smyth et al., 2006; Zhang et al., 2008; Javeed et al., 2009). There is increasing evidence showing the infiltration of Tregs into the tumorigenic site, showing that it might functionally inhibit tumor-specific effector T cells and lead to poor survival (Ichihara et al., 2003; Nishikawa et al., 2005; Ormandy et al., 2005; Zhang et al., 2008). Strikingly, cells receiving Taxol treatment had decrease in the number of Tregs and then enhance the anti-tumor immunity (Zhang et al., 2008; Javeed et al., 2009; Kampan et al., 2015).

In an analysis of peripheral blood samples from non-small cell lung cancer (NSCLC) patients undergoing Taxol-based chemotherapy, it was observed that this treatment selectively reduced the population of Tregs in the patients and suppressed the immune-inhibitory function of Tregs (Zhang et al., 2008). The Tregs that received Taxol treatment were noted to have lesser inhibition on Teff as compared to Treg without Taxol treatment (Zhang et al., 2008). Apoptosis was seen in Tregs post-Taxol treatment but other lymphocytic cells like effector T cells, CD8[+] T cells were not affected (Zhang et al., 2008). Moreover, Vicari et al. also observed reduction in Tregs in mice tumor models which is TLR4-independent (Vicari et al., 2009). Both studies indicated that the inhibition of Tregs by Taxol is possibly correlated to FoxP3 expression as they found decrease in FoxP3 expression alongside Tregs level in the cells after Taxol treatment (Zhang et al., 2008; Vicari et al., 2009).

Hence it is accountable that Taxol plays a certain role in modulating the immune cells and by gaining insight into the immunomodulation mechanism of Taxol could provide aids in developing additional therapeutic strategies to improve the anti-tumor immunity. For example, an inhibitor for IDO (indoleamine 2,3-dioxygenase) which is an enzyme that promotes Tregs proliferation and suppress Teff production, in combined with Taxol treatment effectively regressed tumor size (Javeed et al., 2009).

3.5.2 Taxol and macrophages

Macrophages belong to innate immunity which is involved in immune surveillance and defence against tumors (Nielsen and Schmid, 2017). Macrophages are one of the immune cells found in tumor microenvironment which could destroy tumor cells by several mechanisms for example through the release of lysosomal enzymes and nitric oxide (Javeed et al., 2009). However, there are reports saying that infiltration of macrophages in the tumor site is associated to poor survival in breast, lung, pancreatic cancer and so on (Bingle et al., 2002; Chen et al., 2005; Nielsen and Schmid, 2017). The macrophages located at the tumor microenvironment are known as tumor-associated macrophages (TAM) and are proposed to be polarized into two categories: M1 macrophages and M2 macrophages (Nielsen and Schmid, 2017). M1 macrophages are the typical and classic active macrophages which responds to interferon gamma (IFNγ) and lipopolysaccharide (LPS), they produce cytokines, ROS and are able to remove pathogens (Nielsen and Schmid, 2017). On the other hand, M2 macrophages regulates tissue remodeling, promotes angiogenesis, and suppresses the adaptive immunity (Yamaguchi et al., 2017). Hence, M1 macrophages are suggested to be anti-tumor while M2 macrophages are pro-tumor (Nielsen and Schmid, 2017; Yamaguchi et al., 2017).

3.5.3 Taxol and TLR4-dependent pathway

Taxol was reported to have LPS-like effect on murine macrophages like secretion of cytokines and chemokines, release of ROS and nitrogen species and anti-tumor activity (Bingle et al., 2002; Szajnik et al., 2009; Yamaguchi et al., 2017). The effects of Taxol and LPS on macrophages are toll-like receptor 4 (TLR4) dependent, both of them share a receptor and similar signaling cascade despite lacking structural similarity between the two ligands, thus Taxol was identified as a TLR4 ligand in macrophages (Byrd-Leifer et al., 2001; Szajnik et al., 2009). TLRs that are expressed on immune cells like macrophages serve as a cell-surface receptors for pathogens and the activation of immune responses (Szajnik et al., 2009; Kim et al., 2014). The activation of TLR4 ligand will lead to the activation of downstream signaling cascade, myeloid differentiation primary response gene 88 (MyD88)-dependent or TRIF-dependent pathways (Szajnik et al., 2009; Kampan et al., 2015). Taxol binds to TLR4 receptor on macrophages will trigger MyD88-dependent pathway which in turn leads to activation of mitogen-activated protein kinase (MAPK) and transcription of nuclear factor (NFκB) pathways (Kampan et al., 2015). The activation of MAPK pathways is somehow associated with cell cycle arrest and Bcl-2 proteins phosphorylation, similar with JNK/SAPK pathway activation as mentioned previously (Kampan et al., 2015). Hence the binding of TLR4 with Taxol eventually causes cell apoptosis via cell cycle arrest and Bcl-2 regulations (Kampan et al., 2015). Additionally, Taxol also stimulates macrophages in the tumor site to induce the release of cytokines to activate dendritic cells, and tumor-specific cytotoxic T cells to cause direct tumor cell lysis and anti-tumor responses (Javeed et al., 2009; Kampan et al., 2015).

However, the effects of Taxol on macrophages in the human tumor cells remain ambiguous as there were reports which observed insignificant effects on human macrophages (Kawasaki et al., 2000; Wang et al., 2002). The role of Taxol and the TLR4/MyD88 signaling remains uncertain in the cancer progression. Kelly et al. (2006) reported that the TLR4 signaling in ovarian cancer cells promotes tumor growth and Taxol resistance (Kelly et al., 2006). It is speculated that pro-inflammatory cytokines are released upon TLR4/MyD88 signaling activation and the cytokines allow cancer cells to grow, metastasize and develop resistance to chemotherapy (Szajnik et al., 2009). A study found that Taxol enhance cancer cells survival by releasing cytokines and activating TLR4 pathway, and they promote resistance toward Taxol via NFκB pathway in breast cancer (Rajput et al., 2013). A detailed schematic diagram, which shows the signaling pathways regulated by Taxol and TLR4 is shown in Fig. 3.3. TLR4 inhibitors were

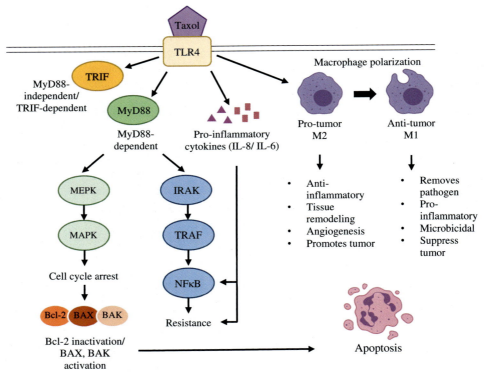

Fig. 3.3 Signaling pathways and macrophage polarization regulated by Taxol. Taxol exerts LPS-mimetic effect on TLR4 receptor, upon binding with TLR4, several downstream signaling pathways will be activated including TLR4/MyD88 pathway which leads to induction of MAPK or NFκB signaling. Activation of MAPK pathway eventually causes cell cycle arrest and Bcl-2 proteins changes and cell apoptosis, whereas NFκB pathway leads to drug resistance. Taxol also regulates macrophage polarization, altering M2 macrophage into anti-tumor M1 macrophage via TLR4 signaling. TLR4: Toll-like receptor 4; MEPK/MAPK: mitogen activated protein kinase; TRIF: TIR-domain-containing adapter-inducing interferon-β; MyD88: Myeloid differentiation primary response 88; IRAK: IL-1 receptor associated kinase; TRAF: TNFR associated factor;NFκB: nuclear factor kappa B; IL-6/8: Interleukin 6/8.

shown to significantly increase the Taxol sensitivity in tumor cells, this strongly proves the importance of this receptor in conferring chemoresistance (Rajput et al., 2013).

3.5.4 Taxol affects macrophages polarization

In contrast with the previous findings, TLR4 is suggested to be an important inducer for M1 macrophage and possibly a suppresser for M2 macrophages (Yamaguchi et al., 2017). It is noteworthy that the induction of M1 macrophage results in an immunogenic and anti-tumor effect, thus Yamaguchi et al. (2017) hypothesized that Taxol is able to suppress M2 macrophage and activates M1 macrophage in gastric cancer (Yamaguchi et al., 2017). Additionally, they found that the expression of NOS2 (M1 macrophage marker) was upregulated while expression of CD204 (M2 macrophage marker) was suppressed in the gastric cells following low-dose Taxol treatment, suggesting that Taxol could alter the macrophages phenotype in the cells (Yamaguchi et al., 2017). They also indicated that the M1 macrophage polarization involves TLR4/NFκB pathway as intranuclear translocation of NFκB p65 subunit was observed in the cells incubated with low-dose of Taxol (Yamaguchi et al., 2017).

The immunomodulation by Taxol on the macrophage polarization was further proven in breast cancer and melanoma tumor models. The study found that Taxol reprogrammed pro-tumor M2 macrophages into anti-tumor M1 macrophages in in vivo tumor models (Wanderley et al., 2018). Clinical samples from Taxol-treated patients showed an increase in the M1 macrophages profile and the M2 macrophages were observed to be polarized in M1-like macrophages with a less immunotolerant status via TLR4 signaling after being treated with Taxol (Wanderley et al., 2018). These results provide a novel finding which suggests macrophage polarization by chemo-drug as an effective control for tumor progression by improving the immune surveillance in the tumor microenvironment (Wanderley et al., 2018).

However, the contribution of TLR4 activation in the cancer setting remains questionable as it could be a promoting factor for chemoresistance and tumor-survival too, hence further investigations are much needed to incorporate TLR4 into the therapeutic strategy. In fact, compound A (CpdA) or 2-(4-acetoxyphenyl)-2- chloro-N-methylethylammonium chloride which is a glucocorticoid receptor ligand that disrupts NFκB activation helps to alleviate the Taxol resistance (Sootichote et al., 2018). It was observed that CpdA efficiently reduce the TLR4-mediated Taxol resistance in breast cancer and melanoma cells by reducing the cytokines (interleukin-6 and interleukin-8) and XIAP in a synergistic manner, this provided proof that TLR4 activated by Taxol was associated with NFκB pathway (Sootichote et al., 2018). Thus they suggest using CpdA as a synergistic drug with Taxol to minimize chemoresistance in Taxol-based cancer treatment (Sootichote et al., 2018).

3.6 Resistance mechanisms of Taxol

Chemoresistance either in intrinsic or acquired manner remains as a major problem for chemotherapy in cancer treatment and with no surprise, Taxol treatment develops drug resistance after prolonged treatment too (Abu Samaan et al., 2019). Taxol resistance is multifactorial and complicated as it involves several different mechanisms which different proteins or pathways are incorporated (Kampan et al., 2015).

3.6.1 Increased drug efflux by multi-drug resistance (MDR-1)

The most common resistance mechanism of Taxol is the overexpression of drug efflux protein, multi-drug resistance (MDR-1) or P-glycoprotein (P-gp) from the drug efflux protein ATP-binding cassette (ABC) superfamily (Abu Samaan et al., 2019; Yusuf et al., 2003). P-gp is known to export drug out of the cells thus affecting the internal drug concentration and chemosensitivity (Abu Samaan et al., 2019). Taxol is a known substrate for P-gp while the upregulated expression of MDR-1 or P-gp in Taxol-resistant cells was positively correlated with drug efflux and negatively correlated with Taxol sensitivity, thus this provided the evidence that overexpression of MDR-1 results in Taxol resistance (Sparreboom et al., 1997; Yusuf et al., 2003; Sherman-Baust et al., 2011; Abu Samaan et al., 2019).

3.6.2 Altered expression of β-tubulins isotypes

β-tubulins in human are classified into several categories from class I β-tubulin to class VI, each encoded by different genes and have different expressions which eventually would affect the overall microtubule dynamics (Yusuf et al., 2005). Several studies had implied that the altered expression of β-tubulin isotypes is correlated to Taxol sensitivity and resistance (Orr et al., 2003). For instance, class III and IVa β-tubulins were found to increase in their expression in Taxol-resistant non-small lung carcinoma cells (Kavallaris et al., 1997). Similarly, a twofold increase in class IVa β-tubulin was observed inTaxol-resistant leukemia cell lines too, but differential β-tubulin was noted to affect Taxol-resistant cell lines in vitro but not in vivo as no correlation was found between their expression with Taxol-resistant human ovarian carcinoma xenografts obtained from patients' samples (Yusuf et al., 2003). It was suggested that the altered expression of tubulin isotypes could modulate microtubule stability in a way that it affects the efficacy of Taxol's action on microtubules (Kavallaris et al., 1997).

3.6.3   Upregulation of Bcl-2 family proteins

The aforementioned apoptosis-related proteins, Bcl-2 family proteins also play an important role in Taxol resistance. The changes in either the anti- or pro-apoptotic members of Bcl-2 family could easily affect the Taxol sensitivity. Bcl-2 and Bcl-XL which both are anti-apoptotic proteins were found to mediate Taxol resistance as their high expressions were observed in resistant hepatocellular carcinoma cells (Chun and Lee, 2004). The upregulation of Bcl-XL into ovarian carcinoma cells by transfection was showed to significantly increase resistance toward Taxol treatment as compared to non-transfected cells, whereas breast cancer patients with over-expressed Bcl-2 also showed poor prognosis (Liu et al., 1998). As apoptosis is a necessary pathway for most of the chemotherapeutic drugs, it is expected that changes in the ratio of anti- or pro-apoptotic proteins would result in changes in drug sensitivity too (Yusuf et al., 2003). However, with the understanding of resistance mechanism, it is possible to develop targeted therapeutic measure for an improved survival outcome. Recently, a pan-cancer genomic data from a batch of cancer patients was conducted to find that certain single nucleotide variants in the Bcl-2 sequence is related to Taxol resistance and could be used as a predictor for patient's response toward the drug (Ben-Hamo et al., 2019) (Table 3.1).

3.7   Conclusion

Taxol is an effective antimitotic chemotherapeutic agent that is widely used in treating several types of cancers. Not only it is a microtubule stabilizer, but it also acts on apoptotic proteins like Bcl-2 family proteins, controls mitochondria and ER calcium store and moreover it was found to exert modulations on the immune system as well. From here it can be concluded that Taxol serves a multidimensional function as a chemo-drug via several mechanisms although the significance of each mechanism could be different between cell lines and dosages applied. However, the battle for minimizing the drug resistance remains as an issue to be tackled thus continuous investigations on this aspect would provide a more comprehensive insight and understanding into the mechanism of Taxol in action. Macrophage polarization by Taxol is a novel insight on its mode of action and provides a potential therapy option by combining chemotherapy and immunotherapy for cancer patients. The drug dosage and treatment time also

Table 3.1 A summarized table of the cancer cell lines, drug concentrations involved in each mechanism of actions for Taxol.

Mechanisms of action		Cancer cell line model(s)	Dosage of Taxol	Reference(s)
Mitotic arrest	G2/M arrest	Human cervical cancer cells (HeLa)	1 or 10 μM	Schiff and Horwitz (1980)
	G2/M arrest	Human lung cancer cells (A549, H1299)	0.002–0.05 mM	Das et al. (2001)
	G1 arrest	Human colon carcinoma cells (HCT116)	25–100 nM	Long and Fairchild (1994)
	G2/M arrest	Human kidney carcinoma cells (A 498), Epithelial-like human ovarian adenocarcinoma cells (Caov-3)	30–100 nM	Yvon et al. (1999)
	G2/M arrest	Human leukemia cells (HL-60, U937)	1–10 μM	Gangemi et al. (1995)
	Multinucleated cells	Human colon carcinoma cells (HCT116)	8–16 nM	Long and Fairchild (1994)
Cell signaling pathway	Bcl-2 phosphorylation—Taxol phosphorylates Bcl-2 proteins and induces cell apoptosis	Human prostate cancer cells (DU145, PC-3, LNCaP)	5–10 μM	Haldar et al. (1996)
		Multiple myeloma cell lines (8226, IM-9, ARH-77)	100 nM	Kroning and Lichtenstein (1998)
		Human hepatocellular carcinoma cells (QGY-7703)	5 nM	Cheng et al. (2001)
		Human lung (H460), Human Colon (RKO), Breast carcinoma cell lines (BT-474)	10–100 nM	Scatena et al. (1998)
	BIM-dependent apoptosis—inhibits anti-apoptotic Bcl-2 proteins or activate pro-apoptotic BAX/ BAK proteins	Human non-small-cell lung cancer cell lines (LuCSF1, LT-46, LT-30, LT-36, CRI-702, and CRI-619)	10–80 nM	Li et al. (2005)
		Breast cancer cell lines (BrCa-101, BccMcGee, MDA-MB-231 and SK-Br-3)	10–200 nM	Li et al. (2005)
		Prostate cancer cell lines (PC3 and Du145)	10–200 nM	Li et al. (2005)

Table 3.1 A summarized table of the cancer cell lines, drug concentrations involved in each mechanism of actions for Taxol—cont'd

Mechanisms of action		Cancer cell line model(s)	Dosage of Taxol	Reference(s)
	Displacement of BIM induces apoptosis	Breast cancer cell lines (MCF-7, T47D, BT20, and MDA-MB-468 cells)	100 nM	Kutuk and Letai (2010)
	JNK/SAPK pathway—Taxol activates JNK pathway which induces down-stream signaling protein activation and causes cell apoptosis	Human ovarian cancer cells (OVCA 420)	30 μM	Lee et al. (1998)
		Human ovarian carcinoma cells (SKOV3)	100 nM	Jo et al. (2016)
		Human cervical carcinoma cell line (KB-3)	30 nM	Kolomeichuk et al. (2008)
ER-calcium dependent apoptosis	Taxol induces calcium changes in ER which in turn attenuates Bcl-2 inhibition and leads to cell apoptosis	Breast carcinoma cell line (M468)	2.5 μM ($\sim 10^{-6}$ M)	Pan and Gollahon (2013)
Immunomodulation	Macrophages polarization—Taxol suppresses M2 macrophage and alter its pro-tumor phenotype to M1 anti-tumor phenotype	Gastric cancer cell lines (TMK-1, MKN45)	Undefined	Yamaguchi et al. (2017)
		Breast cancer cells (4T1)	30 μM	Wanderley et al. (2018)
		Melanoma cancer cells (B16)	30 μM	Wanderley et al. (2018)

need to be taken into consideration as Taxol's anti-cancer action could be dose-dependent.

Acknowledgment

This work was supported by Fundamental Research Grant Scheme grants (FRGS/1/2019/SKK08/MUSM/02/4 and FRGS/1/2019/WAB09/MUSM/02/1) from the Ministry of Higher Education Malaysia.

References

Abu Samaan, T.M., Samec, M., Liskova, A., Kubatka, P., Busselberg, D., 2019. Paclitaxel's mechanistic and clinical effects on breast cancer. Biomolecules 9 (12). https://doi.org/10.3390/biom9120789.

Akimov, S.S., Belkin, A.M., 2001. Cell surface tissue transglutaminase is involved in adhesion and migration of monocytic cells on fibronectin. Blood 98 (5), 1567–1576. https://doi.org/10.1182/blood.v98.5.1567.

Attalla, H., Westberg, J.A., Andersson, L.C., Adlercreutz, H., Mäkelä, T.P., 1998. 2-Methoxyestradiol-induced phosphorylation of Bcl-2: uncoupling from JNK/SAPK activation. Biochem. Biophys. Res. Commun. 247 (3), 616–619. https://doi.org/10.1006/bbrc.1998.8870.

Baggott, R.R., Mohamed, T.M., Oceandy, D., Holton, M., Blanc, M.C., Roux-Soro, S.C., et al., 2012. Disruption of the interaction between PMCA2 and calcineurin triggers apoptosis and enhances paclitaxel-induced cytotoxicity in breast cancer cells. Carcinogenesis 33 (12), 2362–2368. https://doi.org/10.1093/carcin/bgs282.

Ben-Hamo, R., Zilberberg, A., Cohen, H., Bahar-Shany, K., Wachtel, C., Korach, J., et al., 2019. Resistance to paclitaxel is associated with a variant of the gene BCL2 in multiple tumor types. NPJ Precis. Oncol. 3, 12. https://doi.org/10.1038/s41698-019-0084-3.

Bingle, L., Brown, N.J., Lewis, C.E., 2002. The role of tumour-associated macrophages in tumour progression: implications for new anticancer therapies. J. Pathol. 196 (3), 254–265. https://doi.org/10.1002/path.1027.

Boehmerle, W., Splittgerber, U., Lazarus, M.B., McKenzie, K.M., Johnston, D.G., Austin, D.J., et al., 2006. Paclitaxel induces calcium oscillations via an inositol 1,4,5-trisphosphate receptor and neuronal calcium sensor 1-dependent mechanism. Proc. Natl. Acad. Sci. U. S. A. 103 (48), 18356–18361. https://doi.org/10.1073/pnas.0607240103.

Bonneau, B., Prudent, J., Popgeorgiev, N., Gillet, G., 2013. Non-apoptotic roles of Bcl-2 family: the calcium connection. Biochim. Biophys. Acta (BBA) Mol. Cell Res. 1833 (7), 1755–1765. https://doi.org/10.1016/j.bbamcr.2013.01.021.

Byrd-Leifer, C.A., Block, E.F., Takeda, K., Akira, S., Ding, A., 2001. The role of MyD88 and TLR4 in the LPS-mimetic activity of Taxol. Eur. J. Immunol. 31 (8), 2448–2457. https://doi.org/10.1002/1521-4141(200108)31:8<2448::aid-immu2448>3.0.co;2-n.

Campbell, K.J., Tait, S.W.G., 2018. Targeting BCL-2 regulated apoptosis in cancer. Open Biol. 8 (5). https://doi.org/10.1098/rsob.180002.

Chen, J.J., Lin, Y.C., Yao, P.L., Yuan, A., Chen, H.Y., Shun, C.T., et al., 2005. Tumor-associated macrophages: the double-edged sword in cancer progression. J. Clin. Oncol. 23 (5), 953–964. https://doi.org/10.1200/jco.2005.12.172.

Cheng, B., Crasta, K., 2017. Consequences of mitotic slippage for antimicrotubule drug therapy. Endocr. Relat. Cancer 24 (9), T97–T106. https://doi.org/10.1530/ERC-17-0147.

Cheng, S.C., Luo, D., Xie, Y., 2001. Taxol induced Bcl-2 protein phosphorylation in human hepatocellular carcinoma QGY-7703 cell line. Cell Biol. Int. 25 (3), 261–265. https://doi.org/10.1006/cbir.2000.0619.

Chun, E., Lee, K.Y., 2004. Bcl-2 and Bcl-xL are important for the induction of paclitaxel resistance in human hepatocellular carcinoma cells. Biochem. Biophys. Res. Commun. 315 (3), 771–779. https://doi.org/10.1016/j.bbrc.2004.01.118.

Colina, C., Flores, A., Rojas, H., Acosta, A., Castillo, C., Garrido Mdel, R., et al., 2005. Ceramide increase cytoplasmic Ca2+ concentration in Jurkat T cells by liberation of calcium from intracellular stores and activation of a store-operated calcium channel. Arch. Biochem. Biophys. 436 (2), 333–345. https://doi.org/10.1016/j.abb.2005.02.014.

Czernick, M., Rieger, A., Goping, I.S., 2009. Bim is reversibly phosphorylated but plays a limited role in paclitaxel cytotoxicity of breast cancer cell lines. Biochem. Biophys. Res. Commun. 379 (1), 145–150. https://doi.org/10.1016/j.bbrc.2008.12.025.

Das, G.C., Holiday, D., Gallardo, R., Haas, C., 2001. Taxol-induced cell cycle arrest and apoptosis: dose-response relationship in lung cancer cells of different wild-type p53 status and under isogenic condition. Cancer Lett. 165 (2), 147–153. https://doi.org/10.1016/s0304-3835(01)00404-9.

Demeckova, V., Solar, P., Hrckova, G., Mudronova, D., Bojkova, B., Kassayova, M., et al., 2017. Immodin and its immune system supportive role in paclitaxel therapy of 4T1 mouse breast cancer. Biomed. Pharmacother. 89, 245–256. https://doi.org/10.1016/j.biopha.2017.02.034.

Fan, W., 1999. Possible mechanisms of paclitaxel-induced apoptosis. Biochem. Pharmacol. 57 (11), 1215–1221. https://doi.org/10.1016/s0006-2952(99)00006-4.

Ferlini, C., Cicchillitti, L., Raspaglio, G., Bartollino, S., Cimitan, S., Bertucci, C., et al., 2009. Paclitaxel directly binds to Bcl-2 and functionally mimics activity of Nur77. Cancer Res. 69 (17), 6906–6914. https://doi.org/10.1158/0008-5472.CAN-09-0540.

Gangemi, R.M., Tiso, M., Marchetti, C., Severi, A.B., Fabbi, M., 1995. Taxol cytotoxicity on human leukemia cell lines is a function of their susceptibility to programmed cell death. Cancer Chemother. Pharmacol. 36 (5), 385–392. https://doi.org/10.1007/bf00686187.

Giannakakou, P., Robey, R., Fojo, T., Blagosklonny, M.V., 2001. Low concentrations of paclitaxel induce cell type-dependent p53, p21 and G1/G2 arrest instead of mitotic arrest: molecular determinants of paclitaxel-induced cytotoxicity. Oncogene 20 (29), 3806–3813. https://doi.org/10.1038/sj.onc.1204487.

Gogada, R., Yadav, N., Liu, J., Tang, S., Zhang, D., Schneider, A., et al., 2013. Bim, a proapoptotic protein, up-regulated via transcription factor E2F1-dependent mechanism, functions as a prosurvival molecule in cancer. J. Biol. Chem. 288 (1), 368–381. https://doi.org/10.1074/jbc.M112.386102.

Haldar, S., Chintapalli, J., Croce, C.M., 1996. Taxol induces bcl-2 phosphorylation and death of prostate cancer cells. Cancer Res. 56 (6), 1253–1255.

Haldar, S., Jena, N., Croce, C.M., 1995. Inactivation of Bcl-2 by phosphorylation. Proc. Natl. Acad. Sci. U. S. A. 92 (10), 4507–4511. https://doi.org/10.1073/pnas.92.10.4507.

Huang, H.C., Shi, J., Orth, J.D., Mitchison, T.J., 2009. Evidence that mitotic exit is a better cancer therapeutic target than spindle assembly. Cancer Cell 16 (4), 347–358. https://doi.org/10.1016/j.ccr.2009.08.020.

Ichihara, F., Kono, K., Takahashi, A., Kawaida, H., Sugai, H., Fujii, H., 2003. Increased populations of regulatory T cells in peripheral blood and tumor-infiltrating lymphocytes in patients with gastric and esophageal cancers. Clin. Cancer Res. 9 (12), 4404–4408.

Janssen, K., Pohlmann, S., Janicke, R.U., Schulze-Osthoff, K., Fischer, U., 2007. Apaf-1 and caspase-9 deficiency prevents apoptosis in a Bax-controlled pathway and promotes clonogenic survival during paclitaxel treatment. Blood 110 (10), 3662–3672. https://doi.org/10.1182/blood-2007-02-073213.

Javeed, A., Ashraf, M., Riaz, A., Ghafoor, A., Afzal, S., Mukhtar, M.M., 2009. Paclitaxel and immune system. Eur. J. Pharm. Sci. 38 (4), 283–290. https://doi.org/10.1016/j.ejps.2009.08.009.

Jo, H., Ahn, H.J., Lee, J.H., Min, C.K., 2016. Roles of JNK and P53 in taxol-induced apoptotic signaling in SKOV3 human ovarian cancer cells. Elyns J. Cancer Res. 01 (01). https://doi.org/10.19104/ejcr.2015.101.

Kampan, N.C., Madondo, M.T., McNally, O.M., Quinn, M., Plebanski, M., 2015. Paclitaxel and its evolving role in the management of ovarian cancer. Biomed. Res. Int. 2015, 413076. https://doi.org/10.1155/2015/413076.

Kavallaris, M., Kuo, D.Y., Burkhart, C.A., Regl, D.L., Norris, M.D., Haber, M., et al., 1997. Taxol-resistant epithelial ovarian tumors are associated with altered expression of specific beta-tubulin isotypes. J. Clin. Invest. 100 (5), 1282–1293. https://doi.org/10.1172/JCI119642.

Kawasaki, K., Akashi, S., Shimazu, R., Yoshida, T., Miyake, K., Nishijima, M., 2000. Mouse toll-like receptor 4.MD-2 complex mediates lipopolysaccharide-mimetic signal transduction by Taxol. J. Biol. Chem. 275 (4), 2251–2254. https://doi.org/10.1074/jbc.275.4.2251.

Kelly, M.G., Alvero, A.B., Chen, R., Silasi, D.A., Abrahams, V.M., Chan, S., et al., 2006. TLR-4 signaling promotes tumor growth and paclitaxel chemoresistance in ovarian cancer. Cancer Res. 66 (7), 3859–3868. https://doi.org/10.1158/0008-5472.CAN-05-3948.

Kim, J.E., Jang, M.J., Jin, D.H., Chung, Y.H., Choi, B.S., Park, G.B., et al., 2014. Paclitaxel-exposed ovarian cancer cells induce cancerspecific CD4+ T cells after doxorubicin exposure through regulation of MyD88 expression. Int. J. Oncol. 44 (5), 1716–1726. https://doi.org/10.3892/ijo.2014.2308.

Kolomeichuk, S.N., Terrano, D.T., Lyle, C.S., Sabapathy, K., Chambers, T.C., 2008. Distinct signaling pathways of microtubule inhibitors—vinblastine and Taxol induce JNK-dependent cell death but through AP-1-dependent and AP-1-independent mechanisms, respectively. FEBS J. 275 (8), 1889–1899. https://doi.org/10.1111/j.1742-4658.2008.06349.x.

Kroning, R., Lichtenstein, A., 1998. Taxol can induce phosphorylation of BCL-2 in multiple myeloma cells and potentiate dexamethasone-induced apoptosis. Leuk. Res. 22 (3), 275–286. https://doi.org/10.1016/s0145-2126(97)00170-7.

Kutuk, O., Letai, A., 2010. Displacement of Bim by Bmf and Puma rather than increase in Bim level mediates paclitaxel-induced apoptosis in breast cancer cells. Cell Death Differ. 17 (10), 1624–1635. https://doi.org/10.1038/cdd.2010.41.

Lee, L.F., Li, G., Templeton, D.J., Ting, J.P., 1998. Paclitaxel (Taxol)-induced gene expression and cell death are both mediated by the activation of c-Jun NH2-terminal kinase (JNK/SAPK). J. Biol. Chem. 273 (43), 28253–28260. https://doi.org/10.1074/jbc.273.43.28253.

Lee, W.-L., Shiau, J.-Y., Shyur, L.-F., 2012. Taxol, camptothecin and beyond for cancer therapy. In: Recent Trends in Medicinal Plants Research, pp. 133–178.

Li, R., Moudgil, T., Ross, H.J., Hu, H.M., 2005. Apoptosis of non-small-cell lung cancer cell lines after paclitaxel treatment involves the BH3-only proapoptotic protein Bim. Cell Death Differ. 12 (3), 292–303. https://doi.org/10.1038/sj.cdd.4401554.

Liu, J.R., Fletcher, B., Page, C., Hu, C., Nunez, G., Baker, V., 1998. Bcl-xL is expressed in ovarian carcinoma and modulates chemotherapy-induced apoptosis. Gynecol. Oncol. 70 (3), 398–403. https://doi.org/10.1006/gyno.1998.5125.

Long, B.H., Fairchild, C.R., 1994. Paclitaxel inhibits progression of mitotic cells to G1 phase by interference with spindle formation without affecting other microtubule functions during anaphase and telephase. Cancer Res. 54 (16), 4355–4361.

Martin, V., 1993. Overview of Paclitaxel (Taxol*). Semin. Oncol. Nurs. 9 (4), 2–5. https://doi.org/10.1016/s0749-2081(16)30035-3.

Miller, A.V., Hicks, M.A., Nakajima, W., Richardson, A.C., Windle, J.J., Harada, H., 2013. Paclitaxel-induced apoptosis is BAK-dependent, but BAX and BIM-independent in breast tumor. PLoS One 8 (4). https://doi.org/10.1371/journal.pone.0060685, e60685.

Mironov, S.L., Ivannikov, M.V., Johansson, M., 2005. [Ca2+]i signaling between mitochondria and endoplasmic reticulum in neurons is regulated by microtubules. From mitochondrial permeability transition pore to Ca2+-induced Ca2+ release. J. Biol. Chem. 280 (1), 715–721. https://doi.org/10.1074/jbc.M409819200.

Mukhtar, E., Adhami, V.M., Mukhtar, H., 2014. Targeting microtubules by natural agents for cancer therapy. Mol. Cancer Ther. 13 (2), 275–284. https://doi.org/10.1158/1535-7163.MCT-13-0791.

Nielsen, S.R., Schmid, M.C., 2017. Macrophages as key drivers of cancer progression and metastasis. Mediat. Inflamm. 2017, 9624760. https://doi.org/10.1155/2017/9624760.

Nishikawa, H., Kato, T., Tawara, I., Takemitsu, T., Saito, K., Wang, L., et al., 2005. Accelerated chemically induced tumor development mediated by CD4+CD25+ regulatory

T cells in wild-type hosts. Proc. Natl. Acad. Sci. U. S. A. 102 (26), 9253–9257. https://doi.org/10.1073/pnas.0503852102.

Nutt, L.K., Chandra, J., Pataer, A., Fang, B., Roth, J.A., Swisher, S.G., et al., 2002a. Bax-mediated Ca2+ mobilization promotes cytochrome c release during apoptosis. J. Biol. Chem. 277 (23), 20301–20308. https://doi.org/10.1074/jbc.M201604200.

Nutt, L.K., Pataer, A., Pahler, J., Fang, B., Roth, J., McConkey, D.J., et al., 2002b. Bax and Bak promote apoptosis by modulating endoplasmic reticular and mitochondrial Ca2+ stores. J. Biol. Chem. 277 (11), 9219–9225. https://doi.org/10.1074/jbc.M106817200.

Oakes, S.A., Scorrano, L., Opferman, J.T., Bassik, M.C., Nishino, M., Pozzan, T., et al., 2005. Proapoptotic BAX and BAK regulate the type 1 inositol trisphosphate receptor and calcium leak from the endoplasmic reticulum. Proc. Natl. Acad. Sci. U. S. A. 102 (1), 105–110. https://doi.org/10.1073/pnas.0408352102.

Ormandy, L.A., Hillemann, T., Wedemeyer, H., Manns, M.P., Greten, T.F., Korangy, F., 2005. Increased populations of regulatory T cells in peripheral blood of patients with hepatocellular carcinoma. Cancer Res. 65 (6), 2457–2464. https://doi.org/10.1158/0008-5472.Can-04-3232.

Orr, G.A., Verdier-Pinard, P., McDaid, H., Horwitz, S.B., 2003. Mechanisms of taxol resistance related to microtubules. Oncogene 22 (47), 7280–7295. https://doi.org/10.1038/sj.onc.1206934.

Pan, Z., Avila, A., Gollahon, L., 2014. Paclitaxel induces apoptosis in breast cancer cells through different calcium—regulating mechanisms depending on external calcium conditions. Int. J. Mol. Sci. 15 (2), 2672–2694. https://doi.org/10.3390/ijms15022672.

Pan, Z., Gollahon, L., 2011. Taxol directly induces endoplasmic reticulum-associated calcium changes that promote apoptosis in breast cancer cells. Breast J. 17 (1), 56–70. https://doi.org/10.1111/j.1524-4741.2010.00988.x.

Pan, Z., Gollahon, L., 2013. Paclitaxel attenuates Bcl-2 resistance to apoptosis in breast cancer cells through an endoplasmic reticulum-mediated calcium release in a dosage dependent manner. Biochem. Biophys. Res. Commun. 432 (3), 431–437. https://doi.org/10.1016/j.bbrc.2013.01.130.

Pandya, P.H., Murray, M.E., Pollok, K.E., Renbarger, J.L., 2016. The immune system in cancer pathogenesis: potential therapeutic approaches. J Immunol Res 2016, 4273943. https://doi.org/10.1155/2016/4273943.

Pinton, P., Giorgi, C., Siviero, R., Zecchini, E., Rizzuto, R., 2008. Calcium and apoptosis: ER-mitochondria Ca2+ transfer in the control of apoptosis. Oncogene 27 (50), 6407–6418. https://doi.org/10.1038/onc.2008.308.

Rajput, S., Volk-Draper, L.D., Ran, S., 2013. TLR4 is a novel determinant of the response to paclitaxel in breast cancer. Mol. Cancer Ther. 12 (8), 1676–1687. https://doi.org/10.1158/1535-7163.MCT-12-1019.

Rao, R.V., Ellerby, H.M., Bredesen, D.E., 2004. Coupling endoplasmic reticulum stress to the cell death program. Cell Death Differ. 11 (4), 372–380. https://doi.org/10.1038/sj.cdd.4401378.

Rathore, R., McCallum, J.E., Varghese, E., Florea, A.M., Busselberg, D., 2017. Overcoming chemotherapy drug resistance by targeting inhibitors of apoptosis proteins (IAPs). Apoptosis 22 (7), 898–919. https://doi.org/10.1007/s10495-017-1375-1.

Rong, Y.P., Bultynck, G., Aromolaran, A.S., Zhong, F., Parys, J.B., De Smedt, H., et al., 2009. The BH4 domain of Bcl-2 inhibits ER calcium release and apoptosis by binding the regulatory and coupling domain of the IP3 receptor. Proc. Natl. Acad. Sci. U. S. A. 106 (34), 14397–14402. https://doi.org/10.1073/pnas.0907555106.

Scatena, C.D., Stewart, Z.A., Mays, D., Tang, L.J., Keefer, C.J., Leach, S.D., et al., 1998. Mitotic phosphorylation of Bcl-2 during normal cell cycle progression and Taxol-induced growth arrest. J. Biol. Chem. 273 (46), 30777–30784. https://doi.org/10.1074/jbc.273.46.30777.

Schiff, P.B., Horwitz, S.B., 1980. Taxol stabilizes microtubules in mouse fibroblast cells. Proc. Natl. Acad. Sci. U. S. A. 77 (3), 1561–1565. https://doi.org/10.1073/pnas.77.3.1561.

Sherman-Baust, C.A., Becker, K.G., Wood Iii, W.H., Zhang, Y., Morin, P.J., 2011. Gene expression and pathway analysis of ovarian cancer cells selected for resistance to cisplatin, paclitaxel, or doxorubicin. J. Ovarian Res. 4 (1), 21. https://doi.org/10.1186/1757-2215-4-21.

Smyth, M.J., Dunn, G.P., Schreiber, R.D., 2006. Cancer immunosurveillance and immunoediting: the roles of immunity in suppressing tumor development and shaping tumor immunogenicity. Adv. Immunol. 90, 1–50. https://doi.org/10.1016/s0065-2776(06)90001-7.

Sootichote, R., Thuwajit, P., Singsuksawat, E., Warnnissorn, M., Yenchitsomanus, P.T., Ithimakin, S., et al., 2018. Compound A attenuates toll-like receptor 4-mediated paclitaxel resistance in breast cancer and melanoma through suppression of IL-8. BMC Cancer 18 (1), 231. https://doi.org/10.1186/s12885-018-4155-6.

Sorger, P.K., Dobles, M., Tournebize, R., Hyman, A.A., 1997. Coupling cell division and cell death to microtubule dynamics. Curr. Opin. Cell Biol. 9 (6), 807–814. https://doi.org/10.1016/s0955-0674(97)80081-6.

Sparreboom, A., van Asperen, J., Mayer, U., Schinkel, A.H., Smit, J.W., Meijer, D.K.F., et al., 1997. Limited oral bioavailability and active epithelial excretion of paclitaxel (Taxol) caused by P-glycoprotein in theintestine. Proc. Natl. Acad. Sci. 94 (5), 2031–2035. https://doi.org/10.1073/pnas.94.5.2031.

Srivastava, R.K., Mi, Q.S., Hardwick, J.M., Longo, D.L., 1999. Deletion of the loop region of Bcl-2 completely blocks paclitaxel-induced apoptosis. Proc. Natl. Acad. Sci. U. S. A. 96 (7), 3775–3780. https://doi.org/10.1073/pnas.96.7.3775.

Sunters, A., Fernandez de Mattos, S., Stahl, M., Brosens, J.J., Zoumpoulidou, G., Saunders, C.A., et al., 2003. FoxO3a transcriptional regulation of Bim controls apoptosis in paclitaxel-treated breast cancer cell lines. J. Biol. Chem. 278 (50), 49795–49805. https://doi.org/10.1074/jbc.M309523200.

Szajnik, M., Szczepanski, M.J., Czystowska, M., Elishaev, E., Mandapathil, M., Nowak-Markwitz, E., et al., 2009. TLR4 signaling induced by lipopolysaccharide or paclitaxel regulates tumor survival and chemoresistance in ovarian cancer. Oncogene 28 (49), 4353–4363. https://doi.org/10.1038/onc.2009.289.

Tan, T.T., Degenhardt, K., Nelson, D.A., Beaudoin, B., Nieves-Neira, W., Bouillet, P., et al., 2005. Key roles of BIM-driven apoptosis in epithelial tumors and rational chemotherapy. Cancer Cell 7 (3), 227–238. https://doi.org/10.1016/j.ccr.2005.02.008.

Tsavaris, N., Kosmas, C., Vadiaka, M., Kanelopoulos, P., Boulamatsis, D., 2002. Immune changes in patients with advanced breast cancer undergoing chemotherapy with taxanes. Br. J. Cancer 87 (1), 21–27. https://doi.org/10.1038/sj.bjc.6600347.

Varghese, E., Samuel, S.M., Sadiq, Z., Kubatka, P., Liskova, A., Benacka, J., et al., 2019. Anti-cancer agents in proliferation and cell death: the calcium connection. Int. J. Mol. Sci. 20 (12). https://doi.org/10.3390/ijms20123017.

Vicari, A.P., Luu, R., Zhang, N., Patel, S., Makinen, S.R., Hanson, D.C., et al., 2009. Paclitaxel reduces regulatory T cell numbers and inhibitory function and enhances the anti-tumor effects of the TLR9 agonist PF-3512676 in the mouse. Cancer Immunol. Immunother. 58 (4), 615–628. https://doi.org/10.1007/s00262-008-0586-2.

Wanderley, C.W., Colon, D.F., Luiz, J.P.M., Oliveira, F.F., Viacava, P.R., Leite, C.A., et al., 2018. Paclitaxel reduces tumor growth by reprogramming tumor-associated macrophages to an M1 profile in a TLR4-dependent manner. Cancer Res. 78 (20), 5891–5900. https://doi.org/10.1158/0008-5472.CAN-17-3480.

Wang, J., Kobayashi, M., Han, M., Choi, S., Takano, M., Hashino, S., et al., 2002. MyD88 is involved in the signalling pathway for Taxol-induced apoptosis and TNF-alpha expression in human myelomonocytic cells. Br. J. Haematol. 118 (2), 638–645. https://doi.org/10.1046/j.1365-2141.2002.03645.x.

Wang, T.H., Wang, H.S., Ichijo, H., Giannakakou, P., Foster, J.S., Fojo, T., et al., 1998. Microtubule-interfering agents activate c-Jun N-terminal kinase/stress-activated protein kinase through both Ras and apoptosis signal-regulating kinase pathways. J. Biol. Chem. 273 (9), 4928–4936. https://doi.org/10.1074/jbc.273.9.4928.

Wang, T.H., Wang, H.S., Soong, Y.K., 2000. Paclitaxel-induced cell death: where the cell cycle and apoptosis come together. Cancer 88 (11), 2619–2628. https://doi.org/10.1002/1097-0142(20000601)88:11<2619::aid-cncr26>3.0.co;2-j.

Weaver, B.A., 2014. How Taxol/paclitaxel kills cancer cells. Mol. Biol. Cell 25 (18), 2677–2681. https://doi.org/10.1091/mbc.E14-04-0916.

Whitaker, R.H., Placzek, W.J., 2019. Regulating the BCL2 family to improve sensitivity to microtubule targeting agents. Cells 8 (4). https://doi.org/10.3390/cells8040346.

Yamaguchi, T., Fushida, S., Yamamoto, Y., Tsukada, T., Kinoshita, J., Oyama, K., et al., 2017. Low-dose paclitaxel suppresses the induction of M2 macrophages in gastric cancer. Oncol. Rep. 37 (6), 3341–3350. https://doi.org/10.3892/or.2017.5586.

Yamamoto, K., Ichijo, H., Korsmeyer, S.J., 1999. BCL-2 is phosphorylated and inactivated by an ASK1/Jun N-terminal protein kinase pathway normally activated at G(2)/M. Mol. Cell. Biol. 19 (12), 8469–8478. https://doi.org/10.1128/mcb.19.12.8469.

Yusuf, R.Z., Duan, Z., Lamendola, D.E., Penson, R.T., Seiden, M.V., 2003. Paclitaxel resistance: molecular mechanisms and pharmacologic manipulation. Curr. Cancer Drug Targets 3 (1), 1–19. https://doi.org/10.2174/1568009033333754.

Yvon, A.M., Wadsworth, P., Jordan, M.A., 1999. Taxol suppresses dynamics of individual microtubules in living human tumor cells. Mol. Biol. Cell 10 (4), 947–959. https://doi.org/10.1091/mbc.10.4.947.

Zhang, L., Dermawan, K., Jin, M., Liu, R., Zheng, H., Xu, L., et al., 2008. Differential impairment of regulatory T cells rather than effector T cells by paclitaxel-based chemotherapy. Clin. Immunol. 129 (2), 219–229. https://doi.org/10.1016/j.clim.2008.07.013.

Zhu, L., Chen, L., 2019. Progress in research on paclitaxel and tumor immunotherapy. Cell. Mol. Biol. Lett. 24, 40. https://doi.org/10.1186/s11658-019-0164-y.

4

Application of nanocarriers for paclitaxel delivery and chemotherapy of cancer

Saloni Malla[a], Rabin Neupane[a], Sai H.S. Boddu[b,c], Mariam Sami Abou-Dahech[a], Mariah Pasternak[a], Noor Hussein[a], Charles R. Ashby, Jr[d], Yuan Tang[e], R. Jayachandra Babu[f], and Amit K. Tiwari[a]

[a]*Department of Pharmacology and Experimental Therapeutics, College of Pharmacy & Pharmaceutical Sciences, University of Toledo, Toledo, OH, United States,* [b]*Department of Pharmaceutical Sciences, College of Pharmacy and Heath Sciences, Ajman University, Ajman, United Arab Emirates,* [c]*Center of Medical and Bio-allied Health Sciences Research, Ajman University, Ajman, United Arab Emirates,* [d]*Department of Pharmaceutical Sciences, College of Pharmacy and Health Sciences, St. John's University, Queens, NY, United States,* [e]*Department of Bioengineering, College of Engineering, University of Toledo, Toledo, OH, United States,* [f]*Department of Drug Discovery & Development, Harrison School of Pharmacy, Auburn University, Auburn, AL, United States*

4.1 Introduction

Paclitaxel (PTX) was the first taxane-derived antineoplastic drug used to treat certain types of cancer (Jordan and Wilson, 2004). PTX was identified after screening 35,000 medicinal plants for cytotoxic efficacy by the U.S. National Cancer Institute in 1958 and later was extracted from the bark of the Pacific Yew tree *Taxus brevifolia*, in 1971 (Farina, 1995; Wani et al., 1971). It is a polyoxygenated, cyclic diterpenoid containing a taxane nucleus. PTX produces its efficacy by targeting and stabilizing microtubules during the G2-M phase of cell cycle, thereby inhibiting the depolymerization of microtubules into soluble tubulin, which inhibits cancer cell growth (Jennewein and Croteau, 2001; Surapaneni et al., 2012).

Although PTX is considered as a gold standard chemotherapeutic drug with a broad range of anti-cancer efficacy in various cancers such as breast, ovarian, urothelial, head and neck, and non-small cell lung carcinoma and Kaposi's sarcoma (Kim et al., 2001; Bajorin, 2000),

Paclitaxel. https://doi.org/10.1016/B978-0-323-90951-8.00004-7
Copyright © 2022 Elsevier Inc. All rights reserved.

its clinical use has been limited due to its poor solubility and bioavailability at the target tissue (Surapaneni et al., 2012). Paclitaxel is a class IV drug (low solubility, low permeability) based on the biopharmaceutical classification system (BCS) (Ghadi and Dand, 2017). The presence of several hydrophobic moieties and the absence of ionizable functional groups decreases its aqueous solubility to less than 0.01 mg/mL (Straubinger, 1995; Surapaneni et al., 2012). Therefore, the solubilization of PTX requires organic solvents (1:1, v/v mixture), including polyoxyethylated castor oil, yielding Cremophor EL°, and dehydrated ethanol, which is commercially available as Taxol°, which is diluted before parenteral administration. However, a major drawback with Taxol° is the use of Cremophor EL which has a few adverse effects including hypersensitivity reactions, nephrotoxicity and neurotoxicity in a certain patient population (Singla et al., 2002). To overcome poor solubility and permeability issues, various nano-formulation strategies have been used. The nanoparticle (NP) formulation of PTX has been shown to (1) decrease the toxicity due to Cremophor EL° and (2) reduce the precipitation of PTX upon dilution (Paál et al., 2001; Surapaneni et al., 2012). Moreover, of the targetable properties of the nanocarriers specifically at the tumor sites enable decreasing the dose and thereby the dose-related side effects in the cancer therapy (Freitas and Müller, 1999). In 2005, Abraxane°, a NP formulation of PTX complexed human serum albumin (HSA), was approved by the FDA for the treatment of metastatic breast cancer and was later approved for the treatment of metastatic pancreatic adenocarcinoma or locally advanced or metastatic non-small cell lung cancer (Green et al., 2006). HSA is the most abundant protein that is a natural depot of hydrophobic molecules as it has several domains for hydrophobic binding that can also accommodate fatty acids (Stillwell, 2016). The formulation of PTX with HSA increased the maximum tolerated dose (MTD) of PTX, and thus, decreased the administration time of Abraxane° (30 min) compared to Taxol° (3 or 24 h) (Gradishar et al., 2005). In addition, Abraxane° improved PTX's pharmacokinetic profile, eliminated hypersensitivity reactions associated with the use of Cremophor EL° (as an excipient), increased tolerance and the overall response rate (Gradishar et al., 2005). Lipusu° is another nano-formulation of PTX approved for the treatment of breast, ovarian and non-small cell lung cancers (Bernabeu et al., 2017; Koudelka and Turánek, 2012). It is formulated using cholesterol, lecithin and PTX, and this significantly reduced the incidence and magnitude of the adverse effects compared to Taxol° (Xu et al., 2013).

Nanocarriers offer a promising therapeutic platform to directly deliver drugs, such as PTX, to cancer cells, where they are taken into the cells by endocytosis (Markman et al., 2013; Qi et al., 2017). These nanocarriers improve PTX delivery within the tumor site through passive targeting, i.e., via the enhanced permeability and retention (EPR) effect or active targeting, i.e., tumor-site targeting by targeting ligands, thus circumventing P-glycoprotein (P-gp or MDR1 or ABCB1; paclitaxel is effluxed from certain cancer cells by P-gp) mediated drug efflux, greatly decreasing the likelihood of drug resistance (Davis et al., 2008; Ahmad et al., 2016). Recently, several nanocarrier-based drug delivery systems, such as NPs, liposomes and micelles incorporated with PTX alone or in combination with other chemotherapeutic drugs, have been evaluated to determine their potential to overcome PTX resistance. This chapter will discuss various nanocarrier systems and their application, in pre-clinical and clinical studies related to the effective delivery of PTX and circumvention of PTX resistance and recent patents for PTX formulations.

4.2 Nanoparticles

(i) Polymers that are optimal for preparation of NPs should be biodegradable, biocompatible, have high chemical stability and versatility and be non-immunogenic (Choudhury et al., 2019). Consequently, polymeric NPs are more frequently used in drug delivery than inorganic and metallic NPs (Rai et al., 2019; Vahed et al., 2019). The drug is either encapsulated or embedded in the polymeric matrix of NPs (Marin et al., 2013). The release of drugs from NPs can be controlled by the polymer composition, as well as by an external stimuli (Kamaly et al., 2016). The NPs can be designed such that the microenvironment specific to a disease condition, such as change in pH, higher levels of free radicals, etc., can induce drug release from these particles, facilitating their site-specific delivery (Kamaly et al., 2016). NPs are typically transported into cells by endocytosis (Iversen et al., 2011). Therefore, their size and surface characteristics determine the rate of uptake (Iversen et al., 2011). The surface modification of NPs with an appropriate ligand makes them suitable for targeted delivery to specific receptors in the body (Vahed et al., 2019). Furthermore, the surface modification of particles can enhance the half-life of NPs. For example, PEGylated NPs can escape endocytosis by the reticuloendothelial system (RES), if the particle size is < 400 nm. In addition, the nano-size of the particles produces enhanced permeability and retention effect (EPR) in tumor tissues (Tee et al., 2019).

(ii) Advantage of PTX NPs/nano-formulations includes
   1. Targeted delivery can be achieved for the PTX NPs < 200 nm, which accumulate in tumors by the EPR effect (Acharya and Sahoo, 2011)

2. Improved aqueous solubility of PTX can be achieved after cross-linking with hydrophilic polymers or enclosed within NPs that have both hydrophobic and hydrophilic moieties (Guo et al., 2020; Ma and Mumper, 2013)
3. Nanoformulations enable combination therapy of PTX with other drugs (Phung et al., 2020)
4. Decreased incidence of the development of drug resistance compared to the free PTX can be accomplished (Diab et al., 2020; Phung et al., 2020)
5. By choosing an appropriate ligands based on the tumor biology, targeted delivery to single or multiple targets in the tumor cells can be achieved (Houdaihed et al., 2020)
6. Nanoformulations can circumvent using certain excipients such as Cremophor EL° in the currently marketed PTX formulation, which cause severe adverse reactions in some patients (Guo et al., 2020)
7. Overall, a better pharmacokinetic profile, tumor-specific delivery leading to better efficacy of the nano-formulations cancer management can be achieved. For example, Abraxane° (albumin NPs of PTX) has greater efficacy compared to Taxol° (micelle-based formulation of PTX) in the treatment of metastatic breast cancer (Dranitsaris et al., 2016; Yuan et al., 2020). Although the plasma concentration and tissue distribution of Abraxane° are lower than that of Taxol°, the former produces higher tumor/plasma concentration than the latter. Yuan et al. reported that the NP-based formulation of PTX had a greater permeation in cancer stem-like cells compared to a micelle-based formulation (Yuan et al., 2020).

4.2.1 Pharmaceutical aspects of polymeric NPs preparation

(i) There are different methods used for the preparation of polymeric NPs. The choice of a method depends on the physicochemical properties of the drug to be encapsulated as well as the choice of polymer for the preparation of NPs. The characteristics common to these methods include (1) the preparation of a nano-dispersion of polymer solution where after dispersion, the polymers are precipitated as particles, (2) freeze-drying of particles, and (3) suspension of NPs in an aqueous solution (Murakami et al., 2000; Rao et al., 2017; Rajput, 2015).

(ii) Crucho and Barros (2017) summarized the different process parameters that influence the particle size based on the method used. In their review, important process parameters, which are common in different methods of NP preparation, were discussed.

The most frequently used approach for NPs preparation is the emulsification of the polymers dissolved in an organic phase, followed by precipitation of the polymers as an aqueous NP suspension (Deshmukh et al., 2016). Other popular methods include nanoprecipitation and dialysis. Once the NPs are formed, the final product is freeze-dried to obtain a solid final product that is suitable for long-term storage. The lyophilization of the NPs should be carried out using appropriate cryoprotectants, such as trehalose, mannitol, etc., to avoid NP aggregation (Umerska et al., 2018).

(iii) Emulsions, such as oil-in-water (o/w) or water-in-oil-in-water (w/o/w), can be used in NP preparation, depending on the physicochemical characteristics of the drug that is to be incorporated. Hydrophilic drugs can be constituted in a water phase, whereas lipophilic drugs can be incorporated in the oily phase of the emulsion system. The emulsion is then subjected to high shear sonication or homogenization to decrease the size of the dispersed globules. Finally, the organic solvent is evaporated, followed by the precipitation of polymers as NPs. Important process parameters that contribute to particle size control include polymer concentration, surfactant concentration, choice of solvent, etc. (Crucho and Barros, 2017). Dichloromethane, chloroform, and ethyl acetate are the most commonly used organic solvents in the solvent evaporation method, and acetone is most frequently used in the salting-out method (Crucho and Barros, 2017). The presence of organic solvents in NPs, even in trace amounts, could produce toxic effects (Yerlikaya et al., 2013). Therefore, it is best to use solvents known to have a better toxicity profile, such as methylene chloride, acetone, among others (Crucho and Barros, 2017). For the solvent evaporation method of NP preparation, the polymer and surfactant concentration are two important factors that determine the effectiveness of the emulsification process (Crucho and Barros, 2017). Higher polymer concentrations increase the viscous resistance in the emulsion formation process, leading to larger particles size if the energy of ultra-homogenization and sonication is not proportionally increased (Crucho and Barros, 2017). The polymer concentration in the external phase affects the particle size in the emulsion salting-out method (Zweers et al., 2003). Similarly, in nanoprecipitation, increasing the polymer concentration increases particle size (Chorny et al., 2002). In addition, the polymer concentration, the injection rate, and the needle gauge used in injection are determinant parameters of particle size in the nanoprecipitation method (Şimşek et al., 2013). A higher rate of injection and/or smaller sized gauge favors the generation of smaller sized particles during nanoprecipitation (Şimşek et al., 2013).

(iv) The surfactant plays a pivotal role in stabilizing the dispersed phase in the emulsion. Commonly used surfactants are PVA, DMAB (didodecyldimethylammonium bromide) and Pluronic F-68 (Zambaux et al., 1998). Small globules of the dispersed phase in the emulsion require a higher concentration of surfactant for stabilization compared to the larger globules (Crucho and Barros, 2017). Therefore, smaller sized particles can be obtained by increasing the concentration of the surfactants (Zambaux et al., 1998). Depending on the method used to prepare the NPs, the choice of surfactant could affect the particle size. For example, in the solvent diffusion method, DMAB has been reported to produce a better result compared to PVA (Jain et al., 2011; Sahana et al., 2008). However, nanoprecipitation does not necessarily require the use of surfactant. Sonication and homogenization are the most common techniques used to decrease the globule size in emulsion-based NP preparations (Crucho and Barros, 2017). However, the emulsion solvent diffusion and emulsion reverse salting out methods can circumvent the need for high shear forces, such as sonication and homogenization (Crucho and Barros, 2017). When using the solvent evaporation technique, the particles produced by sonication have a more restricted particle distribution and better polydispersity index compared to the homogenization method (Budhian et al., 2007). The use of microfluidics can be a possible alternative to regulate the size and distribution of the droplets formed during emulsion, which, in turn, will produce a better polydispersity index of the NPs (Perez et al., 2015; Karnik et al., 2008). For NPs prepared using a w/o/w emulsion, the particle size is mainly determined by the duration of the sonication in the second step (to achieve w/o/w), rather than the first step where a w/o emulsion is used (Bilati et al., 2003; Iqbal et al., 2015). Different salts, such as magnesium chloride, calcium chloride, magnesium acetate, sodium chloride, etc., are used in the salting-out process, and this requires extensive washing to remove the salt in the NPs (Song et al., 2008; Allémann et al., 1992).

4.2.2 Recent preclinical studies on paclitaxel NPs

(i) The cellular membranes of bacteria, erythrocytes, platelets, and macrophages have been used to modify the surface of NPs (Li et al., 2018). Specific cells are collected, and the membrane is harvested using techniques such as hypotonic lysis and mechanical membrane fragmentation, followed by differential centrifugation (Li et al., 2018). Furthermore, techniques such as extrusion can be used to obtain membrane vesicles (Cao et al., 2020). Utilizing the extrusion technique, Cao et al. (2020) reported the formulation of NPs composed of the membranes from macrophages. After harvesting the membrane from macrophages, extrusion was conducted by passing the preparation through a 400 nm porous polycarbonate to obtain the NPs (Cao et al., 2020). Self-assembled NPs, without excipients, is an approach where a carrier molecule is conjugated with the drug to form the NPs without using metals or polymers (Feng et al., 2020). Feng et al. (2020) reported the use of indocyanine green (ICG) to form NPs with small molecules via self-assembly. This novel approach yielded up to 100% drug loading without the use of excipients. The ICG-templated self-assembly of PTX NPs (ISPN) was efficacious in eliciting antitumor immunity and facilitating the intra-tumoral infiltration of cytotoxic T lymphocytes (CTLs) by photo dynamic-induced immunogenic cell death (ICD) of the tumor cells in vitro and in vivo. The anatomy of the tumor environment, which enhances tumor permeability and retention effect (EPR), contributes to the high efficacy of this delivery system. Overall, it is a novel approach where the antitumor efficacy is produced by two distinct mechanisms using a single carrier system (Feng et al., 2020). Guo et al. formulated RNA four-way junction NPs of paclitaxel that was thermodynamically stable, and this produced an increase in the water solubility of the paclitaxel molecules by 32,000-fold. Twenty-four molecules of paclitaxel could be covalently loaded on to each RNA NP (Guo et al., 2020). This carrier system was shown to have an "X" shaped structure based on cryo-electron microscopy. Upon conjugation of the anti-EGFR aptamer on these particles, specific inhibition of the proliferation of triple negative breast cancer cells was achieved. Furthermore, the systemic administration 8 of $mg\,kg^{-1}$ i.v., every 2 days for a total of five doses of the RNA NPs produced extremely low or undetectable toxicity in mice. The weight of the mice 2 weeks after the last injection was used as indicator of toxicity profile (Guo et al., 2020). Diab et al. (2020) compared the efficacy of paclitaxel- loaded poly lactic-*co*-glycolic acid (PLGA) NPs to a paclitaxel solution, in the hepatoma cell line, HepG2, and the breast adenocarcinoma cells line, MCF-7. The IC_{50} values, expression of PTX resistance gene, Trx1, and the expression of CYP 3A4 and CYP 2C8, which are involved in metabolism of PTX, were determined and compared to cells incubated with paclitaxel solution and the nano-formulation. The IC_{50} value of the NPs treated cells was much lower than that of the cells incubated with the PTX solution, indicating that the nanoformulation had a higher anticancer efficacy. Furthermore, there was a lower expression level of the paclitaxel resistance gene, Txr1, and a higher level of CYP3A4 and CYP2C8 expression in the cells incubated with the NP formulation compared to cells incubated with the PTX. This in vitro study suggests that NPs can decrease resistance to PTX that can occur during chemotherapy (Diab et al., 2020).

(ii) The surface modification of NPs facilitates the targeting and increases the half-life of the delivery system (Madan et al., 2013; Vahed et al., 2019). Targeting can be achieved by conjugating antibodies or peptides onto the surface of the NPs that specifically bind to those receptors which are highly expressed on the targets (Fay and Scott, 2011; Ruoslahti, 2012). Furthermore, the NPs can be coated with polymers, such as PEG, to increase the half-life by avoiding sequestration by the reticuloendothelial cells (Owens III and Peppas, 2006). Recently, there has been an increase in research related to the development of surface coated NPs with different membranes. The membranes from cancer cells, erythrocyte-platelet hybrids, macrophages, etc., have been utilized to increase the circulation time of the NPs (Cao et al., 2020; Fang et al., 2014; Dehaini et al., 2017). Overall, surface modification can produce therapeutic efficacy at a lower dose of the drug, thereby reducing toxicity, which is more frequent and severe at higher doses. In breast cancer, the tumor cells heterogeneously express cell surface receptors, which is the primary cause reason why targeted delivery approaches fail when only one receptor is targeted (Norton et al., 2015). Houdaihed et al. (2020) designed nano-carrier systems for targeting multiple surface receptors to overcome receptor heterogeneity and increase drug efficacy. They formulated surface-modified NPs conjugated with the Fab fragments from trastuzumab and panitumumab to target HER2 and EGFR receptors, respectively, and determined their in vitro efficacy in breast cancer cells with HER2 negative and low EGFR expression (MCF-7, MDA-MB-436) and with high HER2 and moderate EGFR expression (SKBR3). The NPs were loaded with PTX and everolimus (an mTOR inhibitor) to produce synergistic efficacy. The in vitro cytotoxicity of the dual-targeted NPs was superior to HER2 mono-targeted NPs (T-NPs) and untargeted NPs (UT-NPs) and comparable to free PTX-everolimus combination in SKBR3 monolayer cells. However, in case of in vitro SKBR3 spheroids, the cytotoxicity exerted by Dual-NPs and T-NPs were comparable to that of free PTX and everolimus combination. This disparity in results of 2D and 3D cytotoxicity experiments could be due to differential gene expression between 2D and 3D models. However, in case of MCF cells and MDA-MB-436 spheroids, there was no significant difference in spheroid size upon exposure to Dual-NPs, T-NPs, and UT-NPs when compared to free

Chapter 4 Paclitaxel-based formulations in cancer chemotherapy **81**

PTX-everolimus combination. This superior inhibitory activity of free PTX-everolimus compared to NPs is due to facilitated diffusion of free PTX throughout the spheroids due to its smaller size. This study could be further extended in in-vivo models where the targeted delivery could possibly minimize the side effect without compromising the efficacy of the combination (Houdaihed et al., 2020).

(iii) Polyethylene glycol (PEG) is commonly used in the preparation of NPs as it helps to avoid the sequestration of the NPs by the reticuloendothelial system (Owens III and Peppas, 2006). In addition, drugs conjugated with PEG have increased solubility in aqueous solution (D'Souza and Shegokar, 2016). Phung et al. (2020) exploited the inherent character of PEG and conjugated PEG with paclitaxel (PTX) and dihydroartemisinin (DHA). The polymer-drug complex (PTX-PEG-DHA) was further formulated into NPs using the solvent evaporation method. Subsequently, the efficacy of the PTX-PEG-DHA NP formulation was compared to aqueous solutions of PTX and DHA, in HT-29 colon cancer cells. The results indicated that the pegylated nanoparticle formulation of PTX-PEG-DHA was more efficacious in decreasing HT-29 cell proliferation compared to aqueous PTX or DHA Furthermore, the expression of Bcl-2, an anti-apoptotic protein (Belka and Budach, 2002) was significantly decreased in the HT-29 cells incubated with the PTX-PEG-DHA NPs compared to HT-29 cells incubated with aqueous PTX and DHA. Thus, the combination of PTX and PEG in a NP formulation significantly decreased resistance compared to the aqueous formulations of PTX and DHA. In mice xenografted with HT-29 cells, the i.v. administration of the PTX-PEG-DHA formulation (at an equivalent dose of 5 mg/kg of PTX and 10 mg/kg of DHA) significantly decreased tumor volume at 19 days post treatment compared to the mice administered aqueous formulations of 5 mg/kg of PTX and 10 mg/kg of DHA.

(iv) Cao et al. (2020) evaluated the efficacy of macrophage membrane-coated albumin NPs and uncoated albumin NPs of PTX in B16F10 melanoma cells and in in vivo melanoma models. The NP formulation of paclitaxel was significantly more efficacious in inhibiting the growth of B16F10 cell lines compared to an aqueous solution of paclitaxel or uncoated albumin NPs of paclitaxel. In the in vivo experiments, mice were injected subcutaneously in the right flank with B16F10 melanoma cells and were used 1 week later to determine the efficacy of the treatments. Paclitaxel, at a dose equivalent to 1 mg/kg, was injected intravenously as either a paclitaxel solution, uncoated albumin NPs of paclitaxel and macrophage membrane coated albumin

NPs of paclitaxel via the tail vein every other day for 4 dosing cycles. The tumor volume was significantly smaller in animals that were given the macrophage-coated albumin NPs of paclitaxel compared to other treatment groups (paclitaxel solution treated, and uncoated albumin NPs of paclitaxel treated). In the same group (macrophage membrane coated albumin NPs of the paclitaxel-treated group), there was no significant change in a mouse's body weight after the treatment, which indicates a lower toxicity level of the formulation. In addition, the pharmacokinetic experiment indicated that the macrophage-coated albumin NPs of paclitaxel had long half-life and selective accumulation at the tumor site. Yan et al. (2019) modified the NPs with hydroxypropyl-β-cyclodextrin (HP-β-CD) to increase the target specificity of the NPs. The compound, HP-β-CD, is widely used to increase the aqueous solubility of hydrophobic molecules (Pitha et al., 1986). A modified form of HP-β-CD, consisting of biotin and arginine, was used to formulate PTX NPs. Biotin is an essential micronutrient for cell growth and is highly expressed in various types of tumor cells (Ren et al., 2015). In addition, the use of arginine facilitates the internalization of the particles into the cell (Futaki and Nakase, 2017). The modified HP-β-CD NPs were more efficacious than PTX + Cremophor EL® or PTX alone in inhibiting the proliferation of MCF-7 cells.

(v) NPs that are formulated to only release the drug in the target area in a controlled fashion would be ideal for treating various cancers. An external stimulus (such as magnetic field, ultrasound, light rays, etc.) and internal stimulus (pH, redox potential, etc.) are used to induce the release of drugs at the target site. The tumor site has a unique redox gradient that consists of an oxidative extracellular matrix with high levels of ROS, whereas the cytoplasm is highly reductive due to high levels of glutathione (GSH) (Thakkar et al., 2020). Chen et al. (2018) designed thioketal NPs (TKNs) that were responsive to GSH and the ROS, H_2O_2). PTX-TKNs were efficacious in inhibiting the in vitro proliferation of PC-3 prostate cancer cells. Furthermore, the effect of the i.v. administration of TX-TKNs or PTX alone was determined in mice bearing PC-3 tumors. Over a 20-day treatment period, Tumor size was PTX in an aqueous solution decreased tumor growth by 77%, compared to 65% in animals treated with PTX-TKNs. However, the plasma levels of alanine aminotransferase (ALT) and aspartate aminotransferase (AST) were significantly greater in the PTX alone group compared to the PTX-TKNs treated group. Similarly, the use of superparamagnetic iron oxide (SPIOs)-loaded polymeric NPs can be a potential strategy for drug targeting in the presence of an external magnetic field. Ganipineni et al. (2019)

compared the in vivo efficacy of passive targeting of PTX-loaded NPs (without SPIO or RGD), active targeting of PTX NPs using RGD tripeptides which bind to $\alpha_v\beta_3$ integrin, and active targeting of PTX NPs with RGD as well as the SPIO for magnetic targeting in a glioblastoma (GBM) model. The highly proliferating and infiltrating glioblastoma cells express $\alpha_v\beta_3$ integrin, which can be targeted using the RGD peptide. In vitro cytotoxicity assays indicated that in U87MG cells, the IC 50 value for the paclitaxel NPs was 1 ng/mL. The SPIO-loaded NPs and NPs without SPIO had similar cytotoxicity profiles, which were significantly different compared to blank NPs without PTX, indicating that cytotoxicity is solely dependent on the PTX in the NPs. Furthermore, the in vivo efficacy of targeting using a magnetic field was evaluated. Six i.v. doses equivalent to 5 mg/kg of PTX were given to mice bearing orthotopic U87MG tumors over 3 weeks to the following groups: (1) PTX-SPIO NPs without magnet-passive targeting; (2) PTX-SPIO NPs with magnet-active targeting; and (3) PTX-SPIO-RGD with a magnet. Similarly, the control group of mice was administered with six i.v. doses of normal saline. The median survival period and the decrease in tumor volume were used as indicators of efficacy. The passive targeting showed a poor survival period due to an insufficient quantity of NPs reaching the tumor site. The use of active targeting significantly decreased tumor growth and increased the median survival period. Moreover, the animals treated with PTX-SPIO NPs with a magnet had the smallest tumor volumes. Interestingly, the combination of RGD functionalization and magnetic targeting did not significantly increase the median survival period compared to the PTX-SPIO NPS with a magnet. When using a magnet as an external stimulus, the efficacy may be dependent on the proximity of the external magnetic to the target site, the time and strength of the applied magnetic field, and efficacy could be decreased or abrogated as the distance from the magnetic field increases (Ganipineni et al., 2019).

4.3 Liposomes

(i) Liposomes, also known as phospholipid vesicles, are amphiphilic molecules consisting of one or more lipid bilayers with an aqueous interior (Akbarzadeh et al., 2013; Çağdaş et al., 2014). Liposomes can enclose hydrophilic/lyophobic drugs in its aqueous space, and hydrophobic/lipophilic drugs are sequestered in the lipid bilayer (Sercombe et al., 2015). Due to their size, biocompatibility, and the incorporation of hydrophilic and hydrophobic drugs, liposomes has been extensively investigated as a drug carrier to increase the therapeutic efficacy and index of various drugs (Akbarzadeh et al., 2013). Liposomes can be categorized according to their size: small ($\leq$ 100 nm), intermediate (100–250 nm), large ($\geq$ 250 nm), and giant (> 1 μm) (Maja et al., 2020). Liposomes can also be categorized based on lamellarity such as unilamellar and multilamellar types, or surface charge, such as cationic, anionic, or neutral types (Zylberberg and Matosevic, 2016). The alteration of these parameters, along with lipid composition, preparation method, and type of surface ligands, can produce changes in the properties of the liposomes (Sercombe et al., 2015). The delivery of drugs to cells by liposomes is due to cell membrane adsorption, endocytosis, liposome—lipid bilayer fusion with the lipoidal cell membrane, and lipid exchange (Yadav et al., 2017).

(ii) Advantages of liposomal drug encapsulation includes
1. Similar to NPs, PTX liposomes can provide a higher therapeutic index and efficacy in treating various cancers (Anwekar et al., 2011).
2. Improved pharmacokinetic and pharmacodynamic profile of PTX can be achieved (Yadav et al., 2017).
3. Liposomal formulations enable a decreased frequency and magnitude of adverse and toxic drug effects for PTX (Yadav et al., 2017; Anwekar et al., 2011).
4. Active targeting due to surface modifications of liposomes can be achieved (Akbarzadeh et al., 2013).
5. Efficient delivery of poorly water-soluble anticancer drugs such as PTX can be achieved by increasing their solubility, increasing the likelihood of sustained-release, protecting the drug from biodegradation (Le et al., 2021).

4.3.1 Pharmaceutical aspects of liposome preparation

(i) Liposomes have to be specifically prepared based on the specific drug to be encapsulated. Drug encapsulation can be achieved either by passive loading, where the incorporation of the drug occurs during liposome formation or active loading, where drug loading occurs after liposome formation (Akbarzadeh et al., 2013). One of the most commonly used methods of passive loading is solvent dispersion. The first step of solvent dispersion is lipid solubilization, which involves the homogeneous mixing of organic lipids, followed by the addition of drug molecules. Hydrophobic drug molecules are added to the organic solution, whereas hydrophilic drug molecules are added to the aqueous solution. The second step is the evaporation of the organic solvent, typically done using a rotary evaporator. The phospholipids will be dried on the wall of the round bottom glass flask, resulting in a thin lipid film that is then exposed to a vacuum overnight to remove any residual solvent. Also, lipids can be subjected to an alternative freeze-dry method, where organic solvent at a temperature above the freeze-drying condenser temperature is used. The third step involves lipid hydration, where an appropriate buffer, with its temperature above the phospholipid phase-transition temperature, is required. A milky suspension of lipids is obtained after the solution is sonicated or vortex until complete incorporation occurs. The complete swelling of the suspension results in the formation of multi-lamellar large vesicles (MLVs), which are > 500 nm in size (Shashi et al., 2012). Discarding the upper layer of MLVs after centrifugation yields large unilamellar vesicles (LUVs). The final step yields a small, unilamellar vesicle (20–100 nm) from MLVs. Since MLVs are heterogeneous and large in size, strategies such as extrusion, ultra-sonication or micro-emulsification are used. Extrusion is the phenomenon of passing the liposomes through a membrane filter of defined pore size such as 0.1 0.2, 0.5, 2 μM, etc., that produces uniform size, followed by freeze–thaw sonication, ultra-sonication, sonication and homogenization (Olson et al., 1979; Ong et al., 2016). Similarly, the encapsulation efficiency (EE) or lamellarity of liposomes can be modified with processes such as freeze-thawing, freeze-drying, ion gradient, or pH gradient. Freeze-thawing involves applying rapid thawing and freezing cycles to SUVs to form LUVs (Llu and Yonetani, 1994; Pick, 1981). Freeze-drying, also known as lyophilization, removes water from the final product. It is used for thermolabile dry compounds that are susceptible to damage from heat. However, this process may produce leakage of the drug during reconstitution (Anwekar et al., 2011).

4.3.2 Recent preclinical studies on liposomes

(i) Liposomes are formulated as one or more lipid bilayers enclosing an aqueous core. Soy lecithin (SL), obtained from Soybean, can be used to form lipid bilayers of the liposome. It is an unsaturated, natural phospholipid that provides better stability due to the presence of fewer polyunsaturated fatty acids and has a lower chance of proteins or pathogen contamination than saturated phospholipids of animal origin (Le et al., 2019).

(ii) Le et al. (2021) formulated PTX-incorporated SL liposomes where the surface was partially modified by methoxy polyethylene glycol conjugated with cholesterol (mPEG-Chol) to increase the solubility PTX (Le et al., 2021). Liposomal surface modification is one of the techniques that efficiently deliver therapeutic drugs to the target sites as a result of specific modifications, such as PEGylation or covalent bonding of tissue-specific antibodies (Büyükköroğlu et al., 2016), peptides (Luo et al., 2013) or proteins (Wang et al., 2017), to the surface of the liposomes. Using MCF-7 breast cancer cells, the results indicated that formulating PTX liposomes using SLP and mPEG PTX produced sustained and controlled delivery of PTX as compared to Taxol® (Le et al., 2021). Similar findings were reported by Yoshizawa et al. (2011) in C26-colon tumor-bearing mice. A lower deposition of PTX was observed with PTX-liposomes formulated with hydrogenated soybean phosphatidylcholine (HSPC), cholesterol (Chol), and distearoylphosphatidylethanolamine-N-[methoxy poly (ethylene glycol)-2000] (PEG-DSPE) (HSPC:Chol:PEG-DSPE:PTX=90:10:5:8, molar ratio), into cells of the reticuloendothelial system, indicating an efficient enhanced permeability and retention (EPR) effect in the tumors. This resulted in a significant delivery of PTX to tumors, as well as increased anti-tumor efficacy in mice, compared to PTX-naked (without PEG) liposomes (HSPC:Chol:PTX=90:10:8, molar ratio) (Yoshizawa et al., 2011). PEGylated PTX liposomes, consisting of N,N-dimethyl-1,3-propanediamine with an amido linkage to DSPE-PEG$_{2000}$ (PTX-Lip2N), had significantly decreased the tumor volume (mean tumor volume of $0.129 \pm 0.022\,cm^3$ for PTX-Lip2N, $P < 0.01$ vs $0.221 \pm 0.036\,cm^3$ for PTX vs $0.331 \pm 0.043\,cm^3$ for control group) and the tumor proliferation rate ($24.2 \pm 6.65\%$ for PTX-Lip2N vs $56.7 \pm 8.16\%$, $P < 0.001$ for PTX vs $78.3 \pm 8.75\%$, $P < 0.05$ for control-treated group) in a patient-derived, mouse orthotopic xenograft (PDOX) model (Chen et al., 2020). This PEG conjugated cationic liposomes (CL) packed with PTX increased the cellular uptake of PTX and increased cytotoxicity in the prostate cancer cell line, CRL1435, and the melanoma cell line, M21, compared to CL-PTX. The positive charges allowed for the compounds to passively accumulate in tumor tissue (Campbell et al., 2002; Schmitt-Sody et al., 2003; Thurston et al., 1998), in combination with PEG to delay recognition and clearance by RES and increase circulation time (Allen and Cullis, 2013; Allen et al., 1995), produced anti-tumor efficacy. Cationic liposomes have been shown to target tumor blood vessels and disrupt tumor microvasculature (Denekamp, 1982). The administration of cationic liposomes (containing 5 mg/kg of paclitaxel; LipoPac) by continuous i.v. infusion resulted in a fourfold higher selectivity in amelanotic hamster melanoma A-Mel-3 tumors compared to the surrounding normal tissues (Schmitt-Sody et al., 2003). In the same in vivo hamster melanoma (using A-Mel-3 cells) model, tumor growth was also significantly inhibited (tumor volumes of $1.7 \pm 0.3 \, cm^3$ for LipoPac vs $10.7 \pm 1.7 \, cm^3$ for Taxol° vs $17.7 \pm 1.9 \, cm^3$ for the control group) and the prolonged absence of metastasis, compared to PTX alone (Schmitt-Sody et al., 2003). The positive charge in the cationic PTX-liposomes (DLD/PTX-Lips) was due to the use of the pH sensitizer, DLD (DSPE conjugated by L-lysine with 2,3-dimethylmaleic anhydride (DMA). This formulation not only increased the accumulation of PTX in 4T1-murine breast tumors (pH 6.8) compared to normal tissues (pH 7.4) after 2.5 h of incubation, but it also facilitated the endosomal/lysosomal escape of cationic DLD/PTX-Lips after 4.5 h of incubation. Furthermore, DLD/PTX-Lips had the highest 4T1-tumor inhibition rate (57.4%), compared to free PTX (25.1%) and liposomes without a charged conversion function, i.e., without DLD (30.4%) (Jiang et al., 2016). In a humanized SCID melanoma (A-375) murine model, liposomes formulated with 1,2-dioleoyl-3-trimethylammonium propane and 1,2-dioleoyl-sn-glycero-3-phosphocholine produced a 2.5-fold greater inhibition of cancer cell growth in mice, compared to mice treated PTX or glucose alone. PTX liposomes significantly decreased tumor invasiveness as indicated by distinct borders around the tumor and prolonged survival compared to that of the PTX or the vehicle treatment group (Kunstfeld et al., 2003).

(iii) Generally, the pH of healthy tissues (typically pH 7.4) differs from tumor sites (pH ~ 6.8) and endosomes (pH ~ 5.5). Liposomes that are pH-controllable utilize the charge conversion in the surrounding acidic environment, which increases its cellular uptake and facilitates its endosomal escape (Mo et al., 2012). The pH sensitivity of liposomes can be achieved by utilizing pH sensitizers such as Diolyl phosphatidylethanolamine (DOPE) (Monteiro et al., 2019) and pH-sensitive PEG derivatives (such as PEG_{5k} linked to phosphatidylethanolamine (PE) by hydrazone bond) (Zhang et al., 2015) produces liposome instability under acidic conditions. PTX-loaded liposomes consist of pH-responsive DOPE, with cholesterylhemisuccinate (CHEMS), DSPE-PEG$_{2000}$ and DSPE-PEG2000-folate, at a molar ration of 5.7:3.8:0.45:0.05, yielding cytotoxicity efficacy in MDA-MB-231 breast cancer cell line. This liposomal formulation increased the percentage of apoptotic cells by fourfold compared to control and increased the number of cells in G0/G1 phase compared to Taxol° and pH-sensitive PTX-liposomes without folate, suggesting DNA fragmentation (hallmark of apoptosis (Bortner et al., 1995)) in cells incubated with the PTX-folate-liposomal formulation. In an in vivo MDA-MB-231 tumor model, tumor growth was inhibited by 80%, compared to the PTX dispersion and PTX-liposomes without folate conjugation (Monteiro et al., 2019).

(iv) Utilizing the pH-responsive property of $H_7K(R_2)_2$, Zheng et al. (2018) formulated PTX-containing inorganic material and superparamagnetic iron oxide (SPIO)-loaded sterically on stabilized liposome (SSL) (PTX/SPIO-SSL-$H_7K(R_2)_2$). The i.v. administration of PTX/SPIO-SSL- $H_7K(R_2)_2$ (15 mg/kg, i.v., every fourth day) produced a 90% decrease in the growth of tumors in mice xenografted with MDA-MB-231 breast cancer cells, with mean tumor sizes of 177 ± 85 mm^3 compared to the PTX-SSL (70% and 531 ± 128 mm^3), PTX/SPIO-SSL (70% and 493 ± 154 mm^3) and the control groups (70% and 1768 ± 614 mm^3). In addition, PTX/SPIO-SSL-$H_7K(R_2)_2$ significantly increased the number of apoptotic cells (determined by the fluorescence area (about $0.14 \, \mu m^2$, $P < 0.01$), compared to mice administered PTX-SPIO-SSL and PTX-SSL (about $0.04 \, \mu m^2$) and the control group (about $0.01 \, \mu m^2$). Similarly, Bao et al. (2014) developed an inorganic liposomal system encapsulating PTX-conjugated gold NPs, using PEG$_{400}$ (PTX-PEG$_{400}$-GNP-Lips). When PTX-PEG$_{400}$-GNP-Lips was administered via tail vein injection at 7 mg/kg every other day to male ICR mice, the results of this eight-day of treatment indicated that the relative tumor volume was lowest (0.5 cm^3) in the PTX-PEG$_{400}$-GNP-Lips group compared to Taxol® (1.2 cm^3) and the control group (2.1 cm^3) (Bao et al., 2014). The administration of losartan (40 mg/kg via intraperitoneal injection), an angiotensin II receptor antagonist, followed by the administration of cleavable, pH triggered PTX liposomes (5 mg/kg every 3 days via caudal vein injection), formulated using PEG$_{5000}$-Hydrazone-PE and DSPE-PEG$_{2000}$-R8, increased the depletion of collagen and inhibited the 4T1 (murine mammary carcinoma) tumor growth by 59.8%, compared to 37.8% for the PTX-cleavable-liposomes (Zhang et al., 2015). Similarly, the incorporation of weak-acid molecules, such as succinic acid, in PTX-liposomes (PTX-SA LPs, containing 8 mg/kg of PTX), produced an encapsulation efficiency of 97.2%, a fourfold increase in its half-life, a greater tumor penetration, and a significant inhibition ($P < 0.001$) of tumor growth (tumor volume of 400 mm^3 vs 700 mm^3 for Taxol® and 1300 mm^3 for the control group) in a mouse 4T1 xenograft model, as compared to Taxol® (Yu et al., 2020).

(v) The i.v. infusion of multilamellar liposomes, containing concentric phospholipid layers separated by a water layer (Shaheen et al., 2006) and loaded with PTX, also called L-pac and composed of phosphatidylcholine (PC) and phosphatidyl glycerol (PG) (PC:PG:PTX=9:1:3), at a dose of 40 mg/kg equivalent to $22 \, \mu Ci$ of ^3H-paclitaxel, in male Sprague–Dawley rats, produced a delayed release of PTX from plasma to peripheral tissues. Additionally, the apparent tissue partition coefficient value for

L-pac was fivefold lower in bone marrow, a major site of toxicity of Taxol®, compared to paclitaxel formulated in Cremophor EL® (Fetterly and Straubinger, 2003). PTX-coated liposomes, containing soya lecithin, cholesterol, stearyl amine (SA), and PTX, were formulated into multilayer liposomes by coating the cationic surface with anionic polyacrylic acid (PAA), which was re-coated with a cationic chitosan aqueous solution (chitosan-PAA-PTX-liposomes) (Chen et al., 2014). In this study, the results indicated that the chitosan layer slows down the release of PTX from the liposomes and increases the uptake of PTX in human cervical cancer cells (HeLa), compared to PAA-PTX-liposomes and PTX-liposomes.

(vi) One challenge in the development of PTX-liposomes is their instability (Hong et al., 2016; Kan et al., 2011). The asymmetric and bulky nature of PTX, in combination with its marked hydrophobicity, makes the PTX loading difficult. The molar PTX to phospholipid ratio for PTX is 4% (Kan et al., 2011). The loading of PTX beyond this range will result in abrupt PTX release due to the clustering of the liposomal layers or precipitation of the drug during storage (Hong et al., 2016). Interestingly, the PEGylation of liposomes causes premature drug release (Kannan et al., 2015). This rapid drug leakage and the incorporation of lipids, such as DSPE-PEG$_{2000,}$ occurs due to the micelle-forming property of PEGylated lipids (Fujie and Yoshimoto, 2019). Saturated phosphatidylcholine (PC), with the absence of bonds that are susceptible to oxidation and high phase-transition temperature, can be used in liposomes to increase the retention of compounds (Mattjus and Slotte, 1996; Payton et al., 2013). However, some studies have reported that saturated PC in a liposome preparation increases the membrane rigidity, thus, limiting the penetration of PTX into the hydrophobic bilayers (Sharma and Straubinger, 1994; Kannan et al., 2015; Hong et al., 2015). One approach to increase the stability of liposomes based on PEGylated/saturated PC, is the incorporation of medium-chain triglyceride (MCT) or long-chain triglyceride (LCT) (Hong et al., 2016). The incorporation of triglyceride in PTX-containing DMPC:CHOL:PE:PEG (1,2-dimyristoyl-sn-glycero-3-phosphocholine:cholesterol:N-(Carbonyl-methoxypolyethyleneglycol 2000)-1,2-distearoylsn-glycero-3-phospho-ethanolamine) liposomes resulted in the formation of multilamellar liposomes that prevented time-dependent PTX leakage, and the formulation was stabilized, which yielded an injectable PTX formulation.

(vii) In an in vivo SCC7 squamous tumor bearing-mouse model, 15 mg/kg i.v. of either PTX-liposomes and Taxol®, were of 15 mg/kg of PTX, significantly inhibited tumor growth, with mean tumor volumes of $700\,mm^3$ vs $1000\,mm^3$ for Taxol® and $1700\,mm^3$ for the control group). However, Taxol® produced an 11% decrease in body weight, whereas the triglyceride-incorporated PTX-liposomes did not significantly affect body weight. PTX liposomes produced a 5.3-fold decrease in hemolyzed RBC, compared to Taxol® (Hong et al., 2016). Another approach to overcome the adverse effects produced by liposomes is to formulate liposomes in a dry-stable form, known as proliposomes, that can be converted into liposomes with water (Payne et al., 1986). Proliposomes are a mixture of the drug and phospholipid that are consistently distributed throughout carbohydrate carrier particles (Khan et al., 2020). PTX-proliposome tablets, consisting of soya phosphatidylcholine (SPC) and lactose monohydrate (LMH) was reported to decrease the growth (60% inhibition) of human lung fibroblast cancer cell line, MRC-5 SV2, and this formulation was not toxic to normal human lung fibroblast MRC-5 cells (Khan et al., 2020). Coating PTX-loaded liposomes with silica particles is another way of increasing stability. Ingle et al. (2018) formulated PTX liposomes containing PC, cholesterol and stearyl amine, which were then surface-coated with silica to form PTX-liposils. Compared to the conventional liposomes, liposils had significantly greater stability over 6 months and had a longer half-life in rats at a dose of $5\,mg/kg$ i.v., compared to the same doses of PTX liposomes and Taxol®.

(viii) Recently, the use of a bacteriobot, where bacteria are attached to liposomal microcargo to deliver drugs for cancer therapy, has garnered significant attention (Yoo et al., 2011; Akin et al., 2007; Park et al., 2013). Nguyen et al. (2016) formulated *Salmonella typhimurium* bacteriobot, coated with 1% chitosan to adhere to the PTX liposomes, using egg PC and DSPE-PEG$_{2000}$, with biotin displayed on their outer surface proteins. Compared to PTX-loaded liposomes ($IC_{50}=21.91\pm0.74\,\mu g/mL$), PTX-loaded bacteriobots produced a more potent effect ($IC_{50}=16.48\pm0.43\,\mu g/mL$) in mammary 4T1 cells, without altering the growth of normal mouse embryonic fibroblast cells NIH/3T3. This could be due to the longer and faster movement of PTX-loaded bacteriobots toward the tumor sites, as compared to PTX-loaded liposomes, PTX-loaded bacteriobots had a significantly longer displacement ($185.58\pm0.74\,\mu m$ vs $23.99\pm8.48\,\mu m$) and higher velocity ($3.09\pm0.44\,\mu m/s$ vs $0.40\pm80.14\,\mu m/s$) (Nguyen et al., 2016). Another potential approach for efficacious tumor therapy is the use of thermosensitive liposomes (TSL) TSLs release encapsulated drugs upon conversion to the liquid phase from its original gel phase at the phase transition temperature (ranging from 39 to 42°C (Kong et al., 2000; De Smet et al.,

2010). This heat-responsive nanosystem retains drugs in the bloodstream, where the temperature is 37 °C and releases the incorporated drugs at locally—heated tumors, up to 45 °C (Wei et al., 2017).

(ix) Wang et al. (2016) reported the formulation of temperature-sensitive liposomes (DPPC:MSPC:DSPE-PEG$_{2000}$:DSPG (mass ratio) = 83:3:10:4) (TSL) loaded with PTX (PTX-TSL), with a phase-transition temperature of 42 °C. Using an in vivo Lewis lung carcinoma (LLC) mouse model, it was reported that PTX accumulation in tumors was dependent on the temperature. Compared to physiological temperature, mild hyperthermia, at 42 °C, resulted in a higher accumulation of PTX in tumors after PTX-TSL administration, compared to non-TSL liposome PTX or PTX alone. Another study used thermo-reversible Pluronic F127 (Poloxamer 407 (P407)), which exists as a temperature-dependent dual state (i.e., a liquid state at low to ambient temperatures and a gel state at higher temperatures) to formulate PTX-liposomes (PTX-Lip-gel) (Nie et al., 2011). PTX-Lip-gel produced a significantly higher in vitro uptake of PTX in human KB oral cancer cells (Nie et al., 2011). The intratumoral injection of 15 mg/kg i.v. of PTX-Lip-gel produced a significant decrease in the in vivo growth of ascites sarcoma S180 tumors in mice compared to the i.v. administration of 15 mg/kg of PTX-loaded liposomes by 3.5-fold and Taxol® by fourfold. In terms of safety, there was a lower loss of body weight, and the concentration of PTX was < 50 ng/mL of PTX were present in the heart and lungs following the administration of PTX-Lip-gel than compared to Taxol®, suggesting that therapeutic index of PTX can be improved when formulated as a Lip-gel (Mao et al., 2016).

4.4 Dendrimers

(i).Dendrimers have been categorized as cascade polymers (Noriega-Luna et al., 2020). There is a central core in the structure that varies in composition between different dendrimers (Hu et al., 2016). In each core, there are highly ordered and protruding three-dimensional branches (Patravale et al., 2012). The spatial configuration of the branches in the dendrimers defines the physicochemical characteristics of each polymer (Noriega-Luna et al., 2020). These branches also allow the encapsulation of other organic molecules, which increases their solubility, and increases the likelihood of their systemic administration (Noriega-Luna et al., 2020). Polyamidoamine (PAMAM) and polyglycerol are the most commonly used polymers to form dendrimers (Khandare et al., 2006). Drugs can interact with these dendrimers at the surface functional groups by covalent bond, electrostatic interactions, or physical encapsulation within the interior of the molecule (Singh et al., 2016). The encapsulated drugs are released from the dendrimers due to the lysis of the covalent bond and changes in structure, pH, and temperature changes (Singh et al., 2016).

(ii).PTX-loaded dendrimers can increase drug loading, permeability and cytotoxicity compared to conventional chemotherapy. For example, PTX-loaded dendrimers increased the solubility of PTX > 14,000-fold compared to PTX alone, with > 70% of the drug dissolving within 5 min (Xie and Yao, 2019). The same study reported that dendrimers have higher PTX loading and faster release and therefore have the potential to be used for parenteral, topical, and oral formulations. PTX-loaded dendrimers (20 µg/mL) were shown to significantly increase the permeability and apoptosis of HeLa cancer cells (Rompicharla et al., 2018). In Caco-2 and porcine brain endothelial cells, the penetration of PTX was 12-fold greater than that of PTX alone (Teow et al., 2013). In human ovarian carcinoma A2780 cells, PTX-loaded PAMAM dendrimers were 10-fold more cytotoxicity than PTX (Khandare et al., 2006). Overall, the results of the above studies suggest that dendrimers effectively increase the translocation of PTX.

4.5 Micelles

(i). Micelles, or structures with hydrophobic cores and hydrophilic outer shells, are stable systems that transport hydrophobic drugs through an aqueous environment (Xu et al., 2020). They can be formed using lipids or polymers. For this reason, micelles can mimic lipoproteins and other natural transport systems within the body (Shuai et al., 2004). By mimicking these natural transporters, micelles can escape macrophage recognition and accumulate in tumors due to the enhanced permeability and retention (EPR) effect (Soga et al., 2005). These drug delivery systems release the encapsulated drug by the controlled release that is regulated by interactions with the hydrophilic shell or a change in environmental pH (Gao et al., 2020). The advantages of this type of delivery method include high drug loading capacity, surface modifications, and ease of production (Tu et al., 2020).

(ii) The efficacy of PTX-loaded micelles has been evaluated in different dosage forms, including topical and oral formulations (Zhang et al., 2017; Xu et al., 2020). It has been reported that PTX-loaded micelles formulated into hydrogels containing 5 mg PTX for topical delivery were more potent in inhibiting the growth of tumors in vivo in mice xenografted with melanoma B16 cells, compared to Taxol®-treated mice (Xu et al., 2020). This topical micelle gel formulation increased the fluidity of the skin by changing the spatial structure of the skin lipids and keratin (Xu et al., 2020). PTX-loaded micelles formulated into an oral dosage form increased bioavailability of the PTX by 3.8-fold and increased the anti-tumor efficacy in mice bearing A569 lung cancer tumors due to its enhanced intestinal absorption and greater bioavailability (Tu et al., 2020). Preclinical studies have shown that PTX delivered by micelles was mostly restricted to plasma and its accumulation in tissue was due to the EPR effect (Hamaguchi et al., 2007).

4.6 Nanotubes

(i) Nanotubes, specifically carbon nanotubes, can be modified to yield physiochemical requirements for drug delivery, including selectivity, solubility, biocompatibility and loading capability (Hashemzadeh and Raissi, 2017). Therefore, there are many different shapes and forms of these nanocarriers that typically have a very high surface area per unit weight (Naderi et al., 2015). They are good drug carriers due to their binding to drugs by noncovalent and covalent interactions (Hashemzadeh and Raissi, 2017). These carriers can be coupled to a large amount of drug that enters the target cells by endocytosis (Comparetti et al., 2020). Following the entry of the drug into the cells, the endosomes fuse with lysosomes, and the pH is decreased to 4.8, and this induces the release of the drug (Comparetti et al., 2020).

(ii) PTX-loaded nanotubes have been modeled to determine the binding energy produced by the interaction of the drug and its carrier (Al Garalleh and Algarni, 2020). Oxygen carbon nanotubes containing PTX (PTXon) potentiated the in vitro inhibition of MCF-7 and SK-BR-3 breast cancer cell proliferation compared to free paclitaxel by increasing availability of oxygen and thus, the efficacy of PTX (Wang et al., 2014). In vitro, PTX-loaded carbon nanotubes produced greater inhibition of the growth of LNCaP, HCT-116, and CaCo-2 cancer cells compared to Taxol® (Comparetti et al., 2020). In vitro cytotoxicity studies in SKOV3 and A549 cells indicated that paclitaxel containing nanotubes produced a similar magnitude of growth inhibition at 24 h as Taxol® did at 48 h (Sobhani et al., 2011). Overall, nanotubes can effectively deliver paclitaxel to cancer cells (García-Hevia et al., 2015).

4.7 Niosomes

(i) Niosomes are closed spheroidal structures formed by cholesterol and nonionic surfactants, such as Spans and Tweens, in an aqueous environment (Bayindir et al., 2015). They are uni- or multi-lamellar assemblies that can transport hydrophobic and hydrophilic drugs (Ge et al., 2019). Cholesterol is added to produce structural rigidity, whereas charged molecules are commonly used to stabilize the niosomes and prevent aggregation (Bayindir and Yuksel, 2010). Due to their high surface charges, niosomes can be suspended in water, which is advantageous for administration and storage (Bayindir and Yuksel, 2010). This drug delivery system provides multiple advantages compared to alternatives, such as improved stability and bioavailability and lower costs and toxicity (Al Garalleh and Algarni, 2020).

(ii) PTX can be incorporated into niosomes due to its lipophilic nature. Once inside the bilayer membrane, PTX exhibits amorphous properties and is slowly released (Al Garalleh and Algarni, 2020), which reduces potential toxic effects (Bayindir and Yuksel, 2010). Niosomes have variations in the encapsulation efficiency (EE) of PTX and a negative linear relationship between PTX EE and surfactant hydrophobic-lipophilic balance (HLB) value have been reported (Bayindir and Yuksel, 2010). The stability of niosomes in oral dosage forms is affected by bile salts; however, introducing polymers into the niosomes can increase its stability and plasma half-life (Bayindir et al., 2013). In a study using Carbopol 974P coated niosomes of PTX, a 98.7 ± 0.8% encapsulation efficiency of PTX was achieved (Bayindir et al., 2013). These niosomes yielded a higher drug plasma concentration than Taxol® in rats, and bioavailability was increased 3.8- and 1.4-fold compared to uncoated and coated Carbopol 974P niosomes, respectively.

4.8 Proniosomes

(i) Proniosomes are niosome hybrids that can transition into niosomes when hydrated (Mittal et al., 2020). They were developed to increase the stability and shelf life of niosomes (Pandit et al., 2015). Consequently, these dosage forms can transport hydrophobic and hydrophilic drugs. Proniosomes can be formulated as dry sorbitol or maltodextrin-based surfactant coated carriers or nonionic liquid crystals that can be used for transdermal delivery (Ahmad et al., 2017). Due to their marked similarity to niosomes, it can be hypothesized that proniosomes will interact with and deliver PTX to niosomes. Currently, there are no published papers about PTX-loaded proniosomes.

4.9 Ethosomes

(i) Ethosomes are ideal for transdermal dosage forms due to their penetration of the skin (Paolino et al., 2012). They are deformable liposomes that are predominantly composed of phospholipids that contain a high level of water and ethanol (Eskolaky et al., 2015). The ethanol content increases the lipid membrane fluidity of the skin cells, and flexibility of the ethosome increases its skin permeability (Eskolaky et al., 2015). These carriers have a faster penetration rate and greater penetration depth through skin than liposomes and thus have been replacing liposomes in transdermal drug delivery systems (Jadhav et al., 2012). However, the exact mechanism of drug release from ethosomes into the skin remains to be elucidated (Niu et al., 2019).

(ii) PTX is a highly hydrophobic molecule and therefore does not significantly penetrate beyond the stratum corneum (Paolino et al., 2012). Ethosomes can be used for the topical delivery of PTX due to their high rate of penetration into the skin. Studies have shown that PTX-loaded ethosomes are more efficacious in treating various types of skin cancers compared to paclitaxel alone. For example, the incubation of DJM-1 squamous carcinoma cells with PTX-loaded ethosomes elevated the apoptotic rate to 61% than when compared to 36% induced by Taxol® (Paolino et al., 2012). The authors of this study postulated that PTX-loaded ethosomes could be a potential treatment for squamous cell carcinoma disease (Paolino et al., 2012). The EE of PTX from loaded ethosomes is higher than that of other carriers, such as NPs and niosomes (Paolino et al., 2012). In one formulation, a zero-order kinetic profile of PTX was maintained for 20 h using dynamic Franz diffusion cells, and 90% of the drug was released (Paolino et al., 2012). The viability of SK-MEL-3 melanoma cells was 33% lower after incubation with PTX-loaded ethosomes, compared to the aqueous PTX (Eskolaky et al., 2015).

4.10 Microparticles

(i) Microparticles, are carriers with any structure or shape, provided their size is in the micron range (Lengyel et al., 2019). These structures can be created from a wide variety of materials, including, but not limited to, polysaccharides, lipids, proteins, and waxes (Lengyel et al., 2019). Drugs sequestered in microparticles are typically released by erosion, dissolution, diffusion or osmosis (Lengyel et al., 2019). The main advantage of these carriers compared to NPs is that they act locally and, therefore, typically produce a lower incidence of toxicity (Lengyel et al., 2019).

(ii) The loading of PTX into microparticles increases the stability, release kinetics and efficacy of PTX in SKOV-3 ovarian cancer cells in vitro (Han et al., 2019). Smaller PTX-loaded microparticles have been reported to have a greater dose efficiency, compared to larger microparticles, due to their size and larger distribution in the peritoneal cavity (Tsai et al., 2013). However, smaller microparticles also have a rapid drug release, which can produce a lower therapeutic efficacy, as suggested by an in vitro study, where 4T1-Luc tumors were incubated with PTX-loaded microparticles (Chaurasiya et al., 2018). Furthermore, there was a higher incidence of toxicity when smaller microparticles were used, as this would have altered the release kinetics of PTX. Ultimately, the size and composition affect the efficacy of microparticles (Chaurasiya et al., 2018), and this must be considered in tandem with the site of drug action.

4.11 Carbon dots

(i) Carbon dots have drug delivery properties, such as good water solubility and stability, low toxicity, and low cost (Liu et al., 2019). These materials have surface functional groups that facilitate the interaction of the attached drug with the cognate receptor. Carbon dots can be synthesized using different synthesis schemes, and consequently, their composition may differ, producing varying sizes and bioactivity (Pardo et al., 2018). Photoluminescence is one of the unique characteristics of carbon dots, and the emission spectra of photoluminescence is tunable and allows for direct observation in the visible light range (300–700 nm) (Pardo et al., 2018). Carbon dots can deliver, via controlled release, high amounts of hydrophobic drugs, making them promising for tumor detection and anticancer drug delivery (Chowdhuri et al., 2015). PTX-carbon dots, conjugated by ester bonds, increases the water solubility of PTX and produces greater inhibition of the growth of breast (MCF-7, MD-MB-231), lung (A-549, C33-A), prostate (PC3), and cervix cancers (HeLa), compared to the drug alone (Gomez et al., 2018).

4.12 Clinical trials

(i) Although PTX is a commonly used drug to treat various types of cancers, hypersensitivity reactions to components of the formulation, such as polyethylated castor oil (Cremophor EL®) and polysorbate 80, has limited its use (Gradishar et al., 2005). Therefore, various nanocarrier formulations have been developed to eliminate the use of the aforementioned vehicles. The PTX-nanocarrier-based formulations that were subjected to clinical trials are summarized in Table 4.1.

Table 4.1 Summary of PTX-nanocarrier-based formulations under clinical trials application.

Study	Study design	Patients	Treatments	Primary outcomes	Primary outcome results
Awada et al. (2014)	Open-label, phase II, randomized study	Metastatic and/or locally relapsed confirmed triple negative breast cancer	Patients randomized to either weekly EndoTAG-1 (paclitaxel within cationic liposomal membranes) plus paclitaxel group (EndoTag-1 administered at $22\,mg/m^2$ and paclitaxel at $70\,mg/m^2$), twice weekly EndoTag-1 group (EndoTag-1 administered at $44\,mg/m^2$ twice weekly), or weekly paclitaxel group (paclitaxel administered at $90\,mg/m^2$). Patients treated on cycles with 3 weeks of treatment, 1 week of rest, and for 4 cycles unless unacceptable toxicity, a documentation of disease progression, or patient refusal	PFS rate at week 16. PFS was rate of all patients treated without progression and alive 16 weeks after beginning treatment	143 patients were randomized at multiple centers. However, modified intent-to-treat population included 55 patients in EndoTag-1 plus paclitaxel group, 57 patients in EndoTag-1 group, and 28 patients in paclitaxel group. Based on central image evaluation, PFS rate by week sixteen 59.1% (95% CI 45.6–∞) for combination therapy, 34.2% (95% CI 21.6–∞) for EndoTAG-1 group, and 48.0% (95% CI 30.5–∞) for paclitaxel group. When additional analysis was complete with local review and including patients considered to have progression with a lack of assessment, the PFS by week 16 was 47.3% (35.6–∞) in combination group, 29.8% (20.0–∞) in EndoTAG-1 group, and 42.9% (26.9–∞) in paclitaxel group. Since null hypothesis was set for a PFS rate less than or equal to 30%, null hypothesis was rejected only for combination therapy group

Continued

Table 4.1 Summary of PTX-nanocarrier-based formulations under clinical trials application.–cont'd

Study	Study design	Patients	Treatments	Primary outcomes	Primary outcome results
Forero-Torres et al. (2015)	Phase II trial	Confirmed metastatic, triple-negative breast cancer	Randomized 2:1 to nab-paclitaxel with tigatuzumab or only nab-paclitaxel. Nab-paclitaxel administered at $100\,mg/m^2$ IV dose on days 1, 8, and 15 of 28-day cycle. Tigatuzumab administered at $10\,mg/kg$ loading dose followed by $5\,mg/kg$ dose every other week on days 1 and 15 of each cycle	ORR that based on RECIST version 1.1 criteria. ORR was proportion of people who achieved best overall response of complete responses and partial responses.	42 of these patients in combination treatment arm, and 22 patients in nab-paclitaxel arm. 39 patients in combination treatment arm were able to be evaluated for treatment response, and 21 patients in the paclitaxel arm were able to be evaluated for response 11 patients (28%, 95% CI 14.9–45) had objective response in the nab-paclitaxel with tigatuzumab group, and 8 patients (38%, 95% CI 18–61.1) had objective response in the nab-paclitaxel group
He et al. (2018)	Phase II trial	Metastatic gastric cancer	S-1 (oral fluoropyrimidine) administered at 40–60 mg based on BSA twice daily during 21-day cycle on days 1–14 with albumin-bound paclitaxel at $120\,mg/m^2$ days 1 and 8. This combination treatment continued for 6 cycles or until cancer progressed, patients refused, or unacceptable toxicity present	PFS	73 patients enrolled Median PFS9.63 months (95% CI 7.67–11.59). 29% of patient's progression-free at 1 year

Kato et al. (2019)	Single-arm, phase II trial at institution in Tokyo	Confirmed stage IIIB or IV NSCLC	Nab-paclitaxel at 100 mg/m^2 by IV infusion days 1, 8, and 15 of 3-week cycles. For certain toxicities, dose allowed to be reduced to minimum of 50 mg/m^2	ORR with RECIST ver. 1.1 criteria	22 out of 55 patients enrolled treated that met enrollment criteria. Five (22.7%) patients had partial response, 13 (59.1%) patients had stable disease (defined as maintained response for at least 6 weeks following start of treatment), and 3 patients (18.2%) had progressive disease. Overall ORR 22.7% (95% CI 7.8–45.4)
Kumthekar et al. (2020)	Multicenter, open-label, phase II study	Breast cancer and recurrent brain metastases	ANG1005 (paclitaxel trevatide) at 600 mg/m^2 IV once every 3 weeks (1 cycle) until documented disease progression or nonacceptable toxicities	Intracranial ORR (iORR) evaluated by independent radiology facility	72 females with recurrent brain metastases from breast cancer enrolled. Tumor response assessment conducted by local investigators noted overall iORR of 15% for 9 evaluable patients with PR plus 32 evaluable patients with stable disease. The second review had no complete responses and an overall iORR of 8% for 5 evaluable patients with PR plus 41 evaluable patients with stable disease
Murphy et al. (2019)	Phase II, open-label study, three centers in Australia	Stage II or III invasive breast cancer not previously treated	Epirubicin at 90 mg/m^2 and cyclophosphamide at 600 mg/m^2 every 3 weeks for 12 weeks. Following this, nab-paclitaxel administered at 125 mg/m^2 days 1, 8, and 15 every 4 weeks for 12 weeks	PCR defined to be ypT0/is ypN0-3 without invasive tumor cells in the breast sample. However, residual ductal carcinoma in situ was allowed. The non-complete response was the presence of only scattered tumor cells	40 women in the primary cohort made up of 15 patients with HER2-amplified cancer, 15 patients with triple-negative breast cancer, and 10 patients HR-positive but HER2-non-amplified tumors. Overall rate of PCR 55% ($n=22$). 32.5% ($n=13$) considered ypT0 ypN0, 12.5% ($n=5$) ypT0/is ypN0, and 10% ($n=4$) ypT0/is ypN +. PCR differed according to subtype. 80% ($n=12$) patients with PCR had HER2-amplified cancer, 46% ($n=2$) had triple-negative breast cancer, and 30% ($n=3$) had HR-positive but HER2-non-amplified cancer. Another 10% ($n=4$) of patients had non-complete response. Combined PCR and non-complete response 65% ($n=26$) of patients

Continued

Table 4.1 Summary of PTX-nanocarrier-based formulations under clinical trials application.—cont'd

Study	Study design	Patients	Treatments	Primary outcomes	Primary outcome results
Tamura et al. (2020)	Phase II, non-randomized, multicenter	Advanced gastric carcinoma, patients failed more than one regimen having a fluoropyrimidine (HER2 negative) or a fluoropyrimidine and trastuzumab (HER2 positive)	Nab-paclitaxel IV at 220 mg/m^2 day 1 of 21-day cycles. Two dose-reduction regimens could be used: 180 and 150 mg/m^2	Response rate	32 patients included 1 patient had a partial response, giving an overall response rate of 3.1% (95% CI 0–16.2%, $P=0.966$)
Fujiwara et al. (2019)	Multi-national, randomized, non-inferiority, phase III study	Recurrent or metastatic breast cancer with measurable lesion and ECOG of 1 or less	NK105 (nanoparticle formulation with paclitaxel in polymeric micelles) at 65 mg/m^2 infused over 30 min (first arm), paclitaxel at 80 mg/m^2 infused over 1 h (second arm). Both arms dosed days 1, 8, and 15 of 28-day cycles	PFS (time from randomization until first observation of the progression of a lesion or death from any cause)	Total 436 females randomized 1:1, but 422 ultimately included in efficacy analysis ($n=211$ NK105 arm, $n=211$ paclitaxel arm) PFS 8.4 months NK105 and 8.5 months paclitaxel (adjusted HR 1.255, 95% CI 0.989–1.592, did not show non-inferiority of NK105 to paclitaxel)

Gradishar et al. (2005)	International, randomized, open-label, phase III study, 70 sites	Females at least 18 years of age with confirmed metastatic breast cancer with expected survival more than 12 weeks	Random assignment 1:1 to either ABI-007 (albumin-bound paclitaxel with nanometer-sized particle) at 260 mg/m^2 IV over 30 min every 3 weeks without premedication or standard paclitaxel at 175 mg/m^2 IV over 3 h with premedication	Efficacy measure of ORR	Total 460 patients, 454 patients in ITT population ($n=229$ ABI-007, $n=225$ standard paclitaxel) Overall response rate significantly greater for ABI-007 (33%) vs standard paclitaxel (19%), $P=0.001$. Tumor response rate for patients with visceral dominant lesions higher for ABI-007 (34%) vs standard paclitaxel (19%), $P=0.002$. Overall response rate greater for patients with non-visceral dominant lesions for ABI-007 (34%) vs standard paclitaxel (19%)
Park et al. (2017)	Multicenter, randomized, non-inferiority, phase III study	HER2-negative metastatic or advanced breast cancer	Randomized 1:1 to a Genexol-PM (paclitaxel bound to polymeric micelles) administered 260 mg/m^2, or to Cremophor EL-based paclitaxel administered 175 mg/m^2. After first cycle, Genexol-PM dose could be increased to 300 mg/m^2	PFS, which was the patient fraction who had a complete response or partial response with RECIST ver. 1.0 criteria	213 patients were randomly assigned, 105 patients were analyzed in the Genexol-PM arm, and 107 patients were analyzed in the other arm Overall response rate in Genexol-PM group 39.1% (95% CI 31.2–46.9) and 24.3% in Cremophor EL-based paclitaxel group (95% CI 17.5–31.1). Non-inferiority ($P=0.021$) and superiority of Genexol-PM over other arm ($P=0.016$)
Koumarianou et al. (2020)	Multicenter, prospective study	Metastatic breast cancer by histology or cytology	Nab-paclitaxel-containing therapy	ORR. ORR is defined as a proportion of people with the best response of partial response or complete response.	Median age of patients 64.5 years, 40% had received taxane-based therapy before the study for metastatic breast cancer, 36% had de novo metastatic breast cancer, 36% received nab-paclitaxel as second-line therapy, and 53% received nab-paclitaxel as third- or further line-treatment Three patients had a complete response, 37 patients had a partial response, and 55 patients had stable disease. The overall response rate 26.7% (40 patients out of 150 eligible patients; 95% CI 19.6–33.7)

Continued

Table 4.1 Summary of PTX-nanocarrier-based formulations under clinical trials application.—cont'd

Study	Study design	Patients	Treatments	Primary outcomes	Primary outcome results
Slingerland et al. (2013)	Randomized, crossover study	Advanced cancer	All patients in study either received LEP-ETU form of paclitaxel in cycle 1 (at 175 mg/m^2) or the same dose of reference formulation (paclitaxel in dehydrated alcohol and castor oil) during the first cycle. During the second cycle of study, patients switched to the other treatment	Bioequivalency of two treatments and assess LEP-ETU	58 patients enrolled, but only 38 patients able to complete treatment per the protocol. The mean $AUC_{0-\infty}$ was 15,853.8 ng h/mL for LEP-ETU and 18,550.8 ng h/mL for other treatments. Bioavailability of LEP-ETU compared with reference formulation was 84%, and C_{max} ratio 97%. LEP-ETU is deemed to be bioequivalent to the reference formulation. A total of 23 adverse events experienced by 15 patients, the most common event reported was neutropenia. 11 out of 58 patients (19%) had neutropenia all during the first cycle. Five of those patients had grade 4 neutropenia, and another five of those patients had grade 3 neutropenia. One patient had grade 3 febrile neutropenia with grade 2 neutropenia
Wang et al. (2020b)	Retrospective study	Advanced, stage IIIB–IV NSCLC	IV nab-paclitaxel 125 or 130 mg/m^2 days 1 and 8 every 3 weeks, commonly 4–6 cycles, the dose could be adjusted 10%–20%	ORR (ORR, includes complete and partial response rates), disease control rate (DCR, includes a complete response, partial response, and stable disease rates), and PFS (initiation of treatment to disease progression or death (first to occur)	71 patients available for efficacy evaluation. ORR 14.5%, DCR 69.7%, and median PFS 5.2 months (95% CI 4.4–6.0)

PFS, progression-free survival; *ORR*, overall response rate; *PCR*, pathological complete response.

(ii) Some clinical studies have reported that the NP forms of PTX are efficacious in patients with metastatic or advanced breast cancer (Slingerland et al., 2013; Gradishar et al., 2005; Park et al., 2017; Murphy et al., 2019). In a phase III study, ABI-007 (Abraxane) was administered as albumin-bound PTX NPs with 130 nm particles as a colloidal suspension (Gradishar et al., 2005). This formulation of paclitaxel uses albumin as a natural carrier of lipophilic substances, such as paclitaxel, to avoid the toxicities produced by polyethylated castor oil. In this study, in women with metastatic breast cancer, the overall response rate for ABI-007 was significantly greater than standard PTX (33% vs 19%, $P=0.001$) (Gradishar et al., 2005). A phase III trial evaluated the efficacy of Genexol-PM, which is PTX bound to lyophilized, polymeric micelles and Cremophor EL-based PTX, in patients with HER2 negative metastatic or advanced breast cancer (Park et al., 2017). The overall response rate in the Genexol-PM group was 39.1% (95% CI 31.2–46.9) and 24.3% in the Cremophor EL-based PTX group (95% CI 17.5–31.1) (Park et al., 2017). These results indicated non-inferiority ($P=0.021$) and the superiority of Genexol-PM compared to Cremophor EL-based PTX ($P=0.016$) (Park et al., 2017). The NEONAB (NEOadjuvant epirubicin, cyclophosphamide and Nanoparticle Albumin-Bound paclitaxel) trial, consisting of stage II or III invasive breast cancer patients, reported that 55% of the patients had a complete pathological response when receiving epirubicin with cyclophosphamide, followed with nab-paclitaxel and as-needed trastuzumab (Murphy et al., 2019). Similarly, in a study conducted by Slingerland et al. (2013), 58 patients with advanced cancer (prescribed with Paclitaxel as the standard treatment) were administered liposome-entrapped PTX easy-to-use (LEP-ETU) (equivalent to 30 mg of PTX) at a dose of 175 mg/m^2 PTX via i.v. infusion (over 180 min) as a test formulation in first cycle followed by administration of PTX formulation in polyethoxyated castor oil and dehydrated alcohol at the same dose through same route in second cycle with a washout period of 3 weeks. Blood samples collected from patients throughout and after the infusion were then subjected to pharmacokinetic analysis using HPLC-MS/MS method. The results indicated that LEP-ETU formulation was bioequivalent to the reference formulation of paclitaxel with castor oil (Slingerland et al., 2013).

(iii) However, many of the studies in patients with other types of cancer have not shown that the efficacy of NP-formulated PTX is greater than the non-NP formulations of PTX (Awada et al., 2014; Forero-Torres et al., 2015; Fujiwara et al., 2019; He et al., 2018; Kato et al., 2019; Koumarianou et al., 2020; Kumthekar et al., 2020; Tamura et al., 2020; Wang et al., 2020b). A study by Forero-Torres et al. (2015) in patients with metastatic triple negative breast cancer (TNBC) indicated that 11 patients (28%, 95% CI 14.9–45) had an objective response in the nab-PTX + tigatuzumab group and 8 patients (38%, 95% CI 18–61.1) had an objective response in the nab-PTX group. A phase II study with S-1, an oral fluoropyrimidine, in combination with albumin-bound PTX, produced a median PFS(PFS) of 9.63 months (95% CI 7.67–11.59) in patients with advanced gastric cancer (He et al., 2018). A phase II study in breast cancer patients with brain metastases had an investigator-assessed overall intracranial objective response rate, ORR, (iORR) of 15% for 9 patients with partial responses (PR) and 32 patients with stable disease after treatment with ANG1005. ANG1005 is composed of a proprietary peptide of 19 amino acids called Angiopep-2 linked to three molecules of paclitaxel (Kumthekar et al., 2020). Angiopep-2 can cross the blood-brain-barrier (BBB) when used with low-density lipoprotein receptor-related protein 1 (LRP1). After the second review of the iORR by a central independent radiology facility, there were no complete responses and an overall iORR of 8% in 5 patients with PR and 41 patients with stable disease (Kumthekar et al., 2020). Awada et al. (2014) used EndoTAG-1 in their open-label, phase II, randomized study of patients with triple negative breast cancer. EndoTAG-1 is a paclitaxel embedded in lipids, which is in the formulation of cationic liposomes (Fasol et al., 2012). Their results indicated PFS rates of: (1) 59.1% (95% CI 45.6–∞) by week 16 for patients treated with EndoTAG-1 with PTX; (2) 34.2% (95% CI 21.6–∞) for the EndoTAG-1 group; and (3) 48% (95% CI 30.5–∞) for the PTX group (Awada et al., 2014). A single-arm, phase II trial in NSCLC patients indicated that nab-PTX produced an overall ORR of 22.7% (95% CI 7.8–45.4) (Kato et al., 2019). A multi-national, randomized, non-inferiority, phase III study was done to compare the efficacy of NK105 to standard PTX in patients with recurrent or metastatic breast cancer (Fujiwara et al., 2019). NK105 is a formulation that incorporates paclitaxel into polymeric micelles (Hamaguchi et al., 2005). The PFS was 8.4 months for NK105 and 8.5 months for PTX (adjusted HR of 1.255, 95% CI 0.989–1.592) (Fujiwara et al., 2019). In a study conducted by Tamura et al. (2020), only one patient had a partial response after treatment with nab-PTX treatment, which produced an overall response rate of 3.1% (95% CI 0–16.2%). A multicenter, prospective study conducted by Koumarianou et al. (2020) in patients with metastatic breast cancer, reported an overall response rate of 26.7% (40 patients out of 150 eligible patients: 95% CI 19.6–33.7). Although Wang et al. reported that nab-PTX produced a good response in patients with stage IIIB–IV NSCLC, more than half of the patients had stable disease after nab-PTX treatment (Wang et al., 2020b).

4.13 Overcoming paclitaxel resistance by using Nanocarriers

(i) Although significant advances have been made in cancer therapy, such as surgery, radiotherapy or chemotherapy, multi-drug resistance (MDR) is one of the primary factors that attenuates or abrogates the efficacy of anticancer drugs (Persidis, 1999; Wang et al., 2019). MDR can be defined as the resistance of cancer cells to drugs that are structurally and functionally distinct and MDR can be either be intrinsic (i.e., inherent) or acquired after a certain period of chemotherapy treatment (Harris and Hochhauser, 1992). MDR can decrease the intracellular concentration of drugs due the overexpression of certain efflux transporters on the surface of the resistant cancer cells, thereby decreasing or abolishing drug efficacy (Choi and Yu, 2014). These efflux transporters belong to the ATP-binding cassette (ABC) transporters superfamily, and the most commonly overexpressed transporters are P-gp, breast cancer resistance protein (BCRP/ABCG2) or multidrug resistance-associated protein 1 (MRP1/ABCC1) (Glavinas et al., 2004; Choi and Yu, 2014). The overexpression of these transporters can result from the amplification and mutation, enhanced transcription and an alteration in the translational efficiency of genes that code for these transporters (Yusuf et al., 2003).

(ii) Other mechanisms that can produce MDR include evasion of apoptosis and immune system, increased DNA repair, the biotransformation of a drug to an inactive metabolite, increased tolerance to the stressful tumor microenvironment, decreased drug uptake, mutation in the target protein and drug sequestration by organelles (Mansoori et al., 2017; Gottesman et al., 2002). One of the main mechanisms that mediates resistance to paclitaxel is the overexpression of P-gp, as paclitaxel is a substrate of P-gp (Yusuf et al., 2003). The accumulation of paclitaxel in the brain and small intestine of P-gp knockout mouse suggests that P-gp plays a significant role in preventing the entry of paclitaxel across the blood-brain barrier (BBB) and intestinal mucosa (Mayer et al., 1997; Sparreboom et al., 1997). Several studies have reported that paclitaxel resistance can be surmounted by giving paclitaxel with compounds that inhibit the efflux function and/ or the expression of P-gp (Nanayakkara et al., 2018, 2019; Shukla et al., 2011). However, this approach has not been successful for the treatment of cancer patients primarily due to the occurrence of severe toxic effects and problematic drug–drug interactions (Lee et al., 2014; Berg et al., 1995; Callaghan et al., 2014). Recently, several strategies have been developed to overcome MDR in cancer therapy, such as the use of nanocarriers for tumor-targeted delivery (Ahmad et al., 2016), targeting cancer stem cells (Nunes et al., 2018) and using compounds that augment the efficacy of anticancer drugs (i.e., chemosensitizers) (Ozben, 2006). The strategies to overcome PTX resistance by the use of different nanocarriers have been summarized in Table 4.1.

(iii) The resistance of cancer cells to paclitaxel has been overcome by using different types of paclitaxel NPs (Reshma et al., 2019), which can bypass P-gp. Reshma et al. (2019) formulated paclitaxel as polysaccharide galactoxyloglucan (PST001) (obtained from *Tamarindus indica* seed) NPs by epichlorohydrin cross-linking. These PTX-NPs were more efficacious ($IC_{50} = 66.30\,\mu g/mL$) than paclitaxel alone ($IC_{50} = 91.63\,\mu g/mL$) in the paclitaxel resistant lung carcinoma cell line, A549R. In addition, PST-PTX NPs, in vitro, produced a greater number of dead A549R cells (79.30%), compared to PTX alone (68.30%). PST-PTX NPs produced a 25-fold increase in the expression of the pro-apoptotic protein, BAX, in a paclitaxel resistant cancer cell line, A549R, compared to the paclitaxel sensitive cell line, A549R. Although, paclitaxel resistance is not mediated by the overexpression of the BCRP transporter (Mao and Unadkat, 2015; Yuan et al., 2009), BCRP was downregulated in A549R cells after incubation with PST-PTX, compared to paclitaxel alone. These results suggest that PST-PTX could be used to circumvent the MDR phenotype regulated by P-gp and BCRP (Reshma et al., 2019).

(iv) The use of conjugated linoleic acid (CLA) to formulate paclitaxel NPs (PTX-CLA) resulted in a formulation that produced a significant inhibition in the growth of C6 glioblastoma cells, compared to PTX alone (Ke et al., 2010). The uptake of PTX-CLA ($10\,\mu M$) by C6 glioma cell line was increased by twofold, threefold and fourfold after 2, 4, and 6 h of incubation, respectively, compared to PTX. In rat brain tissue, the concentration reached 1000 nmol/kg using PTX-CLA, whereas paclitaxel levels were undetectable after free drug delivery. This effect could be due to the targeting effect of CLA to brain cancer cells, which could help circumvent paclitaxel resistance (Ke et al., 2010). Similarly, Koziara et al. (2004) developed cetyl alcohol/polysorbate NPs using microemulsion precursors. In situ brain uptake of PTX, produced by the NPs, was significantly greater than Taxol®.

(v) Surfactants, commonly used in the NP formulation, were found to be useful in inducing MDR reversal due to effect on P-gp and

MRP2 activity (Bogman et al., 2003). Surfactants, such as Brij 78 and D-α-tocopherol polyethylene glycol 1000 succinate (TPGS), can increase intracellular calcein fluorescence in P-gp overexpressing cells (Liu et al., 2010). Calcein is a fluorescent and cell membrane impermeable molecule obtained following the biotransformation of non-fluorescent Calcein AM molecule (a P-gp substrate) by cytosolic esterase enzymes (Holló et al., 1994). Thus, an increase in Calcein fluorescence indicates the inhibition of P-gp-mediated efflux. P-gp efflux is dependent upon the binding and hydrolysis of ATP (Sauna and Ambudkar, 2007) and the depletion of ATP by Brij78in resistant cancer cell lines mediates its inhibition of P-gp. Paclitaxel nanocrystals formulated using TPGS (PTX/TGPS), induced apoptosis in 95% of NCI/ADR-RES cells (resistant human ovarian cancer cell line), compared to Taxol® (19.4%, 7.1% of early and late apoptosis) and TGPS (11.4% and 18.6%) at a dose of 5 μm PTX. Similarly, in an in vivo NCI/ADR-RES xenograft model, the i.v. administration of PTX/TGPS, at a dose of 10 mg/kg, produced significant tumor regression (mean tumor volume of 220 mm^3) compared to Taxol® (460 mm^3) and the control group (490 mm^3, $P < 0.01$). These results indicated a reversal of paclitaxel resistance in Taxol resistant-NCI/ADR-RES cells (Liu et al., 2010).

(vi) HP-β-CD incorporated PTX liposomes have been developed to overcome MDR (Shen et al., 2020). It has been reported that HP-β-CD can decrease the activity of the P-gp efflux pump, suggesting that it could circumvent MDR, although this remains to determined (Shi et al., 2015). In a study conducted by Shen et al. (2020), PTX/HP-β-CD liposomes produced a significantly greater inhibition of the growth of paclitaxel-resistant human lung adenocarcinoma cells (A549/T), compared to Taxol® and PTX-liposomes, both in vitro and in vivo. Apoptotic cell death in A549/T cells was threefold higher following incubation with HP-β-CD-loaded PTX liposomes, compared to Taxol®. Interestingly, the cellular uptake of cyclodextrin-PTX liposomes by A549/T cells increased with time, whereas the uptake of Taxol® decreased after 4h of incubation. Similar results were achieved using mitochondria-targeting, pH responsive and smart PTX liposomes (Jiang et al., 2015). The PTX-liposomes consisted of DSPE, $_D$[KLAKLAK]$_2$ (KLA) peptide conjugated with DMA (DKD-PTX-Lips), which is selectively internalized by acidic tumor tissues (pH 6.8). DKD-PTX-Lips were 20-fold efficacious than Taxol® in decreasing the growth of A549/T cells. This was likely due to the activation of the mitochondrial-dependent apoptotic pathway as indicated by an increase in caspase-9 and caspase-3 activity and the release of cytochrome c (markers of apoptosis (Matapurkar and Lazebnik, 2006; Brentnall et al., 2013)). These results were further confirmed in vivo using an A549/T cells-bearing nude mouse model. The administration of DKD-PTX-Lips (at a dose of 7.5 mg/kg administered through tail vein injection) produced a significantly greater inhibition of tumor growth (86.7%) compared to Taxol* (61.7%), without causing systemic toxicity.

(vii) Another approach for overcoming MDR is the formulation of PTX/P-gp inhibitor dual-loaded liposomes. Zhang et al. (2016) developed such a formulation by incorporating 1,2-dioleoyl-3-trimethylammonium-propane (DOTAP), PTX and Tariquidar (XR-9576), a third-generation P-gp inhibitor, to form LP(XR,PCT). LP(XR,PCT) (containing equimolar amount (1%, w/w ratio) of PTX and tariquidar to the total lipid), produced synergistic cytotoxicity in SKOV3-TR and HeyA8-MDR ovarian cancer cells (which overexpress P-gp) by inducing mitotic arrest and apoptosis induction, compared to paclitaxel alone and PTX liposome without Tariquidar LP(PCT). Furthermore, the i.p. administration of LP(XR,PCT), at a dose of 1.5 mg/kg of paclitaxel in mice with xenografted with paclitaxel-resistant ovarian cancer Hey8-MDR cells, decreased tumor growth by 43.2% and decreased metastases (44.4%) compared to liposomes containing only PTX LP, where the decrease in tumor growth and metastases was 16.9% and 2.8%, respectively.

(viii) Several studies have reported that use of folic acid to design a PTX-nanocarrier drug delivery system to overcome drug resistance mediated by folate-receptor regulated FPF-PTX endocytosis (Zhang et al., 2008, 2011; Tong et al., 2014; Dai et al., 2019). Liposomes loaded with paclitaxel and folic acid produced a greater inhibition of the in vitro growth of paclitaxel-sensitive, SKOV3 ovarian cancer cells ($IC_{50} = 5.67 \mu g/mL$) compared to PTX-liposomes without FA ($IC_{50} = 50.2 \mu g/mL$). In the PTX-resistant ovarian carcinoma cells, SKOV3/T, FA-PTX produced a much greater decrease in cell growth ($IC_{50} = 0.38 \mu g/mL$), compared to ~ 200 $\mu g/mL$ for PTX-LP without FA. Compared to Taxol*, FA-PTX-liposomes produced a greater increase in SKOV3/T cells apoptosis (35.2% vs 22.4%) in an in vivo murine PTX-resistant ovarian carcinoma xenograft model (Tong et al., 2014). The non-ionic surfactants, such as Solutol HS 15, cremophor EL*, and Pluronic F68, can overcome MDR when encapsulated into liposomes loaded with paclitaxel (Ji et al., 2012). Following the incubation of A549/T cells with Rh123-loaded non-ionic surfactant-liposomes (NLRs), an increase in Rhodamine 123 (Rh123) concentration (a substrate of P-gp), was observed, as compared to free Rh123 or RH123 liposomes with no surfactant. This result suggests that non-ionic surfactants decrease P-gp-mediated drug efflux, a finding that is congruent with in vitro data indicating that non-ionic surfactant paclitaxel liposomes (NLPs) significantly increased the apoptosis of PTX resistant A549/T cells, compared to PTX-liposomes without surfactants. Similarly, utilizing bovine serum albumin (BSA) to form the vitamin E and PTX complex, PTX-VE-NPs, produced a significant in vitro increase in the concentration of PTX in the adriamycin-resistant breast cancer cell line, MCF-7/ADR, compared to PTX-NPs. Similarly, in an in vivo MCF-7/ADR tumor bearing Balb/c mouse model, the i.v. administration of PTX-VE-NPs, at a dose of 10 mg/kg, every 3 days for five times, significantly decreased ($P < 0.05$) mean tumor volume ($250\,mm^3$), compared to mice treated with PTX-NPs ($350\,mm^3$), paclitaxel solution ($550\,mm^3$), and vehicle ($900\,mm^3$) (Tang et al., 2018). PTX micelles containing 4 mg of PTX, formulated using folate-modified, multi-functional pluronic P123/F127 (FPF-PTX) produced a threefold greater bioavailability and higher cytotoxicity in the MDR human oral carcinoma cell line (KBv) ($IC_{50} = 1.17 \pm 0.21\,\mu g/mL$) than PTX alone ($IC_{50} = 15.45 \pm 4.69\,\mu g/mL$) (Zhang et al., 2011). The rate of apoptosis (both early and late apoptosis) induced by FPF-PTX was greater ($14.48 \pm 1.69\%$ and $13.11 \pm 0.14\%$, respectively, $P < 0.01$) in KBv cells compared to Taxol° ($6.06 \pm 1.13\%$ and $5.27 \pm 0.53\%$). In addition, the i.v. administration of FPF-PTX (10 mg/kg) in mice with subcutaneous implanted KBv tumors, produced a 65.6% decrease in tumor weight compared to Taxol° (20.58%) and the vehicle group (tumor weight of $2.07 \pm 0.36\,g$) (Zhang et al., 2011). Using PTX liposomes coupled with folic acid (FA-NP), Tong et al. reported a greater decrease in the growth of SKOV3/TAX cells (IC_{50} value = 0.38), compared to PTX-liposomes without FA (NPs) ($IC_{50} = 200\,\mu g/mL$). In a SKOV3/TAX xenograft mouse model, the i.p. administration of FA-NPs resulted in a $35.2 \pm 7.7\%$ higher rate of tumor apoptosis, a lower number of tumor nodules and an increase in survival compared to PTX-liposome without FA treatment (Tong et al., 2014).

(ix) Yuan et al. (2015) reported the formulation of an enzymatic self-assembly into Taxol° NPs at a furin-activated location, i.e., Golgi bodies, inside HCT116 colon cancer cells. Utilizing a biocompatible condensation reaction, PTX was added to lysine (Lys), along with a 2-cyanobenzothiazole (CBT) motif, a disulfide-functionalized cysteine (Cys) motif and furin substrate, Arg-Val-Arg-Arg (RVRR), which when cleaved, yields the end-product, CBT-Taxol. CBT-Taxol was then co-incubated with the 50 µM of the CBT precursor, Ac-Arg-Val-Arg-Arg-Cys(StBu)-Lys-2-cyanobenzothiazole, to prevent its self-condensation with intracellular Cys, thus ensuring its self-assembly inside the cells. In vitro, CBT-Taxol, when co-incubated with the CBT precursor, produced a 4.5-fold increase in the anti-MDR efficacy of Taxol® in Taxol®-resistant HCT-116 cells compared to Taxol®. Furthermore, in nude mice that were transplanted with parental and Taxol®-resistant HCT-116 cells in the left and right thigh, respectively, CBT-Taxol produced a 1.53-fold increase in the anti-MDR ratio (defines as the ratio of the reduction in the tumor volume of Taxol®-resistant HCT-116 tumor divided by the reduction in tumor volume in parental HCT-116 tumors) compared to Taxol® (Yuan et al., 2015). Similarly, using the self-assembly approach, Wang et al. (2020a) formulated PTX and MDR1-siRNA incorporated nanogels (LNGs-PTX-siRNA), with biomimetic polymeric surfaces, using Dextran and coated with lipids. LNGs-PTX-siRNA were more efficacious in inhibiting the growth of the paclitaxel-resistant ovarian cancer cells, OVCAR-3, (drug resistance index of ~ 60) than PTX-nanogels, due to the in vitro knocking down of MDR1 gene expression by siRNA, thus promoting PTX accumulation. Similarly, in an in vivo DROV xenograft mouse model, the i.v. administration of LNGs-PTX-siRNA (12.5 mg/kg of PTX and 312.5 nmol/kg of siRNA) produced a significant decrease in tumor volume (about 700 mm^3) as compared to PTX-loaded lipid nanogels (about 1000 mm^3) and control group (about 1400 mm^3).

4.14 Selected patents for paclitaxel formulations

(i) Paclitaxel liposomal formulations for the treatment of lower and upper urothelial and bladder cancer: Using paclitaxel liposomal formulations as drug delivery systems, alone or in combination with other chemotherapeutic drugs, has been evaluated to determine their potential to overcome paclitaxel resistance. According to the invention, a liposome formulation, which is marketed as Lipusu®, consists of paclitaxel, lecithin, cholesterol, threonine, and glucose. It is recommended that this compound be administered intravesically to the bladder as a suspension in an aqueous solution, such as saline or water. This compound could also be administered via a urethral catheter or nephrostomy catheter, especially for the treatment of upper tract urothelial cancer (Oefelein et al., 2020).

(ii) Preparation of paclitaxel NPs of polymeric micelles: In 1999, Marita et al. invented a paclitaxel formulation consisting of co-polymeric micelles of paclitaxel, which provided an efficient system for the site-specific delivery of PTX. The compound was synthesized to produce a high encapsulation efficiency of paclitaxel (90%, w/v), and to deliver maximum amounts of PTX to the tumors, with minimum distribution to other tissues (Maitra et al., 2001).

(iii) Preparation and evaluation of Capxol™ formulation and its pharmacokinetic profile in rats: Desai and Soon-Shiong (2004) invented Capxol™, a novel, less toxic NP formulation for PTX. Capxol™ is a cremophor-free formulation that should decrease the incidence and severity of hypersensitivity and anaphylaxis reactions caused by cremophor, which is present in the currently approved and marketed formulation of paclitaxel. The Capxol™ formulation consists of paclitaxel, that is stabilized by a polymer which makes it suitable for parenteral administration in an aqueous suspension. At doses of 90 and 120 mg/kg in male and female rats, respectively, there were no overt signs of toxicity following a single i.v. administration of Capxol™, with the exception of the death of an animal after the administration of 90 mg/kg of Capxol™. The pharmacokinetics and tissue distribution of ^{3}H-Capxol and ^{3}H-Taxol were determined by their i.v. administration in 14 male rats. The results indicated that the pharmacokinetic profile of ^{3}H-Capxol was similar to that of ^{3}H-Taxol. However, following the administration of 5 mg/kg i.v. bolus dose of ^{3}H-Capxol and ^{3}H-Taxol, the levels of both total radioactivity were significantly lower in the blood of rats, with the area under the curve (AUC_{0-24}) μg equiv. h/mL values of 6.1 for ^{3}H-Capxol compared to 10 ^{3}H-Taxol, indicating a rapid distribution rate of ^{3}H-Capxol compared to ^{3}H-Taxol. The tissue distribution rate was higher in rats given 5 mg/kg of Capxol compared to Taxol®. The rate of biotransformation was considerably lower for ^{3}H-Capxol compared to the ^{3}H-Taxol-administered group. Twenty-four hours after administration, paclitaxel represented 44.2% of the blood radioactivity, compared with 27.7% for ^{3}H-TAXO.

(iv) Preparation and evaluation of the efficacy and pharmacokinetic profile of gold NP-paclitaxel formulations: Kingston et al. developed gold NP-paclitaxel formulations to obtain targeted delivery of paclitaxel. The attachment of paclitaxel to the metallic gold NPs was mediated by the presence of thiol groups. The release of active paclitaxel occurred by hydrolysis of the carbonate bond by esterase enzymes present in the plasma (Kingston et al., 2013). Gold NP paclitaxel formulations, at concentrations of 0.1, 2.5, and 15 μg/mL, significantly decreased the growth of A2780 human ovarian cancer (Miao et al., 2017) cells. Pharmacokinetics studies were conducted to determine the distribution and delivery of the thiolated paclitaxel analogs II, IV, and XIII and unformulated paclitaxel. The pharmacokinetic experiments indicated that 50 μg of paclitaxel (dissolved in the Cremophor EL diluent) or equal amounts of the invented formulations were intravenously administered to B16/F10 tumor-bearing mice (C57BL/6, $n=3$/time point/formulation). Blood samples were collected from the retro-orbital sinus at selected time points after each injection and were analyzed for paclitaxel concentration using ELISA. The concentration of Taxol® in the blood, 15 h after injection, was 20-fold higher in animals that received the thiolated paclitaxel formulations compared to unformulated paclitaxel. Tumors were collected to determine the intra-tumor paclitaxel content. The gold NP-paclitaxel formulations increased the delivery of paclitaxel to B16/F10 solid tumors. In the in vivo efficacy studies, the formulations were intravenously administered at a dose of 2.5 mg/kg vs 40 mg/kg for unformulated paclitaxel in C57BL/6 mice. The formulations were 16-fold more efficacious than formulated paclitaxel, based on the inhibition of tumor growth (Kingston et al., 2013).

(v) Preparation and in vitro evaluation of paclitaxel-loaded polymeric NP formulation to overcome MDR in cancer: Paclitaxel-loaded polymeric NPs *co*nsisting of paclitaxel and a water-soluble polymer, poly[2-(dimethylamino) ethyl methacrylate-*co*-methacrylic acid (PDM) that coats and surrounds the surface of paclitaxel, were synthesized. Paclitaxel-loaded polymeric NPs were efficacious in the parental, non-drug resistant cell lines, MCF7 and MT3, and their daughter, MDR cell lines, MCF7/ADR andMT3/ADR. The IC_{50} values for MCF7, MCF7/ADR, MT3, and MT3/ADR cells were 0.001, 4.75, 0.59, and 2.92 nM, respectively. Fluorescein isothiocyanate (FITC)-labeled paclitaxel-loaded polymeric NPs (FITC-PP) were used to determine the intracellular accumulation of paclitaxel-loaded polymeric NPs in MCF7/ADR cells. FITC-PP, at a concentration of 0.1 μM, accumulated inside MCF7/ADR cells, and there was no efflux of paclitaxel, suggesting that the FITC-PP compound increases the delivery of paclitaxel and it circumvents P-gp-mediated MDR (Geckeler and Yeonju, 2012).

(vi) Preparation and the in vitro and in vivo evaluation of the efficacy of gelatin and PLGA polymers encapsulating paclitaxel: Au and Wientjes (2011) synthesized a paclitaxel-loaded gelatin NP formulation that can be retained in the bladder cavity, even after voiding, thereby increasing the total intra-tumor delivery of paclitaxel. The formulation was composed of biodegradable gelatin and poly (lactase-*co*-glycoside) (PLGA) polymers that contained paclitaxel. The efficacy of the paclitaxel-loaded NPs was determined in human RT4 bladder cancer cells. The IC_{50} values, after 48 and 96 h of were 11.0 ± 0.4 and 4.0 ± 0.4 nM for Taxol®, 9.6 ± 1.1 and 4.0 ± 0.3 nM for the paclitaxel-loaded NPs, respectively. The urothelial concentration of PTX-loaded gelatin NPs and paclitaxel were determined after the instillation of the paclitaxel-loaded gelatin NPs (containing 1 mg paclitaxel in a total weight of 250 mg) into the bladder of an anesthetized dog, 2 h after administration. The level of paclitaxel in the urothelium following the administration of the paclitaxel-loaded gelatin NPs was 20-fold greater than the aqueous formulation of paclitaxel. The growth of bladder tumors in dogs was decreased ($IC_{50} = 2 \mu M$) by the paclitaxel-loaded gelatin NPs.

(vii) The in vitro evaluation of a phosphate-containing paclitaxel NP formulation: Hwu et al. (2011) developed phosphate-containing paclitaxel NP formulation, using a phosphodiester moiety to link the NPs and the active ingredient to form a paclitaxel prodrug. The aforementioned formulation has an improved hydrophilicity profile and specificity in tumor cells compared to the unformulated paclitaxel. The paclitaxel-Fe-NP 6 formulation is dephosphorylated specifically in cancer cells because dephosphorylation often occurs more easily in cancer cells than in normal cells. The release of paclitaxel from the NPs is due to the hydrolysis of the phosphodiester moieties by phosphodiesterase. Drug release experiments indicated that 91% of the paclitaxel-containing ligand in the paclitaxel-Fe-NP 6 was hydrolyzed. In vitro data indicated that paclitaxel-Fe-NP 6 significantly inhibited the growth of human OECM1 cancer cells ($IC_{50} = 5.03 \times 10^{-7} \mu g/mL$), and it inhibited the growth of human HUVEC cells at a much higher concentration ($IC_{50} = 3.58 \times 10^{-3} \mu g/mL$).

(viii) The in vitro and in vivo evaluation of the efficacy of galactosamine conjugated Gal-NPs: Sung et al. (2014) developed a self-assembled NP formulation consisting of γ-PGA-PLA (poly(Y-glutamic acid)-poly(lactide)) block copolymers conjugated with galactosamine and paclitaxel NP. The in vitro efficacy of 0.25–8 μg/mL of galactosamine conjugated Gal-NPs was evaluated in HepG2 cancer cells. Conjugation with galactosamine provided a therapeutic advantage in hepatoma cells as they can identify *N*-acetylgalactosamine and galactose terminated glycoproteins through asialoglycoprotein (ASGP) receptors. As a result, increased fluorescence was observed with exposure of HepG2 cells with rhodamine-123 containing Gal-NPs than compared with rhodamine-123-containing NPs without galactosamine. In addition, Gal-NPs showed comparable in-vitro HepG2 cells growth inhibition to that of clinically available paclitaxel formulation, Phyxol® (Sinphar Pharmaceutical). The i.v. administration of 20 mg/kg of paclitaxel-loaded NPs with galactosamine conjugated or the available paclitaxel formulation, Phyxol® in PBS on days 0, 4, 8, 12 and 16, significantly decreased the growth of hepatoma tumors in mice compared to mice treated with vehicles such as PBS, NPs, and Gal-NPs.

4.15 Conclusion

The application of nanocarriers in cancer research has formerly been very successful and continues to thrive into being a crucial component of chemotherapy. Moreover, the nanocarrier drug delivery system is a promising approach that deserves broad attention from researchers because it can overcome the challenges encountered in the conventional delivery of Paclitaxel along with its associated severe toxicity and drug resistance. This book chapter describes the implementation of several Paclitaxel nanocarrier delivery systems, such as NPs, liposomes, dendrimers, micelles, nanotubes, niosomes, proniosomes, ethosomes, microparticles, and carbon dots that offer a lot of benefits ranging from promoting the permeability and retention effect in tumors via surface modification technique to using the pH difference between healthy and tumor tissue as an advantage for targeted delivery through charge conversion in the acidic tumor environment. Although the nanocarriers-based PTX formulation is progressing, we still encounter many obstacles that need to be actively addressed. In addition to the dynamic and heterogeneous nature of the tumor, PTX nanotherapeutics may be influenced by several factors such as choice of solvents, pharmaceutical excipients, differences in in vitro and in vivo behavior of nanocarriers, including drug transportation, accumulation, and biocompatibility, differences in nanocarrier outcome based on its route of administration and selectivity to tumor tissue. Likewise, more optimizations and investigations are required to ensure uniformity in terms of safety, efficacy, and tumor-targeting ability in both preclinical and clinical studies.

References

Acharya, S., Sahoo, S.K., 2011. PLGA nanoparticles containing various anticancer agents and tumour delivery by EPR effect. Adv. Drug Deliv. Rev. 63, 170–183.

Ahmad, J., Akhter, S., Ahmed Khan, M., Wahajuddin, M., Greig, N.H., Amjad Kamal, M., Midoux, P., Pichon, C., 2016. Engineered nanoparticles against MDR in cancer: the state of the art and its prospective. Curr. Pharm. Des. 22, 4360–4373.

Ahmad, M.Z., Mohammed, A.A., Mokhtar Ibrahim, M., 2017. Technology overview and drug delivery application of proniosome. Pharm. Dev. Technol. 22, 302–311.

Akbarzadeh, A., Rezaei-Sadabady, R., Davaran, S., Joo, S.W., Zarghami, N., Hanifehpour, Y., Samiei, M., Kouhi, M., Nejati-Koshki, K., 2013. Liposome: classification, preparation, and applications. Nanoscale Res. Lett. 8, 102.

Akin, D., Sturgis, J., Ragheb, K., Sherman, D., Burkholder, K., Robinson, J.P., Bhunia, A.K., Mohammed, S., Bashir, R., 2007. Bacteria-mediated delivery of nanoparticles and cargo into cells. Nat. Nanotechnol. 2, 441–449.

Al Garalleh, H., Algarni, A., 2020. Modelling of paclitaxel conjugated with carbon nanotubes as an antitumor agent for cancer therapy. J. Biomed. Nanotechnol. 16, 224–234.

Allémann, E., Gurny, R., Doelker, E., 1992. Preparation of aqueous polymeric nanodispersions by a reversible salting-out process: influence of process parameters on particle size. Int. J. Pharm. 87, 247–253.

Allen, T.M., Cullis, P.R., 2013. Liposomal drug delivery systems: from concept to clinical applications. Adv. Drug Deliv. Rev. 65, 36–48.

Allen, T.M., Hansen, C.B., de Menezes, D.E.L., 1995. Pharmacokinetics of long-circulating liposomes. Adv. Drug Deliv. Rev. 16, 267–284.

Anwekar, H., Patel, S., Singhai, A., 2011. Liposome-as drug carriers. Int. J. Pharm. Life Sci. 2, 945–951.

Au, J.L., Wientjes, M.G., 2011. Tumor Targeting Drug-Loaded Particles. Google Patents.

Awada, A., Bondarenko, I., Bonneterre, J., Nowara, E., Ferrero, J., Bakshi, A., Wilke, C., Piccart, M., Group, C. S, 2014. A randomized controlled phase II trial of a novel composition of paclitaxel embedded into neutral and cationic lipids targeting tumor endothelial cells in advanced triple-negative breast cancer (TNBC). Ann. Oncol. 25, 824–831.

Bajorin, D.F., 2000. Paclitaxel in the treatment of advanced urothelial cancer. Oncology 14, 43–52. 57; discussion 58, 61-2.

Bao, Q.Y., Zhang, N., Geng, D.D., Xue, J.W., Merritt, M., Zhang, C., Ding, Y., 2014. The enhanced longevity and liver targetability of paclitaxel by hybrid liposomes encapsulating paclitaxel-conjugated gold nanoparticles. Int. J. Pharm. 477, 408–415.

Bayindir, Z.S., Yuksel, N., 2010. Characterization of niosomes prepared with various nonionic surfactants for paclitaxel oral delivery. J. Pharm. Sci. 99, 2049–2060.

Bayindir, Z.S., Onay-Besikci, A., Vural, N., Yuksel, N., 2013. Niosomes encapsulating paclitaxel for oral bioavailability enhancement: preparation, characterization, pharmacokinetics and biodistribution. J. Microencapsul. 30, 796–804.

Bayindir, Z.S., Beşikci, A., Yüksel, N., 2015. Paclitaxel-loaded niosomes for intravenous administration: pharmacokineticsand tissue distribution in rats. Turk. J. Med. Sci. 45, 1403–1412.

Belka, C., Budach, W., 2002. Anti-apoptotic Bcl-2 proteins: structure, function and relevance for radiation biology. Int. J. Radiat. Biol. 78, 643–658.

Berg, S.L., Tolcher, A., O'Shaughnessy, J.A., Denicoff, A.M., Noone, M., Ognibene, F.P., Cowan, K.H., Balis, F.M., 1995. Effect of R-verapamil on the pharmacokinetics of paclitaxel in women with breast cancer. J. Clin. Oncol. 13, 2039–2042.

Bernabeu, E., Cagel, M., Lagomarsino, E., Moretton, M., Chiappetta, D.A., 2017. Paclitaxel: what has been done and the challenges remain ahead. Int. J. Pharm. 526, 474–495.

Bilati, U., Allémann, E., Doelker, E., 2003. Sonication parameters for the preparation of biodegradable nanocapsulesof controlled size by the double emulsion method. Pharm. Dev. Technol. 8, 1–9.

Bogman, K., Erne-Brand, F., Alsenz, J., Drewe, J., 2003. The role of surfactants in the reversal of active transport mediated by multidrug resistance proteins. J. Pharm. Sci. 92, 1250–1261.

Bortner, C.D., Oldenburg, N.B., Cidlowski, J.A., 1995. The role of DNA fragmentation in apoptosis. Trends Cell Biol. 5, 21–26.

Brentnall, M., Rodriguez-Menocal, L., De Guevara, R.L., Cepero, E., Boise, L.H., 2013. Caspase-9, caspase-3 and caspase-7 have distinct roles during intrinsic apoptosis. BMC Cell Biol. 14, 32.

Budhian, A., Siegel, S.J., Winey, K.I., 2007. Haloperidol-loaded PLGA nanoparticles: systematic study of particle size and drug content. Int. J. Pharm. 336, 367–375.

Büyükköroğlu, G., Şenel, B., Başaran, E., Gezgin, S., 2016. Development of paclitaxel-loaded liposomal systems with anti-her2 antibody for targeted therapy. Trop. J. Pharm. Res. 15, 895–903.

Çağdaş, M., Sezer, A.D., Bucak, S., 2014. Liposomes as potential drug carrier systems for drug delivery. In: Application of Nanotechnology in Drug Delivery. IntechOpen.

Callaghan, R., Luk, F., Bebawy, M., 2014. Inhibition of the multidrug resistance P-glycoprotein: time for a change of strategy? Drug Metab. Dispos. 42, 623–631.

Campbell, R.B., Fukumura, D., Brown, E.B., Mazzola, L.M., Izumi, Y., Jain, R.K., Torchilin, V.P., Munn, L.L., 2002. Cationic charge determines the distribution of liposomes between the vascular and extravascular compartments of tumors. Cancer Res. 62, 6831–6836.

Cao, X., Tan, T., Zhu, D., Yu, H., Liu, Y., Zhou, H., Jin, Y., Xia, Q., 2020. Paclitaxel-loaded macrophage membrane camouflaged albumin nanoparticles for targeted Cancer therapy. Int. J. Nanomedicine 15, 1915–1928.

Chaurasiya, B., Huang, L., Du, Y., Tang, B., Qiu, Z., Zhou, L., Tu, J., Sun, C., 2018. Size-based anti-tumoral effect of paclitaxel loaded albumin microparticle dry powders for inhalation to treat metastatic lung cancer in a mouse model. Int. J. Pharm. 542, 90–99.

Chen, M.X., Li, B.K., Yin, D.K., Liang, J., Li, S.S., Peng, D.Y., 2014. Layer-by-layer assembly of chitosan stabilized multilayered liposomes for paclitaxel delivery. Carbohydr. Polym. 111, 298–304.

Chen, D., Zhang, G., Li, R., Guan, M., Wang, X., Zou, T., Zhang, Y., Wang, C., Shu, C., Hong, H., 2018. Biodegradable, hydrogen peroxide, and glutathione dual responsive nanoparticles for potential programmable paclitaxel release. J. Am. Chem. Soc. 140, 7373–7376.

Chen, Y., Wang, L., Luo, S., Hu, J., Huang, X., Li, P.-W., Zhang, Y., Wu, C., Tian, B.-L., 2020. Enhancement of antitumor efficacy of paclitaxel-loaded PEGylated liposomes by N, N-dimethyl tertiary amino moiety in pancreatic cancer. Drug Des. Dev. Ther. 14, 2945.

Choi, Y.H., Yu, A.-M., 2014. ABC transporters in multidrug resistance and pharmacokinetics, and strategies for drug development. Curr. Pharm. Des. 20, 793–807.

Chorny, M., Fishbein, I., Danenberg, H.D., Golomb, G., 2002. Lipophilic drug loaded nanospheres prepared by nanoprecipitation: effect of formulation variables on size, drug recovery and release kinetics. J. Control. Release 83, 389–400.

Choudhury, H., Gorain, B., Pandey, M., Khurana, R.K., Kesharwani, P., 2019. Strategizing biodegradable polymeric nanoparticles to cross the biological barriers for cancer targeting. Int. J. Pharm. 565, 509–522.

Chowdhuri, A.R., Tripathy, S., Haldar, C., Roy, S., Sahu, S.K., 2015. Single step synthesis of carbon dot embedded chitosan nanoparticles for cell imaging and hydrophobic drug delivery. J. Mater. Chem. B 3, 9122–9131.

Comparetti, E.J., Romagnoli, G.G., Gorgulho, C.M., Pedrosa, V.A., Kaneno, R., 2020. Anti-PSMA monoclonal antibody increases the toxicity of paclitaxel carried by carbon nanotubes. Mater. Sci. Eng. C 116, 111254.

Crucho, C.I.C., Barros, M.T., 2017. Polymeric nanoparticles: a study on the preparation variables and characterization methods. Mater. Sci. Eng. C 80, 771–784.

Dai, Y., Cai, X., Bi, X., Liu, C., Yue, N., Zhu, Y., Zhou, J., Fu, M., Huang, W., Qian, H., 2019. Synthesis and anti-cancer evaluation of folic acid-peptide-paclitaxel conjugates for addressing drug resistance. Eur. J. Med. Chem. 171, 104–115.

Davis, M.E., Chen, Z.G., Shin, D.M., 2008. Nanoparticle therapeutics: an emerging treatment modality for cancer. Nat. Rev. Drug Discov. 7, 771–782.

De Smet, M., Langereis, S., van den Bosch, S., Grull, H., 2010. Temperature-sensitive liposomes for doxorubicin delivery under MRI guidance. J. Control. Release 143, 120–127.

Dehaini, D., Wei, X., Fang, R.H., Masson, S., Angsantikul, P., Luk, B.T., Zhang, Y., Ying, M., Jiang, Y., Kroll, A.V., Gao, W., Zhang, L., 2017. Erythrocyte-platelet hybrid membrane coating for enhanced nanoparticle functionalization. Adv. Mater. 29, 1606209.

Denekamp, J., 1982. Endothelial cell proliferation as a novel approach to targeting tumour therapy. Br. J. Cancer 45, 136.

Desai, N.P., Soon-Shiong, P., 2004. Paclitaxel-Containing Formulations. Google Patents.

Deshmukh, R., Wagh, P., Naik, J., 2016. Solvent evaporation and spray drying technique for micro-and nanospheres/particles preparation: a review. Dry. Technol. 34, 1758–1772.

Diab, T., Alkafaas, S.S., Shalaby, T.I., Hessien, M., 2020. Paclitaxel nanoparticles induce apoptosis and regulate TXR1, CYP3A4 and CYP2C8 in breast cancer and hepatoma cells. Anti Cancer Agents Med. Chem. 20, 1582–1591.

Dranitsaris, G., Yu, B., Wang, L., Sun, W., Zhou, Y., King, J., Kaura, S., Zhang, A., Yuan, P., 2016. Abraxane® versus Taxol® for patients with advanced breast cancer: a prospective time and motion analysis from a Chinese health care perspective. J. Oncol. Pharm. Pract. 22, 205–211.

D'Souza, A.A., Shegokar, R., 2016. Polyethylene glycol (PEG): a versatile polymer for pharmaceutical applications. Expert Opin. Drug Deliv. 13, 1257–1275.

Eskolaky, E.B., Ardjmand, M., Akbarzadeh, A., 2015. Evaluation of anti-cancer properties of pegylated ethosomal paclitaxel on human melanoma cell line SKMEL-3. Trop. J. Pharm. Res. 14, 1421–1425.

Fang, R.H., Hu, C.-M.J., Luk, B.T., Gao, W., Copp, J.A., Tai, Y., O'Connor, D.E., Zhang, L., 2014. Cancer cell membrane-coated nanoparticles for anticancer vaccination and drug delivery. Nano Lett. 14, 2181–2188.

Farina, V., 1995. Chemistry and pharmacology of Taxol® and its derivatives. The. Pharmacochemistry Library, vol. 22 Elsevier Science & Technology.

Fasol, U., Frost, A., Büchert, M., Arends, J., Fiedler, U., Scharr, D., Scheuenpflug, J., Mross, K., 2012. Vascular and pharmacokinetic effects of EndoTAG-1 in patients with advanced cancer and liver metastasis. Ann. Oncol. 23, 1030–1036.

Fay, F., Scott, C.J., 2011. Antibody-targeted nanoparticles for cancer therapy. Immunotherapy 3, 381–394.

Feng, B., Niu, Z., Hou, B., Zhou, L., Li, Y., Yu, H., 2020. Enhancing triple negative breast cancer immunotherapy by ICG-templated self-assembly of paclitaxel nanoparticles. Adv. Funct. Mater. 30, 1906605.

Fetterly, G.J., Straubinger, R.M., 2003. Pharmacokinetics of paclitaxel-containing liposomes in rats. Aaps Pharmsci. 5, 90–100.

Forero-Torres, A., Varley, K.E., Abramson, V.G., Li, Y., Vaklavas, C., Lin, N.U., Liu, M.C., Rugo, H.S., Nanda, R., Storniolo, A.M., 2015. TBCRC 019: a phase II trial of nanoparticle albumin-bound paclitaxel with or without the anti-death receptor 5 monoclonal antibody tigatuzumab in patients with triple-negative breast cancer. Clin. Cancer Res. 21, 2722–2729.

Freitas, C., Müller, R., 1999. Correlation between long-term stability of solid lipid nanoparticles (SLN™) and crystallinity of the lipid phase. Eur. J. Pharm. Biopharm. 47, 125–132.

Fujie, T., Yoshimoto, M., 2019. Rapid leakage from PEGylated liposomes triggered by bubbles. Soft Matter 15, 9537–9546.

Fujiwara, Y., Mukai, H., Saeki, T., Ro, J., Lin, Y.-C., Nagai, S.E., Lee, K.S., Watanabe, J., Ohtani, S., Kim, S.B., 2019. A multi-national, randomised, open-label, parallel, phase III non-inferiority study comparing NK105 and paclitaxel in metastatic or recurrent breast cancer patients. Br. J. Cancer 120, 475–480.

Futaki, S., Nakase, I., 2017. Cell-surface interactions on arginine-rich cell-penetrating peptides allow for multiplex modes of internalization. Acc. Chem. Res. 50, 2449–2456.

Ganipineni, L.P., Ucakar, B., Joudiou, N., Riva, R., Jerome, C., Gallez, B., Danhier, F., Preat, V., 2019. Paclitaxel-loaded multifunctional nanoparticles for the targeted treatment of glioblastoma. J. Drug Target. 27, 614–623.

Gao, M., Yang, Y., Bergfel, A., Huang, L., Zheng, L., Bowden, T.M., 2020. Self-assembly of cholesterol end-capped polymer micelles for controlled drug delivery. J. Nanobiotechnol. 18, 13.

García-Hevia, L., Valiente, R., González, J., Luis Fernandez-Luna, J., Villegas, C.J., Fanarraga, M.L., 2015. Anti-cancer cytotoxic effects of multiwalled carbon nanotubes. Curr. Pharm. Des. 21, 1920–1929.

Ge, X., Wei, M., He, S., Yuan, W.-E., 2019. Advances of non-ionic surfactant vesicles (niosomes) and their application in drug delivery. Pharmaceutics 11, 55.

Geckeler, K., Yeonju, L., 2012. Paclitaxel-Loaded Polymeric Nanoparticle and Preparation Thereof. WO 2012138013A1.

Ghadi, R., Dand, N., 2017. BCS class IV drugs: highly notorious candidates for formulation development. J. Control. Release 248, 71–95.

Glavinas, H., Krajcsi, P., Cserepes, J., Sarkadi, B., 2004. The role of ABC transporters in drug resistance, metabolism and toxicity. Curr. Drug Deliv. 1, 27–42.

Gomez, I.J., Arnaiz, B., Cacioppo, M., Arcudi, F., Prato, M., 2018. Nitrogen-doped carbon nanodots for bioimaging and delivery of paclitaxel. J. Mater. Chem. B 6, 5540–5548.

Gottesman, M.M., Fojo, T., Bates, S.E., 2002. Multidrug resistance in cancer: role of ATP-dependent transporters. Nat. Rev. Cancer 2, 48–58.

Gradishar, W.J., Tjulandin, S., Davidson, N., Shaw, H., Desai, N., Bhar, P., Hawkins, M., O'Shaughnessy, J., 2005. Phase III trial of nanoparticle albumin-bound paclitaxel compared with polyethylated castor oil-based paclitaxel in women with breast cancer. J. Clin. Oncol. 23, 7794–7803.

Green, M., Manikhas, G., Orlov, S., Afanasyev, B., Makhson, A., Bhar, P., Hawkins, M., 2006. Abraxane®, a novel Cremophor®-free, albumin-bound particle form of paclitaxel for the treatment of advanced non-small-cell lung cancer. Ann. Oncol. 17, 1263–1268.

Guo, S., Vieweger, M., Zhang, K., Yin, H., Wang, H., Li, X., Li, S., Hu, S., Sparreboom, A., Evers, B.M., 2020. Ultra-thermostable RNA nanoparticles for solubilizing and high-yield loading of paclitaxel for breast cancer therapy. Nat. Commun. 11, 1–11.

Hamaguchi, T., Matsumura, Y., Suzuki, M., Shimizu, K., Goda, R., Nakamura, I., Nakatomi, I., Yokoyama, M., Kataoka, K., Kakizoe, T., 2005. NK105, a paclitaxel-incorporating micellar nanoparticle formulation, can extend in vivo antitumour activity and reduce the neurotoxicity of paclitaxel. Br. J. Cancer 92, 1240–1246.

Hamaguchi, T., Kato, K., Yasui, H., Morizane, C., Ikeda, M., Ueno, H., Muro, K., Yamada, Y., Okusaka, T., Shirao, K., 2007. A phase I and pharmacokinetic study of NK105, a paclitaxel-incorporating micellar nanoparticle formulation. Br. J. Cancer 97, 170–176.

Han, S., Dwivedi, P., Mangrio, F.A., Dwivedi, M., Khatik, R., Cohn, D.E., Si, T., Xu, R.X., 2019. Sustained release paclitaxel-loaded core-shell-structured solid lipid microparticles for intraperitoneal chemotherapy of ovarian cancer. Artif. Cells Nanomed. Biotechnol. 47, 957–967.

Harris, A.L., Hochhauser, D., 1992. Mechanisms of multidrug resistance in cancer treatment. Acta Oncol. 31, 205–213.

Hashemzadeh, H., Raissi, H., 2017. The functionalization of carbon nanotubes to enhance the efficacy of the anticancer drug paclitaxel: a molecular dynamics simulation study. J. Mol. Model. 23, 1–10.

He, M.-M., Wang, F., Jin, Y., Yuan, S.-Q., Ren, C., Luo, H.-Y., Wang, Z.-Q., Qiu, M.-Z., Wang, Z.-X., Zeng, Z.I., et al., 2018. Phase II clinical trial of S-1 plus nanoparticle albumin-bound paclitaxel in untreated patients with metastatic gastric cancer. Cancer Sci. 109, 3575–3582.

Holló, Z., Homolya, L., Davis, C.W., Sarkadi, B., 1994. Calcein accumulation as a fluorometric functional assay of the multidrug transporter. Biochim. Biophys. Acta 1191, 384–388.

Hong, S.-S., Kim, S.H., Lim, S.-J., 2015. Effects of triglycerides on the hydrophobic drug loading capacity of saturated phosphatidylcholine-based liposomes. Int. J. Pharm. 483, 142–150.

Hong, S.-S., Choi, J.Y., Kim, J.O., Lee, M.-K., Kim, S.H., Lim, S.-J., 2016. Development of paclitaxel-loaded liposomal nanocarrier stabilized by triglyceride incorporation. Int. J. Nanomedicine 11, 4465–4477.

Houdaihed, L., Evans, J.C., Allen, C., 2020. Dual-targeted delivery of nanoparticles encapsulating paclitaxel and everolimus: a novel strategy to overcome breast cancer receptor heterogeneity. Pharm. Res. 37, 1–10.

Hu, J., Hu, K., Cheng, Y., 2016. Tailoring the dendrimer core for efficient gene delivery. Acta Biomater. 35, 1–11.

Hwu, J.-R., Lin, Y.-S., Yeh, C.-S., Shieh, D.-B., 2011. Phosphate-Containing Nanoparticle Delivery Vehicle. Google Patents.

Ingle, S.G., Pai, R.V., Monpara, J.D., Vavia, P.R., 2018. Liposils: an effective strategy for stabilizing paclitaxel loaded liposomes by surface coating with silica. Eur. J. Pharm. Sci. 122, 51–63.

Iqbal, M., Valour, J.-P., Fessi, H., Elaissari, A., 2015. Preparation of biodegradable PCL particles via double emulsion evaporation method using ultrasound technique. Colloid Polym. Sci. 293, 861–873.

Iversen, T.-G., Skotland, T., Sandvig, K., 2011. Endocytosis and intracellular transport of nanoparticles: present knowledge and need for future studies. Nano Today 6, 176–185.

Jadhav, S.M., Morey, P., Karpe, M., Kadam, V., 2012. Novel vesicular system: an overview. J. Appl. Pharm. Sci. 2, 193–202.

Jain, A.K., Swarnakar, N.K., Godugu, C., Singh, R.P., Jain, S., 2011. The effect of the oral administration of polymeric nanoparticles on the efficacy and toxicity of tamoxifen. Biomaterials 32, 503–515.

Jennewein, S., Croteau, R., 2001. Taxol: biosynthesis, molecular genetics, and biotechnological applications. Appl. Microbiol. Biotechnol. 57, 13–19.

Ji, X., Gao, Y., Chen, L., Zhang, Z., Deng, Y., Li, Y., 2012. Nanohybrid systems of non-ionic surfactant inserting liposomes loading paclitaxel for reversal of multidrug resistance. Int. J. Pharm. 422, 390–397.

Jiang, L., Li, L., He, X., Yi, Q., He, B., Cao, J., Pan, W., Gu, Z., 2015. Overcoming drug-resistant lung cancer by paclitaxel loaded dual-functional liposomes with mitochondria targeting and pH-response. Biomaterials 52, 126–139.

Jiang, L., Li, L., He, B., Pan, D., Luo, K., Yi, Q., Gu, Z., 2016. Anti-cancer efficacy of paclitaxel loaded in pH triggered liposomes. J. Biomed. Nanotechnol. 12, 79–90.

Jordan, M.A., Wilson, L., 2004. Microtubules as a target for anticancer drugs. Nat. Rev. Cancer 4, 253–265.

Kamaly, N., Yameen, B., Wu, J., Farokhzad, O.C., 2016. Degradable controlled-release polymers and polymeric nanoparticles: mechanisms of controlling drug release. Chem. Rev. 116, 2602–2663.

Kan, P., Tsao, C.-W., Wang, A.-J., Su, W.-C., Liang, H.-F., 2011. A liposomal formulation able to incorporate a high content of paclitaxel and exert promising anticancer effect. J. Drug Deliv. 2011, 629234.

Kannan, V., Balabathula, P., Divi, M.K., Thoma, L.A., Wood, G.C., 2015. Optimization of drug loading to improve physical stability of paclitaxel-loaded long-circulating liposomes. J. Liposome Res. 25, 308–315.

Karnik, R., Gu, F., Basto, P., Cannizzaro, C., Dean, L., Kyei-Manu, W., Langer, R., Farokhzad, O.C., 2008. Microfluidic platform for controlled synthesis of polymeric nanoparticles. Nano Lett. 8, 2906–2912.

Kato, Y., Okuma, Y., Watanabe, K., Yomota, M., Kawai, S., Hosomi, Y., Okamura, T., 2019. A single-arm phase II trial of weekly nanoparticle albumin-bound paclitaxel (nab-paclitaxel) monotherapy after standard of chemotherapy for previously treated advanced non-small cell lung cancer. Cancer Chemother. Pharmacol. 84, 351–358.

Ke, X.-Y., Zhao, B.-J., Zhao, X., Wang, Y., Huang, Y., Chen, X.-M., Zhao, B.-X., Zhao, S.-S., Zhang, X., Zhang, Q., 2010. The therapeutic efficacy of conjugated linoleic acid–paclitaxel on glioma in the rat. Biomaterials 31, 5855–5864.

Khan, I., Lau, K., Bnyan, R., Houacine, C., Roberts, M., Isreb, A., Elhissi, A., Yousaf, S., 2020. A facile and novel approach to manufacture paclitaxel-loaded proliposome tablet formulations of micro or nano vesicles for nebulization. Pharm. Res. 37, 1–19.

Khandare, J.J., Jayant, S., Singh, A., Chandna, P., Wang, Y., Vorsa, N., Minko, T., 2006. Dendrimer versus linear conjugate: influence of polymeric architecture on the delivery and anticancer effect of paclitaxel. Bioconjug. Chem. 17, 1464–1472.

Kim, S.C., Kim, D.W., Shim, Y.H., Bang, J.S., Oh, H.S., Kim, S.W., Seo, M.H., 2001. In vivo evaluation of polymeric micellar paclitaxel formulation: toxicity and efficacy. J. Control. Release 72, 191–202.

Kingston, D.G., Cao, S., Zhao, J., Paciotti, G.F., Hubta, M.S., 2013. Thiolated Paclitaxels for Reaction with Gold Nanoparticles as Drug Delivery Agents. Google Patents.

Kong, G., Anyarambhatla, G., Petros, W.P., Braun, R.D., Colvin, O.M., Needham, D., Dewhirst, M.W., 2000. Efficacy of liposomes and hyperthermia in a human tumor xenograft model: importance of triggered drug release. Cancer Res. 60, 6950–6957.

Koudelka, Š., Turánek, J., 2012. Liposomal paclitaxel formulations. J. Control. Release 163, 322–334.

Koumarianou, A., Makrantonakis, P., Zagouri, F., Papadimitriou, C., Christopoulou, A., Samantas, E., Christodoulou, C., Psyrri, A., Bafaloukos, D., Aravantinos, G., 2020. ABREAST: a prospective, real-world study on the effect of nab-paclitaxel treatment on clinical outcomes and quality of life of patients with metastatic breast cancer. Breast Cancer Res. Treat. 182, 85–96.

Koziara, J.M., Lockman, P.R., Allen, D.D., Mumper, R.J., 2004. Paclitaxel nanoparticles for the potential treatment of brain tumors. J. Control. Release 99, 259–269.

Kumthekar, P., Tang, S.-C., Brenner, A.J., Kesari, S., Piccioni, D.E., Anders, C., Carrillo, J., Chalasani, P., Kabos, P., Puhalla, S., 2020. ANG1005, a brain-penetrating peptide-drug conjugate, shows activity in patients with breast cancer with leptomeningeal carcinomatosis and recurrent brain metastases. Clin. Cancer Res. 26, 2789–2799.

Kunstfeld, R., Wickenhauser, G., Michaelis, U., Teifel, M., Umek, W., Naujoks, K., Wolff, K., Petzelbauer, P., 2003. Paclitaxel encapsulated in cationic liposomes diminishes tumor angiogenesis and melanoma growth in a "humanized" SCID mouse model. J. Invest. Dermatol. 120, 476–482.

Le, N.T.T., Cao, V.D., Nguyen, T.N.Q., Le, T.T.H., Tran, T.T., Hoang Thi, T.T., 2019. Soy lecithin-derived liposomal delivery systems: surface modification and current applications. Int. J. Mol. Sci. 20, 4706.

Le, N.T.T., Nguyen, D.T.D., Nguyen, N.H., Nguyen, C.K., Nguyen, D.H., 2021. Methoxy polyethylene glycol–cholesterol modified soy lecithin liposomes for poorly water-soluble anticancer drug delivery. J. Appl. Polym. Sci. 138, 49858.

Lee, H.J., Heo, D.-S., Cho, J.-Y., Han, S.-W., Chang, H.-J., Yi, H.-G., Kim, T.-E., Lee, S.-H., Oh, D.-Y., Im, S.-A., Jang, I.-J., Bang, Y.-J., 2014. A phase I study of oral paclitaxel with a novel P-glycoprotein inhibitor, HM30181A, in patients with advanced solid Cancer. Cancer Res. Treat. 46, 234–242.

Lengyel, M., Kállai-Szabó, N., Antal, V., Laki, A.J., Antal, I., 2019. Microparticles, microspheres, and microcapsules for advanced drug delivery. Sci. Pharm. 87, 20.

Li, R., He, Y., Zhang, S., Qin, J., Wang, J., 2018. Cell membrane-based nanoparticles: a new biomimetic platform for tumor diagnosis and treatment. Acta Pharm. Sin. B 8, 14–22.

Liu, Y., Huang, L., Liu, F., 2010. Paclitaxel nanocrystals for overcoming multidrug resistance in cancer. Mol. Pharm. 7, 863–869.

Liu, M.L., Chen, B.B., Li, C.M., Huang, C.Z., 2019. Carbon dots: synthesis, formation mechanism, fluorescence origin and sensing applications. Green Chem. 21, 449–471.

Llu, L., Yonetani, T., 1994. Preparation and characterization of liposome-encapsulated haemoglobin by a freeze-thaw method. J. Microencapsul. 11, 409–421.

Luo, L.-M., Huang, Y., Zhao, B.-X., Zhao, X., Duan, Y., Du, R., Yu, K.-F., Song, P., Zhao, Y., Zhang, X., 2013. Anti-tumor and anti-angiogenic effect of metronomic cyclic NGR-modified liposomes containing paclitaxel. Biomaterials 34, 1102–1114.

Ma, P., Mumper, R.J., 2013. Paclitaxel nano-delivery systems: a comprehensive review. J. Nanomed. Nanotechnol. 4, 1000164.

Madan, J., Pandey, R.S., Jain, V., Katare, O.P., Chandra, R., Katyal, A., 2013. Poly (ethylene)-glycol conjugated solid lipid nanoparticles of noscapine improve biological half-life, brain delivery and efficacy in glioblastoma cells. Nanomed. Nanotechnol. Biol. Med. 9, 492–503.

Maitra, A., Sahoo, S.K., Ghosh, P.K., Burman, A.C., Mukherjee, R., Khattar, D., Kumar, M., Paul, S., 2001. Formulations of Paclitaxel, its Derivatives or its Analogs Entrapped into Nanoparticles of Polymeric Micelles, Process for Preparing Same and the Use Thereof. Google Patents.

Maja, L., Željko, K., Mateja, P., 2020. Sustainable technologies for liposome preparation. J. Supercrit. Fluids 165, 104984.

Mansoori, B., Mohammadi, A., Davudian, S., Shirjang, S., Baradaran, B., 2017. The different mechanisms of cancer drug resistance: a brief review. Adv. Pharm. Bull. 7, 339–348.

Mao, Q., Unadkat, J.D., 2015. Role of the breast cancer resistance protein (BCRP/ABCG2) in drug transport—an update. AAPS J. 17, 65–82.

Mao, Y., Li, X., Chen, G., Wang, S., 2016. Thermosensitive hydrogel system with paclitaxel liposomes used in localized drug delivery system for in situ treatment of tumor: better antitumor efficacy and lower toxicity. J. Pharm. Sci. 105, 194–204.

Marin, E., Briceño, M.I., Caballero-George, C., 2013. Critical evaluation of biodegradable polymers used in nanodrugs. Int. J. Nanomedicine 8, 3071.

Markman, J.L., Rekechenetskiy, A., Holler, E., Ljubimova, J.Y., 2013. Nanomedicine therapeutic approaches to overcome cancer drug resistance. Adv. Drug Deliv. Rev. 65, 1866–1879.

Matapurkar, A., Lazebnik, Y., 2006. Requirement of cytochrome c for apoptosis in human cells. Cell Death Differ. 13, 2062–2067.

Mattjus, P., Slotte, J.P., 1996. Does cholesterol discriminate between sphingomyelin and phosphatidylcholine in mixed monolayers containing both phospholipids? Chem. Phys. Lipids 81, 69–80.

Mayer, U., Wagenaar, E., Dorobek, B., Beijnen, J.H., Borst, P., Schinkel, A.H., 1997. Full blockade of intestinal P-glycoprotein and extensive inhibition of blood-brain barrier P-glycoprotein by oral treatment of mice with PSC833. J. Clin. Invest. 100, 2430–2436.

Miao, F., Zhang, X., Cao, Y., Wang, Y., Zhang, X., 2017. Effect of siRNA-silencing of SALL2 gene on growth, migration and invasion of human ovarian carcinoma A2780 cells. BMC Cancer 17, 838.

Mittal, S., Chaudhary, A., Chaudhary, A., Kumar, A., 2020. Proniosomes: the effective and efficient drug-carrier system. Ther. Deliv. 11, 125–137.

Mo, R., Sun, Q., Xue, J., Li, N., Li, W., Zhang, C., Ping, Q., 2012. Multistage pH-responsive liposomes for mitochondrial-targeted anticancer drug delivery. Adv. Mater. 24, 3659–3665.

Monteiro, L.O., Fernandes, R.S., Castro, L., Reis, D., Cassali, G.D., Evangelista, F., Loures, C., Sabino, A.P., Cardoso, V., Oliveira, M.C., 2019. Paclitaxel-loaded folate-coated pH-sensitive liposomes enhance cellular uptake and antitumor activity. Mol. Pharm. 16, 3477–3488.

Murakami, H., Kobayashi, M., Takeuchi, H., Kawashima, Y., 2000. Further application of a modified spontaneous emulsification solvent diffusion method to various types of PLGA and PLA polymers for preparation of nanoparticles. Powder Technol. 107, 137–143.

Murphy, C., Muscat, A., Ashley, D., Mukaro, V., West, L., Liao, Y., Chisanga, D., Shi, W., Collins, I., Baron-Hay, S., Patil, S., Lindeman, G., Khasraw, M., 2019. Tailored NEOadjuvant epirubicin, cyclophosphamide and nanoparticle albumin-bound paclitaxel for breast cancer: the phase II NEONAB trial-clinical outcomes and molecular determinants of response. PLoS One 14, e0210891.

Naderi, N., Madani, S.Y., Mosahebi, A., Seifalian, A.M., 2015. Octa-ammonium POSS-conjugated single-walled carbon nanotubes as vehicles for targeted delivery of paclitaxel. Nano Rev. 6, 28297.

122 Chapter 4 Paclitaxel-based formulations in cancer chemotherapy

Nanayakkara, A.K., Follit, C.A., Chen, G., Williams, N.S., Vogel, P.D., Wise, J.G., 2018. Targeted inhibitors of P-glycoprotein increase chemotherapeutic-induced mortality of multidrug resistant tumor cells. Sci. Rep. 8, 967.

Nanayakkara, A.K., Vogel, P.D., Wise, J.G., 2019. Prolonged inhibition of P-glycoprotein after exposure to chemotherapeutics increases cell mortality in multidrug resistant cultured cancer cells. PLoS One 14, e0217940.

Nguyen, V.D., Han, J.-W., Choi, Y.J., Cho, S., Zheng, S., Ko, S.Y., Park, J.-O., Park, S., 2016. Active tumor-therapeutic liposomal bacteriobot combining a drug (paclitaxel)-encapsulated liposome with targeting bacteria (*Salmonella yyphimurium*). Sensors Actuators B Chem. 224, 217–224.

Nie, S., Hsiao, W.W., Pan, W., Yang, Z., 2011. Thermoreversible Pluronic® F127-based hydrogel containing liposomes for the controlled delivery of paclitaxel: in vitro drug release, cell cytotoxicity, and uptake studies. Int. J. Nanomedicine 6, 151.

Niu, X.-Q., Zhang, D.-P., Bian, Q., Feng, X.-F., Li, H., Rao, Y.-F., Shen, Y.-M., Geng, F.-N., Yuan, A.-R., Ying, X.-Y., 2019. Mechanism investigation of ethosomes transdermal permeation. Int. J. Pharm. 1, 100027.

Noriega-Luna, B., Godínez, L.A., Rodríguez, F., Rodríguez, A., Zaldívar-Lelo de Larrea, G., Sosa-Ferreyra, C., Mercado-Curiel, R., Manríquez, J., Bustos, E., 2020. Corrigendum to "applications of dendrimers in drug delivery agents, diagnosis, therapy, and detection". J. Nanomater. 2020, 3020287.

Norton, K.-A., Popel, A.S., Pandey, N.B., 2015. Heterogeneity of chemokine cell-surface receptor expression in triple-negative breast cancer. Am. J. Cancer Res. 5, 1295.

Nunes, T., Hamdan, D., Leboeuf, C., El Bouchtaoui, M., Gapihan, G., Nguyen, T.T., Meles, S., Angeli, E., Ratajczak, P., Lu, H., Di Benedetto, M., Bousquet, G., Janin, A., 2018. Targeting cancer stem cells to overcome chemoresistance. Int. J. Mol. Sci. 19, 4036.

Oefelein, M.G., Betageri, G.V., Venkatesan, N., 2020. Liposomal Paclitaxel Formulation for Treating Bladder Cancer. Google Patents.

Olson, F., Hunt, C., Szoka, F., Vail, W., Papahadjopoulos, D., 1979. Preparation of liposomes of defined size distribution by extrusion through polycarbonate membranes. Biochim. Biophys. Acta Biomembr. 557, 9–23.

Ong, S.G.M., Chitneni, M., Lee, K.S., Ming, L.C., Yuen, K.H., 2016. Evaluation of extrusion technique for nanosizing liposomes. Pharmaceutics 8, 36.

Owens III, D.E., Peppas, N.A., 2006. Opsonization, biodistribution, and pharmacokinetics of polymeric nanoparticles. Int. J. Pharm. 307, 93–102.

Ozben, T., 2006. Mechanisms and strategies to overcome multiple drug resistance in cancer. FEBS Lett. 580, 2903–2909.

Paál, K., Müller, J., Hegedûs, L., 2001. High affinity binding of paclitaxel to human serum albumin. Eur. J. Biochem. 268, 2187–2191.

Pandit, P.G., Parakh, D.R., Patil, M.P., 2015. Proniosomal gel for improved transdermal drug delivery: an overview. World J. Pharm. Res. 4, 560–586.

Paolino, D., Celia, C., Trapasso, E., Cilurzo, F., Fresta, M., 2012. Paclitaxel-loaded ethosomes®: potential treatment of squamous cell carcinoma, a malignant transformation of actinic keratoses. Eur. J. Pharm. Biopharm. 81, 102–112.

Pardo, J., Peng, Z., Leblanc, R.M., 2018. Cancer targeting and drug delivery using carbon-based quantum dots and nanotubes. Molecules 23, 378.

Park, S.J., Park, S.-H., Cho, S., Kim, D.-M., Lee, Y., Ko, S.Y., Hong, Y., Choy, H.E., Min, J.-J., Park, J.-O., 2013. New paradigm for tumor theranostic methodology using bacteria-based microrobot. Sci. Rep. 3, 1–8.

Park, I.H., Sohn, J.H., Kim, S.B., Lee, K.S., Chung, J.S., Lee, S.H., Kim, T.Y., Jung, K.H., Cho, E.K., Kim, Y.S., Song, H.S., Seo, J.H., Ryoo, H.M., Lee, S.A., Yoon, S.Y., Kim, C.S., Kim, Y.T., Kim, S.Y., Jin, M.R., Ro, J., 2017. An open-label, randomized, parallel, phase III trial evaluating the efficacy and safety of polymeric micelle-formulated paclitaxel compared to conventional cremophor EL-based paclitaxel for recurrent or metastatic HER2-negative breast cancer. Cancer Res. Treat. 49, 569–577.

Patravale, V., Dandekar, P., Jain, R., 2012. Nanoparticles as drug carriers. In: Nanoparticulate Drug Delivery: Perspectives on the Transition from Laboratory to Market. Elsevier.

Payne, N.I., Browning, I., Hynes, C.A., 1986. Characterization of proliposomes. J. Pharm. Sci. 75, 330–333.

Payton, N., Wempe, M., Betker, J.L., Randolph, T., Anchordoquy, T., 2013. Lyophilization of a triply unsaturated phospholipid: effects of trace metal contaminants. Eur. J. Pharm. Biopharm. 85, 306–313.

Perez, A., Hernandez, R., Velasco, D., Voicu, D., Mijangos, C., 2015. Poly (lactic-co-glycolic acid) particles prepared by microfluidics and conventional methods. Modulated particle size and rheology. J. Colloid Interface Sci. 441, 90–97.

Persidis, A., 1999. Cancer multidrug resistance. Nat. Biotechnol. 17, 94–95.

Phung, C., Le, T.G., Nguyen, V.H., Vu, T.T., Nguyen, H.Q., Kim, J.O., Yong, C.S., Nguyen, C.N., 2020. PEGylated-paclitaxel and dihydroartemisinin nanoparticles for simultaneously delivering paclitaxel and dihydroartemisinin to colorectal cancer. Pharm. Res. 37, 1–11.

Pick, U., 1981. Liposomes with a large trapping capacity prepared by freezing and thawing of sonicated phospholipid mixtures. Arch. Biochem. Biophys. 212, 186–194.

Pitha, J., Milecki, J., Fales, H., Pannell, L., Uekama, K., 1986. Hydroxypropyl-β-cyclodextrin: preparation and characterization; effects on solubility of drugs. Int. J. Pharm. 29, 73–82.

Qi, S.-S., Sun, J.-H., Yu, H.-H., Yu, S.-Q., 2017. Co-delivery nanoparticles of anti-cancer drugs for improving chemotherapy efficacy. Drug Deliv. 24, 1909–1926.

Rai, R., Alwani, S., Badea, I., 2019. Polymeric nanoparticles in gene therapy: new avenues of design and optimization for delivery applications. Polymers 11, 745.

Rajput, N., 2015. Methods of preparation of nanoparticles-a review. Int. J. Adv. Res. Technol. 7, 1806.

Rao, B.G., Mukherjee, D., Reddy, B.M., 2017. Novel approaches for preparation of nanoparticles. In: Nanostructures for Novel Therapy. Elsevier.

Ren, W.X., Han, J., Uhm, S., Jang, Y.J., Kang, C., Kim, J.H., Kim, J.S., 2015. Recent development of biotin conjugation in biological imaging, sensing, and target delivery. Chem. Commun. 51, 10403–10418.

Reshma, P., Unnikrishnan, B., Preethi, G., Syama, H., Archana, M., Remya, K., Shiji, R., Sreekutty, J., Sreelekha, T., 2019. Overcoming drug-resistance in lung cancer cells by paclitaxel loaded galactoxyloglucan nanoparticles. Int. J. Biol. Macromol. 136, 266–274.

Rompicharla, S.V.K., Kumari, P., Ghosh, B., Biswas, S., 2018. Octa-arginine modified poly (amidoamine) dendrimers for improved delivery and cytotoxic effect of paclitaxel in cancer. Artif. Cells Nanomed. Biotechnol. 46, 847–859.

Ruoslahti, E., 2012. Peptides as targeting elements and tissue penetration devices for nanoparticles. Adv. Mater. 24, 3747–3756.

Sahana, D., Mittal, G., Bhardwaj, V., Kumar, M.R., 2008. PLGA nanoparticles for oral delivery of hydrophobic drugs: influence of organic solvent on nanoparticle formation and release behavior in vitro and in vivo using estradiol as a model drug. J. Pharm. Sci. 97, 1530–1542.

Sauna, Z.E., Ambudkar, S.V., 2007. About a switch: how P-glycoprotein (ABCB1) harnesses the energy of ATP binding and hydrolysis to do mechanical work. Mol. Cancer Ther. 6, 13–23.

Schmitt-Sody, M., Strieth, S., Krasnici, S., Sauer, B., Schulze, B., Teifel, M., Michaelis, U., Naujoks, K., Dellian, M., 2003. Neovascular targeting therapy: paclitaxel encapsulated in cationic liposomes improves antitumoral efficacy. Clin. Cancer Res. 9, 2335–2341.

Sercombe, L., Veerati, T., Moheimani, F., Wu, S.Y., Sood, A.K., Hua, S., 2015. Advances and challenges of liposome assisted drug delivery. Front. Pharmacol. 6, 286.

Shaheen, S.M., Shakil Ahmed, F., Hossen, M.N., Ahmed, M., Amran, M.S., Ul-Islam, M., 2006. Liposome as a carrier for advanced drug delivery. Pak. J. Biol. Sci. 9, 1181–1191.

Sharma, A., Straubinger, R.M., 1994. Novel taxol formulations: preparation and characterization of taxol-containing liposomes. Pharm. Res. 11, 889–896.

Shashi, K., Satinder, K., Bharat, P., 2012. A complete review on: liposomes. Int. Res. J. Pharm. 3, 10–16.

Shen, Q., Shen, Y., Jin, F., Du, Y.Z., Ying, X.Y., 2020. Paclitaxel/hydroxypropyl-β-cyclodextrin complex-loaded liposomes for overcoming multidrug resistance in cancer chemotherapy. J. Liposome Res. 30, 12–20.

Shi, Q., Zhang, L., Liu, M., Zhang, X., Zhang, X., Xu, X., Chen, S., Li, X., Zhang, J., 2015. Reversion of multidrug resistance by a pH-responsive cyclodextrin-derived nanomedicine in drug resistant cancer cells. Biomaterials 67, 169–182.

Shuai, X., Merdan, T., Schaper, A.K., Xi, F., Kissel, T., 2004. Core-cross-linked polymeric micelles as paclitaxel carriers. Bioconjug. Chem. 15, 441–448.

Shukla, S., Ohnuma, S., Ambudkar, S.V., 2011. Improving cancer chemotherapy with modulators of ABC drug transporters. Curr. Drug Targets 12, 621–630.

Şimşek, S., Eroğlu, H., Kurum, B., Ulubayram, K., 2013. Brain targeting of atorvastatin loaded amphiphilic PLGA-b-PEG nanoparticles. J. Microencapsul. 30, 10–20.

Singh, J., Jain, K., Mehra, N.K., Jain, N., 2016. Dendrimers in anticancer drug delivery: mechanism of interaction of drug and dendrimers. Artif. Cells Nanomed. Biotechnol. 44, 1626–1634.

Singla, A.K., Garg, A., Aggarwal, D., 2002. Paclitaxel and its formulations. Int. J. Pharm. 235, 179–192.

Slingerland, M., Guchelaar, H.-J., Rosing, H., Scheulen, M.E., van Warmerdam, L.J., Beijnen, J.H., Gelderblom, H., 2013. Bioequivalence of liposome-entrapped paclitaxel easy-to-use (LEP-ETU) formulation and paclitaxel in polyethoxylated castor oil: a randomized, two-period crossover study in patients with advanced cancer. Clin. Ther. 35, 1946–1954.

Sobhani, Z., Dinarvand, R., Atyabi, F., Ghahremani, M., Adeli, M., 2011. Increased paclitaxel cytotoxicity against cancer cell lines using a novel functionalized carbon nanotube. Int. J. Nanomedicine 6, 705.

Soga, O., van Nostrum, C.F., Fens, M., Rijcken, C.J., Schiffelers, R.M., Storm, G., Hennink, W.E., 2005. Thermosensitive and biodegradable polymeric micelles for paclitaxel delivery. J. Control. Release 103, 341–353.

Song, X., Zhao, Y., Wu, W., Bi, Y., Cai, Z., Chen, Q., Li, Y., Hou, S., 2008. PLGA nanoparticles simultaneously loaded with vincristine sulfate and verapamil hydrochloride: systematic study of particle size and drug entrapment efficiency. Int. J. Pharm. 350, 320–329.

Sparreboom, A., Van Asperen, J., Mayer, U., Schinkel, A.H., Smit, J.W., Meijer, D.K., Borst, P., Nooijen, W.J., Beijnen, J.H., Van Tellingen, O., 1997. Limited oral bioavailability and active epithelial excretion of paclitaxel (Taxol) caused by P-glycoprotein in the intestine. Proc. Natl. Acad. Sci. 94, 2031–2035.

Stillwell, W., 2016. Membrane biogenesis, fatty acids. In: An Introduction to Biological Membranes, second ed. Elsevier.

Straubinger, R.M., 1995. Biopharmaceutics of paclitaxel (Taxol): formulation, activity and pharmacokinetics. In: Suffness, M. (Ed.), Taxol: Science and Applications, second ed. CRC Press, pp. 237–258.

Sung, H.-W., Liao, Z.-X., Chung, M.-F., Chen, K.-J., Cheng, P.-Y., Tu, H., 2014. Pharmaceutical Composition of Nanoparticles. Google Patents.

Surapaneni, M.S., Das, S.K., Das, N.G., 2012. Designing paclitaxel drug delivery systems aimed at improved patient outcomes: current status and challenges. ISRN Pharmacol. 2012, 623139.

Tamura, S., Taniguchi, H., Nishikawa, K., Imamura, H., Fujita, J., Takeno, A., Matsuyama, J., Kimura, Y., Kawada, J., Hirao, M., Hirota, M., Fujitani, K., Kurokawa, Y., Sakai,

D., Kawakami, H., Shimokawa, T., Satoh, T., 2020. A phase II trial of dose-reduced nab-paclitaxel for patients with previously treated, advanced or recurrent gastric cancer (OGSG 1302). Int. J. Clin. Oncol. 25, 2035–2043.

Tang, B., Qian, Y., Gou, Y., Cheng, G., Fang, G., 2018. VE-albumin core-shell nanoparticles for paclitaxel delivery to treat MDR breast cancer. Molecules 23, 2760.

Tee, J.K., Yip, L.X., Tan, E.S., Santitewagun, S., Prasath, A., Ke, P.C., Ho, H.K., Leong, D.T., 2019. Nanoparticles' interactions with vasculature in diseases. Chem. Soc. Rev. 48, 5381–5407.

Teow, H.M., Zhou, Z., Najlah, M., Yusof, S.R., Abbott, N.J., D'Emanuele, A., 2013. Delivery of paclitaxel across cellular barriers using a dendrimer-based nanocarrier. Int. J. Pharm. 441, 701–711.

Thakkar, S., Sharma, D., Kalia, K., Tekade, R.K., 2020. Tumor microenvironment targeted nanotherapeutics for cancer therapy and diagnosis: a review. Acta Biomater. 101, 43–68.

Thurston, G., McLean, J.W., Rizen, M., Baluk, P., Haskell, A., Murphy, T.J., Hanahan, D., McDonald, D.M., 1998. Cationic liposomes target angiogenic endothelial cells in tumors and chronic inflammation in mice. J. Clin. Invest. 101, 1401–1413.

Tong, L., Chen, W., Wu, J., Li, H., 2014. Folic acid-coupled nano-paclitaxel liposome reverses drug resistance in SKOV3/TAX ovarian cancer cells. Anti-Cancer Drugs 25, 244–254.

Tsai, M., Lu, Z., Wientjes, M.G., Au, J.L.-S., 2013. Paclitaxel-loaded polymeric microparticles: quantitative relationships between in vitro drug release rate and in vivo pharmacodynamics. J. Control. Release 172, 737–744.

Tu, L., Wang, G., Qi, N., Wu, W., Zhang, W., Feng, J., 2020. Multi-functional chitosan polymeric micelles as oral paclitaxel delivery systems for enhanced bioavailability and anti-tumor efficacy. Int. J. Pharm. 578, 119105.

Umerska, A., Paluch, K.J., Santos-Martinez, M.J., Corrigan, O.I., Medina, C., Tajber, L., 2018. Freeze drying of polyelectrolyte complex nanoparticles: effect of nanoparticle composition and cryoprotectant selection. Int. J. Pharm. 552, 27–38.

Vahed, S., Fathi, N., Samiei, M., Maleki Dizaj, S., Sharifi, S., 2019. Targeted cancer drug delivery with aptamer-functionalized polymeric nanoparticles. J. Drug Target. 27, 292–299.

Wang, Y., Wang, C., Jia, Y., Cheng, X., Lin, Q., Zhu, M., Lu, Y., Ding, L., Weng, Z., Wu, K., 2014. Oxygen-carbon nanotubes as a chemotherapy sensitizer for paclitaxel in breast cancer treatment. PLoS One 9, 1–6.

Wang, Z.Y., Zhang, H., Yang, Y., Xie, X.Y., Yang, Y.F., Li, Z., Li, Y., Gong, W., Yu, F.L., Yang, Z., Li, M.Y., Mei, X.G., 2016. Preparation, characterization, and efficacy of thermosensitive liposomes containing paclitaxel. Drug Deliv. 23, 1222–1231.

Wang, F., Porter, M., Konstantopoulos, A., Zhang, P., Cui, H., 2017. Preclinical development of drug delivery systems for paclitaxel-based cancer chemotherapy. J. Control. Release 267, 100–118.

Wang, X., Zhang, H., Chen, X., 2019. Drug resistance and combating drug resistance in cancer. Cancer Drug Resist. 2, 141–160.

Wang, C., Guan, W., Peng, J., Chen, Y., Xu, G., Dou, H., 2020a. Gene/paclitaxel co-delivering nanocarriers prepared by framework-induced self-assembly for the inhibition of highly drug-resistant tumors. Acta Biomater. 103, 247–258.

Wang, S., Liang, Q., Chi, Y., Zhuo, M., An, T., Duan, J., Wang, Z., Wang, Y., Zhong, J., Yang, X., 2020b. Retrospective analysis of the effectiveness and tolerability of nab-paclitaxel in Chinese elderly patients with advanced non-small-cell lung carcinoma. Thorac. Cancer 11, 1149–1159.

Wani, M.C., Taylor, H.L., Wall, M.E., Coggon, P., McPhail, A.T., 1971. Plant antitumor agents. VI. The isolation and structure of taxol, a novel antileukemic and antitumor agent from *Taxus brevifolia*. J. Am. Chem. Soc. 93, 2325–2327.

Wei, Y., Wang, Y., Xia, D., Guo, S., Wang, F., Zhang, X., Gan, Y., 2017. Thermosensitive liposomal codelivery of HSA-paclitaxel and HSA-ellagic acid complexes for enhanced drug perfusion and efficacy against pancreatic cancer. ACS Appl. Mater. Interfaces 9, 25138–25151.

Xie, Y., Yao, Y., 2019. Incorporation with dendrimer-like biopolymer leads to improved soluble amount and in vitro anticancer efficacy of paclitaxel. J. Pharm. Sci. 108, 1984–1990.

Xu, X., Wang, L., Xu, H.-Q., Huang, X.-E., Qian, Y.-D., Xiang, J., 2013. Clinical comparison between paclitaxel liposome (Lipusu®) and paclitaxel for treatment of patients with metastatic gastric cancer. Asian Pac. J. Cancer Prev. 14, 2591–2594.

Xu, H., Wen, Y., Chen, S., Zhu, L., Feng, R., Song, Z., 2020. Paclitaxel skin delivery by micelles-embedded Carbopol 940 hydrogel for local therapy of melanoma. Int. J. Pharm. 587, 119626.

Yadav, D., Sandeep, K., Pandey, D., Dutta, R.K., 2017. Liposomes for drug delivery. J. Biotechnol. Biomater. 7, 276.

Yan, C., Liang, N., Li, Q., Yan, P., Sun, S., 2019. Biotin and arginine modified hydroxypropyl-β-cyclodextrin nanoparticles as novel drug delivery systems for paclitaxel. Carbohydr. Polym. 216, 129–139.

Yerlikaya, F., Ozgen, A., Vural, I., Guven, O., Karaagaoglu, E., Khan, M.A., Capan, Y., 2013. Development and evaluation of paclitaxel nanoparticles using a quality-by-design approach. J. Pharm. Sci. 102, 3748–3761.

Yoo, J.W., Irvine, D.J., Discher, D.E., Mitragotri, S., 2011. Bio-inspired, bioengineered and biomimetic drug delivery carriers. Nat. Rev. Drug Discov. 10, 521–535.

Yoshizawa, Y., Kono, Y., Ogawara, K., Kimura, T., Higaki, K., 2011. PEG liposomalization of paclitaxel improved its in vivo disposition and anti-tumor efficacy. Int. J. Pharm. 412, 132–141.

Yu, J., Zhou, S., Li, J., Wang, Y., Su, Y., Chi, D., Wang, J., Wang, X., He, Z., Lin, G., 2020. Simple weak-acid derivatives of paclitaxel for remote loading into liposomes and improved therapeutic effects. RSC Adv. 10, 27676–27687.

Yuan, J., Lv, H., Peng, B., Wang, C., Yu, Y., He, Z., 2009. Role of BCRP as a biomarker for predicting resistance to 5-fluorouracil in breast cancer. Cancer Chemother. Pharmacol. 63, 1103–1110.

Yuan, Y., Wang, L., Du, W., Ding, Z., Zhang, J., Han, T., An, L., Zhang, H., Liang, G., 2015. Intracellular self-assembly of Taxol nanoparticles for overcoming multidrug resistance. Angew. Chem. Int. Ed. Engl. 54, 9700–9704.

Yuan, H., Guo, H., Luan, X., He, M., Li, F., Burnett, J., Truchan, N., Sun, D., 2020. Albumin nanoparticle of paclitaxel (Abraxane) decreases while taxol increases breast cancer stem cells in treatment of triple negative breast cancer. Mol. Pharm. 17, 2275–2286.

Yusuf, R.Z., Duan, Z., Lamendola, D.E., Penson, R.T., Seiden, M.V., 2003. Paclitaxel resistance: molecular mechanisms and pharmacologic manipulation. Curr. Cancer Drug Targets 3, 1–19.

Zambaux, M., Bonneaux, F., Gref, R., Maincent, P., Dellacherie, E., Alonso, M., Labrude, P., Vigneron, C., 1998. Influence of experimental parameters on the characteristics of poly (lactic acid) nanoparticles prepared by a double emulsion method. J. Control. Release 50, 31–40.

Zhang, X.G., Miao, J., Dai, Y.Q., Du, Y.Z., Yuan, H., Hu, F.Q., 2008. Reversal activity of nanostructured lipid carriers loading cytotoxic drug in multi-drug resistant cancer cells. Int. J. Pharm. 361, 239–244.

Zhang, W., Shi, Y., Chen, Y., Ye, J., Sha, X., Fang, X., 2011. Multifunctional Pluronic P123/F127 mixed polymeric micelles loaded with paclitaxel for the treatment of multidrug resistant tumors. Biomaterials 32, 2894–2906.

Zhang, L., Wang, Y., Yang, Y., Liu, Y., Ruan, S., Zhang, Q., Tai, X., Chen, J., Xia, T., Qiu, Y., Gao, H., He, Q., 2015. High tumor penetration of paclitaxel loaded pH sensitive cleavable liposomes by depletion of tumor collagen I in breast cancer. ACS Appl. Mater. Interfaces 7, 9691–9701.

Zhang, Y., Sriraman, S.K., Kenny, H.A., Luther, E., Torchilin, V., Lengyel, E., 2016. Reversal of chemoresistance in ovarian cancer by co-delivery of a p-glycoprotein inhibitor and paclitaxel in a liposomal platform. Mol. Cancer Ther. 15, 2282–2293.

Zhang, T., Luo, J., Fu, Y., Li, H., Ding, R., Gong, T., Zhang, Z., 2017. Novel oral administrated paclitaxel micelles with enhanced bioavailability and antitumor efficacy for resistant breast cancer. Colloids Surf. B. Biointerfaces 150, 89–97.

Zheng, X.C., Ren, W., Zhang, S., Zhong, T., Duan, X.C., Yin, Y.F., Xu, M.Q., Hao, Y.L., Li, Z.T., Li, H., Liu, M., Li, Z.Y., Zhang, X., 2018. The theranostic efficiency of tumor-specific, pH-responsive, peptide-modified, liposome-containing paclitaxel and superparamagnetic iron oxide nanoparticles. Int. J. Nanomedicine 13, 1495–1504.

Zweers, M.L., Grijpma, D.W., Engbers, G.H., Feijen, J., 2003. The preparation of monodisperse biodegradable polyester nanoparticles with a controlled size. J Biomed Mater Res B Appl Biomater 66, 559–566.

Zylberberg, C., Matosevic, S., 2016. Pharmaceutical liposomal drug delivery: a review of new delivery systems and a look at the regulatory landscape. Drug Deliv. 23, 3319–3329.

5

Strategies for enhancing paclitaxel bioavailability for cancer treatment

Mina Salehi and Siamak Farhadi

Department of Plant Genetics and Breeding, Faculty of Agriculture, Tarbiat Modares University, Tehran, Iran

5.1   Introduction

Plant active pharmaceutical ingredients are the basis for treating of many human diseases (Firenzuoli and Gori, 2007; Salehi et al., 2018a,b, 2019a,d). Paclitaxel is an effective chemotherapeutic agent with a unique mode of action and wide-spectrum activity against cancers; so that it has been termed the best-selling anticancer drug in the world. The significant anticancer activity of the extract of *Taxus brevifolia* bark tissue was found during a massive screening program of 35,000 plant species, and its structure (Fig. 5.1) was published in 1971 (Wani et al., 1971). Paclitaxel displays a unique antitumor action mode by the inhibition of cell cycle in G2/M phase through microtubule stabilization (Schif et al., 1979). While most anticancer agents interact with DNA of tumor cells, paclitaxel focuses on the microtubular cell system (Florea and Büsselberg, 2011). This powerful anticancer diterpene was originally extracted from *T. brevifolia* bark. Unfortunately, *Taxus* species grow slowly, and also their bark contains an extremely low amount of paclitaxel (0.01%–0.05% DW) (Wheeler et al., 1992). Approximately, 2.5–3 g paclitaxel is needed for each cancer patient, eight mature yew trees (Malik et al., 2011). On the other hand, harvesting the bark for paclitaxel production is destructive for *Taxus* trees. Overharvesting these valuable natural resources have made *Taxus* populations be on the verge of extinction worldwide (Shinwari and Qaiser, 2011; Salehi et al., 2018c, 2020a). Therefore, discovering alternative viable paclitaxel sources is considered to be essential for the protection of these natural limited resources and drug therapy cost decrement. This chapter focuses on methods and developments toward bioprocessing with emphasis on those in plant cell culture.

Paclitaxel. https://doi.org/10.1016/B978-0-323-90951-8.00006-0
Copyright © 2022 Elsevier Inc. All rights reserved.

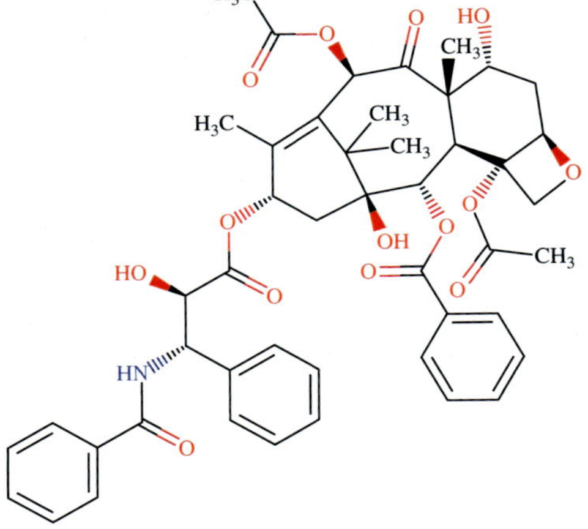

Fig. 5.1 Structure of paclitaxel.

5.2 Alternative paclitaxel sources

Researchers have tried and found alternative sources to natural paclitaxel; (a) semisynthesis, (b) chemical synthesis, (c) nursery cultivated *Taxus*, (d) heterologous expression systems, (e) fungal endophytes, (f) *Corylus avellana*, and (g) plant cell culture (Fig. 5.2).

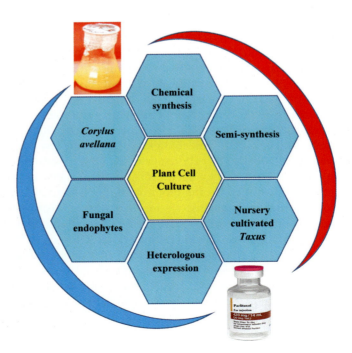

Fig. 5.2 Alternative sources for paclitaxel production.

5.2.1 Semisynthesis

The first relief for paclitaxel supply problems was a semisynthetic approach that was first established in 1986, and then modified to be more efficient (Denis et al., 1988). The semisynthesis of paclitaxel is commercially used by Bristol Myers Squibb, marketed as Taxol, involving the isolation of paclitaxel intermediates, 10-deacetylbaccatin III or baccatin III, from *Taxus* spp. needles. It is noteworthy that needle harvest, unlike bark harvest, is environmentally friendly keeping the tree alive, and paclitaxel semisynthesis was approved by the Food and Drug Administration (FDA) in 1994. However, semisynthesis process of paclitaxel displays some disadvantages including; (a) depending on environmental and epigenetic factors, (b) slow growth rate of *Taxus*, and (c) expensive purification of the intermediates. Generally, paclitaxel semisynthesis is expensive and environmentally unfriendly because a number of used harsh chemical solvents.

5.2.2 Chemical synthesis

Paclitaxel total synthesis was first reported by two different groups, Nicolaou et al. (1994) and Holton et al. (1994a,b) introducing various routes. Several further schemes have since been published, but none of them are industrially feasible by reason of their complex multistep pathways, low yield, and several toxic side products.

5.2.3 Nursery cultivated *Taxus*

Harvesting the bark for paclitaxel production has caused *Taxus* wild species to be at risk of extinction. To meet the increasing demand for paclitaxel and protect this natural finite resources, Canadian Forest Service-Atlantic Forestry Centre has been engaged in a program to minimize overharvesting *T. canadensis* by research in (a) developing the guidelines for ecologically sustainable harvest of *Taxus* in their natural stands, and (b) by a domestication program to convert elite cultivars of wild *Taxus* species into an industrially reared crop since 1997. Nearly simultaneously, China introduced 20,000 seedlings of *Taxus* x *media* (shrub) from TPL Phytogen Inc. (Vancouver, Canada), and established Bei-chuan and Hong-ya seedling bases located in Sichuan province. Ultimately, the seedlings of these seedling bases were introduced to other provinces of China for plantation. Also, *T. chinensis* was industrially cultivated in China. Paclitaxel and its precursor "10-deacetylbaccatin III" are extracted from the twigs and/or needles for keeping the plant alive. This method not only conserves wild *Taxus* species but also present paclitaxel sustainable alternative.

5.2.4 Heterologous expression systems

Some steps of paclitaxel biosynthetic pathway have been transferred into heterologous expression systems, inclusive of *Saccharomyces cerevisiae*, *Escherichia coli*, and plants. Isoprenyl diphosphate (IPP) isomerase, geranyl geranyl diphosphate (GGPP) synthase and taxadiene synthase (TS) expression in *E. coli* resulted in $1.3\,mg\,L^{-1}$ taxadiene production in its culture (Huang et al., 2001). Nevertheless, the further development of paclitaxel biosynthesis pathway is limited because of the problems of cytochromes P450 (CYP) expression in microbial systems. Mostly, CYPs displayed no functionality by reason of incorrect folding, translation, and insertion into cell membrane (Howat et al., 2014). Also, cofactor availability problem and the absence of specific CYP reductases needed for the efficient function of each CYP are significant obstacles (Wilson and Roberts, 2012). Five sequential paclitaxel biosynthetic genes were installed in *S. cerevisiae* to produce taxadien-5a-acetoxy-10b-ol; however, only taxadiene was accumulated (DeJong et al., 2006). The pathway restriction was experienced at the first hydroxylation of CYP, which co-expression of CYP and its cognate "CYP reductase" requires for efficient function (Jennewein et al., 2005). Also, the plant hosts have been used to production paclitaxel intermediates. The expression of GGPP synthase and TS in *Arabidopsis thaliana* resulted in $600\,ng\,g^{-1}DW$ taxadiene production. Nevertheless, introducing the exogenous genes resulted in growth retardation because of taxadiene accumulation (Besumbes et al., 2004). Expression of TS in a tomato line lacking the ability to utilize GGPP for carotenoid synthesis led to $160\,mg\,kg^{-1}$ DW taxadiene production in fruits (Kovacs et al., 2007). Besides, expressing TS in *Physcomitrella patens* led to taxadiene production (0.05% FW) without growth inhibition (Anterola et al., 2009).

There is an evident obstacle in engineering long and highly complex pathway of paclitaxel biosynthesis in heterologous hosts. The transformation of the entire pathway genes is very challenging and also paclitaxel biosynthesis regulation involving epigenetic modulation and signaling crosstalk is still poorly understood. To date, there is no fulfillment on producing final product "paclitaxel" by metabolic engineering or synthetic biology, except for a few intermediates. Since paclitaxel biosynthetic pathway is highly complex and the biosynthetic mechanisms remain poorly understood, the achievement of this final target faces many challenges.

5.2.5 Fungal endophytes

In 1993, the first paclitaxel-producing endophytic fungus was isolated from Pacific yew, *T. brevifolia* (Stierle et al., 1993). This initial finding was followed by plenty of studies reporting various paclitaxel-producing endophytic fungi (Guo et al., 2006; Zhao et al., 2009; Salehi et al., 2018c). To date, continuously more chemical evidence confirms the biosynthesis of paclitaxel and its analogs in some endophytic fungi. It is thought that paclitaxel biosynthetic pathway in the endophytic fungi may include a noticeably distinct evolutionary pattern than that in *Taxus* (Yang et al., 2014). Nevertheless, the fungal paclitaxel biosynthetic pathway has remained elusive, and identification and functional confirmation of corresponding biosynthetic genes will be the trustworthy genetic evidence. In spite of various paclitaxel-producing fungal endophytes (Flores-Bustamante et al., 2010; Salehi et al., 2018c), there has been no key advance about large-scale paclitaxel production using fungal fermentation. Unsustained paclitaxel production by repeated subculture on defined artificial media, and also its low productivity has raised doubts on the possibility of commercializing fungal endophytes as a sustainable platform for paclitaxel production. There are contradictive reports on independent paclitaxel biosynthetic capability of fungal endophytes (Heinig et al., 2013; Kusari et al., 2014). Heinig et al. (2013) found no evidence for the independent biosynthesis of taxanes in none of fungal endophytes described in the various publication, even using a combination of molecular biology, genome sequencing and phytochemistry. Therefore, fungal paclitaxel biosynthesis remains a controversial and ongoing issue.

5.2.6 *Corylus avellana*

Paclitaxel was detected in *C. avellana*, an angiosperm species, by Hoffman et al. (1998). Taxanes including paclitaxel were detected in the shells and leaves (Ottaggio et al., 2008), and also in the branches and stems of *C. avellana* (Hoffman and Shahidi, 2009). Also, several biological activities were reported for *C. avellana* extracts (Hoffman et al., 1998; Bestoso et al., 2006; Ottaggio et al., 2008). It is documented that paclitaxel biosynthesis in *C. avellana* cultures is not caused by fungus contamination (Bestoso et al., 2006).

5.2.7 Plant cell culture

Plant cell factories are a promising eco-friendly alternative platform for commercial paclitaxel production (Salehi et al., 2017; Espinosa-Leal et al., 2018; Salehi et al., 2019a,b). The advantages of paclitaxel production by plant cell culture are (a) independent of seasonal and geographical variations, (b) a limitless, continuous and uniform supply of paclitaxel, (c) possibility of cultivation in large bioreactors, and also paclitaxel biosynthesis induction by manipulating environmental conditions, (d) simpler isolation and purification because of producing a simpler spectrum of compounds than bark or needles (e) faster adaptation to rapid changes in demand as compared to agriculture-based processes, and (f) a renewable and environmentally friendly resource (Bringi and Kadkade, 1995). The first patent of *T. brevifolia* cell suspension culture for paclitaxel production was published in 1991, with recorded yields of $1-3\,mg\,L^{-1}$ (Christen et al., 1991). Currently, Phyton Biotech, Canada and Samyang Genex, South Korea use *Taxus* cell culture for paclitaxel production on a large scale. FDA approved plant cell culture for paclitaxel production in 2004. Presently Phyton Biotech is the largest single supplier of paclitaxel in the world. Also, Samyang Genex produces paclitaxel marketed as Genexol using plant cell culture process. Cell suspension culture of *C. avellana* has been addressed as a promising alternative for producing paclitaxel (Salehi et al., 2017, 2018c, 2019b,c, 2020a,b, 2021; Farhadi et al., 2020a,b). It is noteworthy that paclitaxel production in *C. avellana* cell culture is less than that reported for *Taxus* ones. Nonetheless, a high growth rate of *C. avellana* cells and a series of optimization process may compensate for the lower paclitaxel production in *C. avellana* cells. Recent researches focus on paclitaxel productivity improvement and production variability minimization through a better understanding of paclitaxel biosynthesis regulation on the genetic level, and also the effects of key bioprocess variables. The available strategies for paclitaxel biosynthesis improvement in plant in vitro cultures are discussed here (Fig. 5.3).

5.3 Strategies of paclitaxel biosynthesis improvement in plant cell culture

5.3.1 Selection of high-producing cell lines

The cells in plant cell cultures mostly display the significant variability in the biosynthesis capacity of secondary metabolites (Larkin and Scowcroft, 1981) because of cell genetic variation or heterogeneity in plant cell culture. Indeed, establishing fast-growing and high-producing cell culture is an important stage for paclitaxel production on a large scale. It has been reported that paclitaxel biosynthesis is more influenced by variability in biosynthetic capacity amongst the cultured cells than by any other factors (Bonfill et al., 2006). The callus originated from different parts of the same mother plant or the same explant of different mother plants displayed the considerable variability in growth and paclitaxel biosynthesis (Bruňáková et al., 2004).

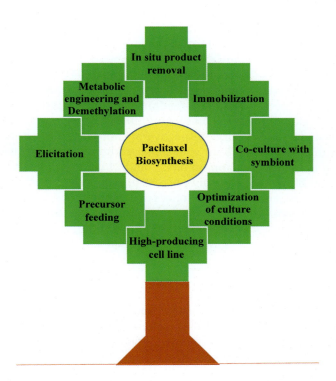

Fig. 5.3 Strategies for enhancing paclitaxel biosynthesis in plant in vitro cultures.

5.3.2 Optimization of culture conditions

Darkness is favorable for cell growth and paclitaxel biosynthesis (Wickremesinhe and Arteea, 1993; Hirasuna et al., 1996). Additionally, optimization of culture medium improves cell growth and paclitaxel biosynthesis in plant in vitro culture (Salehi et al., 2017).

5.3.2.1 Two-stage culture

Paclitaxel is mainly biosynthesized by the plant cells in the stationary growth phase, after exponential growth phase. Accordingly, two-stage system in which the cells are first cultured for growth (biomass) and then transferred to a favorable medium for paclitaxel biosynthesis is an effective strategy for production improvement. Two-stage system improved paclitaxel biosynthesis in *T. baccata* suspension culture (Cusido et al., 1999; Khosroushahi et al., 2006; Palazón et al., 2003; Navia-Osorio et al., 2002).

5.3.2.2 Carbohydrate source

Paclitaxel biosynthesis and cell growth in plant in vitro culture are greatly affected by carbon source and also its concentration. Feeding sucrose, maltose, or a combination of different sugars has been reported to improve paclitaxel biosynthesis in *Taxus* cell culture (Wang et al., 1999). A combination of low initial sucrose concentration with sucrose feeding during culture period improved paclitaxel biosynthesis. Also, Choi et al. (2000) investigated the effects of sucrose feeding to the culture medium at different times. The results showed that sucrose feeding to culture medium on 7th and 21st days of cell culture cycle improved paclitaxel biosynthesis by increasing primary and secondary metabolism, respectively (Choi et al., 2000). It was also found that culture medium with initial carbon source and maltose feeding to culture medium on days 7 and 21 considerably improved paclitaxel biosynthesis (Choi et al., 2000). The various carbohydrate sources led to remarkable differences in paclitaxel biosynthesis in cell cultures, biosynthesis improvement by fructose, and its suppression by glucose. Thus, it is concluded that the limiting step in paclitaxel synthesis is inhibited by glucose and stimulated by fructose (Hirasuna et al., 1996).

5.3.2.3 Phytohormones

The effects of various concentrations and also combinations of auxins (NAA, 2,4-D, 2,4,5-trichlorophenoxyacetic acid or 2,4,5-T, picloram) and cytokinins (kinetin and 6-benzylaminopurine) have been investigated to optimize cell growth and paclitaxel biosynthesis in *T. baccata* cell culture (Moon et al., 1998; Khosroushahi et al., 2006). Picloram enhanced cell growth but decreased paclitaxel biosynthesis (Hirasuna et al., 1996). Therefore, phytohormone type and concentration should be optimized for paclitaxel biosynthesized improvement.

5.3.2.4 Ethylene inhibitors

Another main component of culture conditions is the concentration of dissolved gases in culture flask headspace. Gas components cannot be modulated in the same way as dissolved solids; thus, this item has often been neglected as one of the important culture conditions. The effects of one of the most important dissolved gases, ethylene, was determined on cell growth and paclitaxel production in *Taxus* cell culture (Linden et al., 2001; Zhang and Wu, 2003; Tabata, 2004). Plant cells exposed to biotic and abiotic elicitors activate defense system including secondary metabolite production. Ethylene plays a role, in the regulation of plant defense responses induced with the elicitors (Feys and Parker, 2000). Ethylene production occurs in plant cell culture subjected to the elicitors. It is noteworthy that ethylene displays the positive effects at low concentrations, but the adverse effects at high concentrations on secondary metabolite biosynthesis including paclitaxel and also cell growth in plant in vitro culture (Linden et al., 2001). So, the use of different ethylene biosynthesis inhibitors including $NiCl_2$, $CoCl_2$, and α-amino isobutyric acid, and also an ethylene action inhibitor such as Ag^+ can enhance paclitaxel production (Zhang and Wu, 2003).

5.3.3 Precursor feeding

One of the main limiting factors preventing the optimal biosynthesis level of paclitaxel is precursor availability for use in the corresponding biosynthetic pathway. In precursor feeding, plant cell cultures may convert precursors into products by preexisting enzyme systems (Murthy et al., 2014), and improve desired product biosynthesis. This strategy is effective when the precursors are inexpensive. Phenylalanine is one of the precursors in paclitaxel biosynthetic pathway involving in the synthesis of the side chain of paclitaxel. Phenylalanine feeding enhanced paclitaxel accumulation in *Taxus cuspidata* (Syklowska-Baranek and Furmanowa, 2005) and *C. avellana* (Rahpeyma et al., 2015). Also, mevalonate feeding in *Taxus* cell culture elicited with methyl jasmonate increased paclitaxel biosynthesis (Cusidó et al., 2002).

5.3.4 Elicitation strategy

Elicitation is one of the most impressive strategies used for the dramatic increment of secondary metabolite biosynthesis in plant cell cultures (Wang and Zhong, 2002; Zhong, 2002). Elicitors are defined as compounds stimulating any type of plant defense and increasing secondary metabolism to protect the cell and also the whole plant (Zhao et al., 2005).

Elicitors can be grouped based on their nature as biotic (biological origin) and abiotic (nonbiological origin) elicitors, or based on their origin as exogenous (originated from outside the cell) and endogenous (originated from inside the cell) elicitors. Fig. 5.4 represents the classification of elicitors.

Elicitors have been widely used for enhancing paclitaxel biosynthesis in plant cell cultures since 1990s (Mirjalili and Linden, 1996; Malik et al., 2011; Onrubia et al., 2013; Sabater-Jara et al., 2010; Salehi et al., 2019b,c, 2020a; Farhadi et al., 2020a). Abiotic and biotic elicitors such as vanadyl (IV) sulfate, lanthanum salts, arachidonic acid, salicylic acid, jasmonates, and fungal extracts, alone or in combination, has been successfully used for paclitaxel biosynthesis increment in plant cell culture. Amongst jasmonates, the most widely used one for inducing paclitaxel production in *Taxus* cell cultures is methyl jasmonate (MeJA) (Sabater-Jara et al., 2010).

Since the induction of paclitaxel biosynthesis pathway is part of the plant defense response to extracellular signals, the strategies

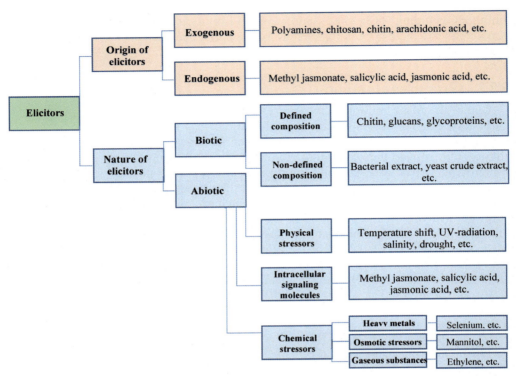

Fig. 5.4 Basic classification of elicitors used for eliciting in vitro biosynthesis of plant secondary metabolites.

of paclitaxel yield improvement in plant in vitro cultures should be based on this mechanism. Fungal endophytes contribute a very remarkable role in producing plant secondary metabolites (Faeth and Fagan, 2002; Jia et al., 2016). Fungal endophytes produce the bioactive compounds that elicit the plant defense system and improve the biosynthesis of secondary metabolites in the plants (Li and Tao, 2009; Ding et al., 2018; Salehi et al., 2019b,c, 2020a; Farhadi et al., 2020a). The fungal elicitors, because of their low toxicity for plant cells and high efficiency, are mostly used for the elicitation of producing plant secondary metabolites (Yuan et al., 2002). Besides, the fungal elicitors were reported to be more efficient than the chemical elicitors for the elicitation of biosynthesizing secondary metabolites in plant in vitro culture (Awad et al., 2014). It was reported that various fungal extracts increased paclitaxel productivity in *Taxus* in vitro culture (Yuan et al., 2002; Badi et al., 2015). Paclitaxel content significantly increased in *T. baccata* cell suspension culture exposed to fungal metabolite (Amini et al., 2014). However, all fungal elicitors are not functional to elicit the biosynthesis of secondary metabolites, several pathogenic fungi lead to cell hypersensitive responses and programmed cell death (DiCosmo et al., 1987). It is noteworthy that endophytic fungi do not result in hypersensitivity reactions in host plant cells and even a number of them can display growth-promoting effects on the plants (Wang et al., 2011; Ming et al., 2013). The various fungal elicitors may display the different effects on the same plant (Zhai et al., 2017). Different fungal elicitors have led to paclitaxel productivity enhancement in *C. avellana* cell culture. *C. avellana* cell culture exposed to culture filtrate (CF) (2.5% (v/v) on 17th day) of *Paraconiothyrium brasiliense* strain HEF_{114} isolated from *C. avellana* exhibited a 3.0-folds enhancement in paclitaxel production (Salehi et al., 2019b). Besides, *C. avellana* cell culture treated with cell extract (CE) (10% (v/v) on 17th day) of *Chaetomium globosum* strain YEF_{20} resulted in a 4.1-folds enhancement in paclitaxel production (Salehi et al., 2019b). In another study, adding 2.5% (v/v) CE of *Epicoccum nigrum* strain YEF_2 to *C. avellana* cell culture led to a 3.6-folds enhancement in paclitaxel production (Salehi et al., 2019c). In another attempt to find the efficient elicitors, it was found that amongst different fungal elicitors, the addition of 5% (v/v) CE and also 2.5 and 5% (v/v) of cell wall (CW) derived from *Coniothyrium palmarum* at day 17 led to highest paclitaxel production (3.6-folds) (Farhadi et al., 2020a). Also, the results showed that a combined treatment of CW (2.5% (v/v) on 17th day) and 50 mM of methyl-β-cyclodextrin (MBCD) synergistically increased paclitaxel production (5.8-fold) in *C. avellana* cell culture (Farhadi et al., 2020a). Salehi et al. (2020a) reported that the joint effects of CF and CE derived from endophytic fungus isolated from *C. avellana* leaf, *Camarosporomyces flavigenus* strain HEF17, resulted in more paclitaxel biosynthesis and secretion as compared to the individual use of them.

Coronatine (Cor) is a plant pathogenic toxin biosynthesized by *Pseudomonas syringae* strains. Cor has recently attracted much attention as an elicitor of plant secondary metabolism. It is reported that Cor is a more effective elicitor than MeJA in paclitaxel biosynthesis improvement in *Taxus* x *media* cell culture (Onrubia et al., 2013).

Cyclodextrin has caught considerable attention not only as an agent eliciting the production of secondary metabolites in plant in vitro cultures, as a result of defense response elicitation but also for its ability to constitute the inclusion complexes with poorly water-soluble apolar compounds and make possible the secretion of metabolites from cell to culture medium, thus considered to be a true elicitor (Ramirez-Estrada et al., 2015; Farhadi et al., 2020a). Several studies have been reported that MBCD increased paclitaxel production as well as its secretion from the cells to culture medium in *C. avellana* (Farhadi et al., 2020a,b) and *Taxus* (Ramirez-Estrada et al., 2015; Sabater-Jara et al., 2014) cell cultures. It is concluded that MBCD, by enhancing paclitaxel secretion, decreased the toxicity and retro-inhibition processes due to paclitaxel accumulation in the cytoplasm and in this way enhanced paclitaxel production (Farhadi et al., 2020a). It is noteworthy that the secretion of paclitaxel from the cells into culture medium assuredly make easy its extraction and purification for paclitaxel production on a large scale.

As mentioned earlier, the combined treatment of various compounds or elicitors that improve paclitaxel biosynthesis may display a prominent effect on paclitaxel productivity because of their interaction with various enzymes of the biosynthesis pathway (Khosroushahi et al., 2006; Salehi et al., 2020a,b, 2021; Farhadi et al., 2020a,b).

5.3.5 In situ product removal and two-phase culture

Low paclitaxel production in plant cell cultures may be due to feedback inhibition, and/or its nonenzymatic or enzymatic degradation in the medium. In these cases, the installation of an artificial site for paclitaxel accumulation, second liquid or solid phase into aqueous culture medium (Cai et al., 2012), can increase paclitaxel production (Fig. 5.5). The inhibitory product removal as soon as it is biosynthesized in cell culture is named in situ product removal (ISPR) (Dafoe and Daugulis, 2014). ISPR can increase secondary metabolite productivity by any of the following means (Freeman et al., 1993): (a) minimizing the inhibitory or toxic effects of the product allowing continuous biosynthesis at the maximal level, (b) the minimization of the product losses caused the degradation or uncontrolled removal by evaporation, etc., and (c) decrement of the total number of downstream processing steps. Such a processing system has been named "two-phase culture." The two-phase culture system is an effective strategy to improve in vitro biosynthesis of secondary metabolites by a promoted release of these compounds from the plant cells. The first phase supports cell growth, the second phase supplies an additional site and acts as a metabolic sink for secondary metabolite accumulation reducing feedback inhibition (Malik et al., 2013). Two-phase culture has been successfully used in *Taxus* for paclitaxel in vitro production (Collins-Pavao et al., 1996; Sykłowska-Baranek et al., 2018).

5.3.6 Immobilization

Immobilization is one of the most effective strategies for in vitro production of secondary metabolites. The advantages of immobilization are easy separation of the biomass and easy product recovery, providing high cell concentrations per unit volume, better cell-cell contact and protection from shear stress, and avoidance of cell washout even at the high dilution rates of the continuous operation mode

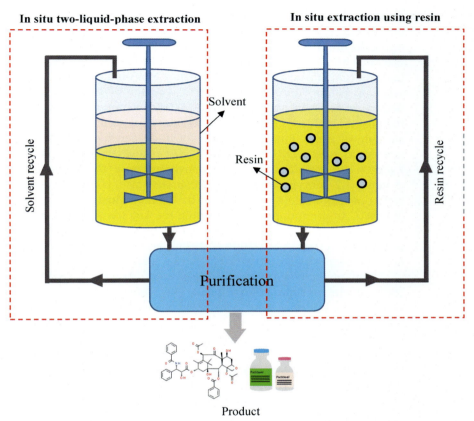

Fig. 5.5 Schematic diagram of in situ product removal system using liquid solvent and solvent impregnated resin extractants.

(Dörnenburg, 2004). Immobilization enhanced paclitaxel biosynthesis as compared to free cells (Bentebibel et al., 2005).

5.3.7 Metabolic engineering

Paclitaxel biosynthesis increment could be achieved using metabolic engineering. Metabolic engineering is rewiring science of cellular metabolism using a modification of regulatory functions and enzyme activities of the cell using recombinant DNA technologies to enhance native metabolite or new products (Bailey, 1991). Metabolic engineering is performed in two approaches including direct and holistic approaches. The direct method is the manipulation of single gene to increase the biosynthesis of the desired or novel compound either by upregulating an intermediate step (rate-limiting enzymes) or inhibiting the undesirable products. Meanwhile, in a holistic method, multiple steps are targeted simultaneously. In the complex pathway of the desired compound such as paclitaxel, a transcription factor is used for simultaneous upregulation of several enzymes, and polycistronic antisense RNA is used for simultaneous suppression of several enzymes in a competing pathway (Capell and Christou, 2004). Three MeJA-inducible MYC transcription factors "TcJAMYC1, TcJAMYC2, and TcJAMYC4" were identified in *Taxus*. Overexpression of TcJAMYC transcription factors made the positive regulation of the promoter of several paclitaxel biosynthesis pathway genes. But the promoters of 10-deacetylbaccatin III-10-*O*-acetyltransferase (*DBAT*), taxane 2α-*O*-benzoyltransferase (*DBBT*), baccatin III: 3-amino, 3-phenylpropanoyltransferase (*BAPT*), and 3′-*N*-debenzoyl-2-deoxypaclitaxel-*N*-benzoyltransferase (*DBTNBT*) genes were downregulated or not affected by the individual action of three TcJAMYC transcription factors. Finding the key transcription factors uncoupled from negative regulators and can particularly simulate MeJA activity by expression activation of specific genes is a key challenge for sustainable metabolite biosynthesis (Lenka et al., 2015).

Regulating related primary metabolism of plant cells can be used for improving paclitaxel biosynthesis. The overexpressing of the neutral/alkaline invertase gene (a key gene of sucrose hydrolysis) significantly enhanced TS expression level in *T. chinensis* cells and improved paclitaxel biosynthesis (Dong et al., 2015). Nevertheless, the metabolic engineering of *Taxus* cell culture for regulating paclitaxel biosynthesis is very challenging because genetic transformation of gymnosperm plant such as *Taxus* is difficult, and also *Taxus* cells are slow-growing (Kundu et al., 2017), and paclitaxel biosynthetic pathway is not fully characterized (Lenka et al., 2015).

It is noteworthy that even plant cells engineered to overexpress biosynthetic pathway key genes still need the elicitation for high production of paclitaxel (Expósito et al., 2010).

5.3.8 Reactivation of paclitaxel biosynthesis pathway

The strategies of paclitaxel biosynthesis improvement including elicitation have been successfully used in plant cell cultures. However, the long-term maintenance of plant in vitro culture results in loss of secondary biosynthesis, a critical handicap for a large-scale production system (Kim et al., 2004; Sanchez-Muñoz et al., 2018). An evident correlation between low biosynthesis of paclitaxel and DNA methylation levels were observed in *Taxus* in vitro cultures (Sanchez-Muñoz et al., 2018). In plants, DNA methylation occurs in three cytosine contexts: CG, CHG, and CHH (H is A, C, or T). Heterochromatin regions enriched with transposons and repetitive DNA sequences are heavily methylated (Zhang et al., 2006, 2018). Also, a remarkable level of

DNA methylation occurs in euchromatic regions (Zhang et al., 2006, 2018). DNA methylation has a main role in the different processes including transcription regulation, genome stability, vernalization or long-term adaption to the environment (Sanchez-Muñoz et al., 2018; Gallego-Bartolomé, 2020). It is suggested that *Taxus* cell lines cultured for long periods in in vitro condition use DNA methylation to regulate paclitaxel biosynthesis and avoid its cytotoxic effects (Sanchez-Muñoz et al., 2018).

DNA methylation inhibitors such as 5-azacytidine and N6-benzyladeninecan have been used to inhibit DNA methylation and reactivate the gene expression and, therefore, improve the biosynthesis of secondary metabolites (Ngernprasirtsiri and Akazawa, 1990; Zeng et al., 2019). 5-Azacytidine is a well-known methylation inhibitor lowering DNA methylation levels (Christman, 2002). 5-Azacytidine is incorporated into DNA and inhibits methylation in the daughter strand during replication resulting in methylation loss in specific gene regions and their activation (Christman, 2002; McClenahan and Rogers, 2018). Adding 5-azacytidine to *Taxus* cell culture reduced methylation levels and improved paclitaxel biosynthesis (Li et al., 2013).

5.3.9 Co-culture of plant cells with paclitaxel-producing fungal endophyte(s)

A friendly relationship has been established between host plants and some fungal endophytes in the course of long-term co-evolution. In this special relationship, fungal endophytes biosynthesize bioactive compounds which protect the host plants against the abiotic and biotic stresses and enhance their growth while the host plants supply the habitat and abundant nutrition nourishment for the fungal endophytes resulting in their survival (Rodriguez et al., 2009; Firáková et al., 2007). The metabolism of the endophytes is affected by the host plants (Kusari et al., 2012) and reciprocally the fungal endophytes increase the transcription of rate-limiting genes in plant paclitaxel biosynthetic pathway (Soliman et al., 2013). It is also observed that the number of endophytic fungi isolated from *Taxus* and its paclitaxel content in old tissue are higher than those in young tissues (Nadeem et al., 2002). It is concluded that the endophytic fungi play key roles in paclitaxel biosynthesis in their host plants.

On the other hand, unsustained paclitaxel production over repeated subculture on artificial culture media (Somjaipeng et al., 2016) is considered to be one of the most key problems for the large-scale production of paclitaxel using the fungal endophytes.

Overall, the interaction of the host plant-fungal endophyte metabolism can occur at several levels: (a) the fungal endophyte elicits host secondary metabolism, (b) the host plant elicits the endophyte secondary metabolism, (c) the fungal endophyte and host plant partially contribute to secondary metabolite biosynthesis by sharing parts of a biosynthesis pathway, (d) the plant host metabolize secondary compounds of the fungal endophyte, and (e) the fungal endophyte metabolizes the host plant products.

The enhancement of paclitaxel production was observed in *E. nigrum* in vitro culture by adding *C. avellana* cell extract (Salehi et al., 2018c). It was stated that *Taxus* extract could provide the essential stimulus to in vitro culture of paclitaxel-producing endophytic fungi (Soliman and Raizada, 2013). It seems that adding plant cell extract to in vitro culture of paclitaxel-producing endophytic fungi or the co-cultivation of these fungal endophyte(s) with *C. avellana* or *Taxus* cells can supply the essential stimulus to fungal endophyte(s) for sustainable and increased paclitaxel production (Salehi et al., 2018c, 2019c). The co-culture has been addressed as an efficient strategy for paclitaxel production improvement in *Taxus* (Li et al., 2009) and *C. avellana* (Salehi et al., 2019c). Indeed, the co-culture strategy can simulate the natural habitat of the fungal endophytes and improve paclitaxel production. Salehi et al. (2019c) described that in a co-culture system, both plant cells and fungal endophyte(s) produce paclitaxel, and also they can reciprocally benefit paclitaxel precursors biosynthesized by another partner, likewise, the elicitation effects of fungal metabolites on plant cells are long-lasting until the end of culture period because fungal mycelia are viable, and vice versa.

Also, the studies (Kusari et al., 2012; Akone et al., 2016; Do Nascimento et al., 2020) have offered the evidence that the interaction between the endophytes can play a major role in quorum-sensing signals and the onset of the biosynthesis of defense metabolites such as paclitaxel. Indeed, the co-culture of the endophytes may trigger the gene expression of the silent biosynthetic pathway (Rateb et al., 2013; Akone et al., 2016). Therefore, the simulation of the natural conditions by fermentation of two endophytes (also called co-culture) may lead to increased production of secondary metabolites including paclitaxel. In light of the available evidence, it seems that multiple co-culture system "endophytes-plant cells" can simulate the natural habitat of the endophytes and host plants. So that it is possible to establish relationships between the endophytes and plant cells in cell suspension culture to increase paclitaxel production.

5.4 Mathematical modeling for paclitaxel biosynthesis optimization

Paclitaxel biosynthesis and its elicitation in plant in vitro culture are complex biological processes as they are influenced by many factors interacting in nonlinear ways (Salehi et al., 2020b, 2021; Farhadi et al., 2020b). The bioprocess optimization by experimenting is high-cost, time-consuming and toilsome (Salehi et al., 2020b, 2021; Farhadi et al., 2020b). Accurate analysis of the effects of the factors affecting bioprocess, and their optimization would pave the way for large-scale production of paclitaxel using the plant cells. The mathematical methods can effectively model and forecast the optimized conditions for a complex multifactorial process. Artificial intelligence technology is the algorithm, designed by the humans, capable of thinking and computing intelligently and humanely (Agatonovic-Kustrin and Beresford, 2000). Mathematical approaches including multilayer perceptron-genetic algorithm (Salehi et al., 2020b), adaptive neuro-fuzzy inference system-genetic algorithm (Farhadi et al., 2020b) and general regression neural network-fruit fly optimization algorithm (Salehi et al., 2021) have been successfully used for modeling and optimizing paclitaxel biosynthesis in *C. avellana* cell culture treated with the fungal elicitors. Fig. 5.6 presents the outline of modeling and optimization of secondary metabolite biosynthesis in plant in vitro culture.

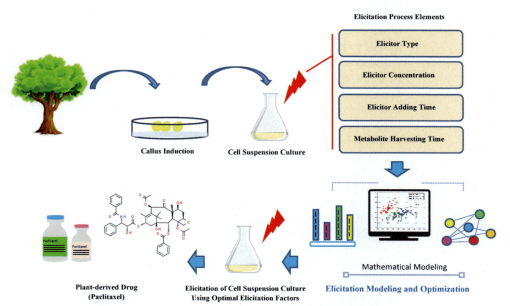

Fig. 5.6 The outline for modeling and optimization of secondary metabolites biosynthesis in plant in vitro culture. Adapted from Farhadi, S., Salehi, M., Moieni, A., Safaie, N., Sabet, M.S., 2020. Modeling of paclitaxel biosynthesis elicitation in *Corylus avellana* cell culture using adaptive neuro-fuzzy inference system-genetic algorithm (ANFIS-GA) and multiple regression methods. PLoS One 15(8), e0237478.

5.5 Concluding remarks and future perspectives

Bioprocessing strategies for paclitaxel production have been discussed in this chapter. Plant cell culture is a promising ecologically sustainable platform for commercial paclitaxel production. Currently, Phyton Biotech, Canada and Samyang Genex, South Korea use *Taxus* cell culture for paclitaxel production on a large scale. Paclitaxel production in *Taxus* cell culture has been enhanced using the selection of high-producing cell lines, optimization of culture conditions including two-stage culture, sugar feeding, optimizing phytohormone type and concentration, adding ethylene inhibitors, precursor feeding, elicitor adding, two-phase culture and immobilization. Nevertheless, these empirical strategies have not been enough to meet the growing world demand for valuable drug "paclitaxel." Most notably, there is an obvious hurdle in engineering long and highly complex pathway of paclitaxel biosynthesis, in its whole, in heterologous hosts. The transformation of the entire pathway genes is very challenging and also paclitaxel biosynthesis regulation involving epigenetic modulation and signaling crosstalk is still poorly understood. The exploitation of interorganismal advantages is promising for paclitaxel production enhancement. This enhancement strategy is co-culture of two or more organisms for eliciting paclitaxel production, biotransformation of paclitaxel pathway-intermediate for using high yield paclitaxel pathway-intermediates, metabolic dead ends, from one organism(s) by another organism(s), and even combinatorial production of paclitaxel by various organisms. The analysis of expression levels of paclitaxel biosynthesis genes in *Taxus* high-producing cell culture exposed to the effective elicitors, and also multiple co-culture of *Taxus* cells and paclitaxel producing endophyte(s) shed light on the bottleneck steps limiting paclitaxel biosynthesis. The overexpression of bottleneck steps of paclitaxel biosynthesis pathway can be promising for enhancing paclitaxel production. Also, the selective manipulation of the chromatin and epigenetic modifications can be applied for engineering paclitaxel biosynthetic pathway. The identification of the transcription factors involved in regulating paclitaxel biosynthesis pathways by functional genomics techniques allows metabolic engineering to increase paclitaxel biosynthesis and reduce production costs in plant cell cultures. Low-cost genome sequencing techniques and user-friendly bioinformatic tools have allowed genome mining of *C. avellana* and paclitaxel-producing endophytic fungi for identifying candidate biosynthetic gene clusters. Evaluating paclitaxel biosynthetic gene clusters is essential for the exploitation of *C. avellana* and paclitaxel-producing endophytic fungi as a source of paclitaxel. Such genome sequence information is crucial to clone missing genes of paclitaxel-biosynthesis pathway in the endophytes, and also detect the bottleneck steps of paclitaxel biosynthesis pathway and then overexpress them in *C. avellana* cells.

References

Agatonovic-Kustrin, S., Beresford, R., 2000. Basic concepts of artificial neural network (ANN) modeling and its application in pharmaceutical research. J. Pharm. Biomed. Anal. 22 (5), 717–727.

Akone, S.H., Mándi, A., Kurtán, T., Hartmann, R., Lin, W., Daletos, G., Proksch, P., 2016. Inducing secondary metabolite production by the endophytic fungus *Chaetomium* sp. through fungal–bacterial co-culture and epigenetic modification. Tetrahedron 72 (41), 6340–6347.

Amini, S.A., Shabani, L., Afghani, L., Jalalpour, Z., Sharif-Tehrani, M., 2014. Squalestatin-induced production of taxol and baccatin in cell suspension culture of yew (*Taxus baccata* L.). Turk. J. Biol. 38, 528–536.

Anterola, A., Shanle, E., Perroud, P.-F., Quatrano, R., 2009. Production of taxa-4(5), 11(12)-diene by transgenic Physcomitrella patens. Transgenic Res. 18, 655–660.

Awad, V., Kuvalekar, A., Harsulkar, A., 2014. Microbial elicitation in root cultures of *Taverniera cuneifolia* (Roth) Arn. For elevated glycyrrhizic acid production. Ind. Crop. Prod. 54, 13–16.

Badi, H.N., Abdoosi, V., Farzin, N., 2015. New approach to improve taxol biosynthetic. Trakia J. Sci. 2, 115–124.

Bailey, J.E., 1991. Toward a science of metabolic engineering. Science 252 (5013), 1668–1675.

Bentebibel, S., Moyano, E., Palazón, J., Cusidó, R.M., Bonfill, M., Eibl, R., Pinol, M.T., 2005. Effects of immobilization by entrapment in alginate and scale-up on paclitaxel and baccatin III production in cell suspension cultures of *Taxus baccata*. Biotechnol. Bioeng. 89 (6), 647–655.

Bestoso, F., Ottaggio, L., Armirotti, A., Balbi, A., Damonte, G., Degan, P., Mazzei, M., Cavalli, F., Ledda, B., Miele, M., 2006. In vitro cell cultures obtained from different explants of *Corylus avellana* produce taxol and taxanes. BMC Biotechnol. 6 (1), 45.

Besumbes, O.´., Sauret-Güeto, S., Phillips, M.A., Imperial, S., Rodríguez-Concepción, M., Boronat, A., 2004. Metabolic engineering of isoprenoid biosynthesis in Arabidopsis for the production of taxadiene, the first committed precursor of taxol. Biotechnol. Bioeng. 88 (2), 168–175.

Bonfill, M., Exposito, O., Moyano, E., Cusidó, R.M., Palazón, J., Pinol, M.T., 2006. Manipulation by culture mixing and elicitation of paclitaxel and baccatin III production in *Taxus baccata* suspension cultures. In Vitro Cell. Dev. Biol. 42 (5), 422–426.

Bringi, V., Kadkade, P.G., 1995. U.S. Patent No. 5,407,816. U.S. Patent and Trademark Office, Washington, DC.

Bruňáková, K., Babincova, Z., Čellárová, E., 2004. Selection of callus cultures of *Taxus baccata* L. as a potential source of paclitaxel production. Eng. Life Sci. 4 (5), 465–469.

Cai, Z., Kastell, A., Knorr, D., Smetanska, I., 2012. Exudation: an expanding technique for continuous production and release of secondary metabolites from plant cell suspension and hairy root cultures. Plant Cell Rep. 31 (3), 461–477.

Capell, T., Christou, P., 2004. Progress in plant metabolic engineering. Curr. Opin. Biotechnol. 15 (2), 148–154.

Choi, H.K., Kim, S.I., Son, J.S., Hong, S.S., Lee, H.S., Chung, I.S., Lee, H.J., 2000. Intermittent maltose feeding enhances paclitaxel production in suspension culture of *Taxus chinensis* cells. Biotechnol. Lett. 22 (22), 1793–1796.

Christen, A.A., Gibson, D.M., Bland, J., 1991. U.S. Patent No. 5,019,504. U.S. Patent and Trademark Office, Washington, DC.

Christman, J.K., 2002. 5-Azacytidine and 5-aza-2′-deoxycytidine as inhibitors of DNA methylation: mechanistic studies and their implications for cancer therapy. Oncogene 21 (35), 5483–5495.

Collins-Pavao, M., Chin, C.K., Pedersen, H., 1996. Taxol partitioning in two-phase plant cell cultures of *Taxus brevifolia*. J. Biotechnol. 49 (1–3), 95–100.

Cusido, R.M., Palazon, J., Navia-Osorio, A., Mallol, A., Bonfill, M., Morales, C., Piñol, M.T., 1999. Production of Taxol® and baccatin III by a selected *Taxus baccata* callus line and its derived cell suspension culture. Plant Sci. 146 (2), 101–107.

Cusidó, R.M., Palazón, J., Bonfill, M., Navia-Osorio, A., Morales, C., Piñol, M.T., 2002. Improved paclitaxel and baccatin III production in suspension cultures of *Taxus media*. Biotechnol. Prog. 18 (3), 418–423.

Dafoe, J.T., Daugulis, A.J., 2014. In situ product removal in fermentation systems: improved process performance and rational extractant selection. Biotechnol. Lett. 36 (3), 443–460.

DeJong, J.M., Liu, Y., Bollon, A.P., Long, R.M., Jennewein, S., Williams, D., et al., 2006. Genetic engineering of taxol biosynthetic genes in *Saccharomyces cerevisiae*. Biotechnol. Bioeng. 93, 212–224.

Denis, J.N., Greene, A.E., Guenard, D., Gueritte-Voegelein, F., Mangatal, L., Potier, P., 1988. Highly efficient, practical approach to natural taxol. J. Am. Chem. Soc. 110 (17), 5917–5919.

DiCosmo, F., Quesnel, A., Misawa, M., Tallevi, S.G., 1987. Increased synthesis of ajmalicine and catharanthine by cell suspension cultures of *Catharanthus roseus* in response to fungal culture-filtrates. Appl. Biochem. Biotechnol. 14, 101–106.

Ding, C.H., Wang, Q.B., Guo, S., Wang, Z.Y., 2018. The improvement of bioactive secondary metabolites accumulation in *Rumex gmelini* Turcz through co-culture with endophytic fungi. Braz. J. Microbiol. 49 (2), 362–369.

Do Nascimento, J.S., Silva, F.M., Magallanes-Noguera, C.A., Kurina-Sanz, M., Dos Santos, E.G., Caldas, I.S., Luiz, J.H.H., de Oliveira Silva, E., 2020. Natural trypanocidal product produced by endophytic fungi through co-culturing. Folia Microbiol. 65 (2), 323–328.

Dong, Y.S., Duan, W.L., He, H.X., Su, P., Zhang, M., Song, G.H., et al., 2015. Enhancing taxane biosynthesis in cell suspension culture of *Taxus chinensis* by overexpressing the neutral/alkaline invertase gene. Process Biochem. 50 (4), 651–660.

Dörnenburg, H., 2004. Evaluation of immobilisation effects on metabolic activities and productivity in plant cell processes. Process Biochem. 39 (11), 1369–1375.

Espinosa-Leal, C.A., Puente-Garza, C.A., García-Lara, S., 2018. *In vitro* plant tissue culture: means for production of biological active compounds. Planta 248 (1), 1–18.

Expósito, O., Syklowska-Baranek, K., Moyano, E., Onrubia, M., Bonfill, M., Palazon, J., et al., 2010. Metabolic responses of *Taxus* media transformed cell cultures to the addition of methyl jasmonate. Biotechnol. Prog. 26, 1145–1153.

Faeth, S.H., Fagan, W.F., 2002. Fungal endophytes: common host plant symbionts but uncommon mutualists. Integr. Comp. Biol. 42, 360–368.

Farhadi, S., Moieni, A., Safaie, N., Sabet, M.S., Salehi, M., 2020a. Fungal cell wall and methyl-β-cyclodextrin synergistically enhance paclitaxel biosynthesis and secretion in *Corylus avellana* cell suspension culture. Sci. Rep. 10 (1), 1–10.

Farhadi, S., Salehi, M., Moieni, A., Safaie, N., Sabet, M.S., 2020b. Modeling of paclitaxel biosynthesis elicitation in *Corylus avellana* cell culture using adaptive neuro-fuzzy inference system-genetic algorithm (ANFIS-GA) and multiple regression methods. PLoS One 15 (8), e0237478.

Feys, B.J., Parker, J.E., 2000. Interplay of signaling pathways in plant disease resistance. Trends Genet. 16 (10), 449–455.

Firáková, S., Šturdíková, M., Múčková, M., 2007. Bioactive secondary metabolites produced by microorganisms associated with plants. Biologia 62 (3), 251–257.

Firenzuoli, F., Gori, L., 2007. Herbal medicine today: clinical and research issues. Evid. Based Complement. Alternat. Med. 4 (S1), 37–40.

Florea, A.M., Büsselberg, D., 2011. Cisplatin as an anti-tumor drug: cellular mechanisms of activity, drug resistance and induced side effects. Cancers 3 (1), 1351–1371.

Flores-Bustamante, Z.R., Rivera-Orduna, F.N., Martínez-Cárdenas, A., Flores-Cotera, L.B., 2010. Microbial paclitaxel: advances and perspectives. J. Antibiot. 63 (8), 460–467.

Freeman, A., Woodley, J.M., Lilly, M.D., 1993. In situ product removal as a tool for bioprocessing. Biotechnology 11 (9), 1007–1012.

Gallego-Bartolomé, J., 2020. DNA methylation in plants: mechanisms and tools for targeted manipulation. New Phytol. 227 (1), 38–44.

Guo, B.H., Wang, Y.C., Zhou, X.W., Hu, K., Tan, F., Miao, Z.Q., Tang, K.X., 2006. An endophytic taxol-producing fungus BT2 isolated from *Taxus chinensis* var. mairei. Afr. J. Biotechnol. 5 (10), 875–877.

Heinig, U., Scholz, S., Jennewein, S., 2013. Getting to the bottom of taxol biosynthesis by fungi. Fungal Divers. 60 (1), 161–170.

Hirasuna, T.J., Pestchanker, L.J., Srinivasan, V., Shuler, M.L., 1996. Taxol production in suspension cultures of *Taxus baccata*. Plant Cell Tissue Organ Cult. 44 (2), 95–102.

Hoffman, A., Shahidi, F., 2009. Paclitaxel and other taxanes in hazelnut. J. Funct. Foods 1 (1), 33–37.

Hoffman, A., Khan, W., Worapong, J., Strobel, G., Grifn, D., Arbogast, B., Barofsky, D., Boone, R.B., Ning, L., Zheng, P., Daley, L., 1998. Bioprospecting for taxol in angiosperm plant extracts—using high performance liquid chromatography thermospray mass spectrometry to detect the anticancer agent and its related metabolites in filbert trees. Spectroscopy 13 (6), 22–32.

Holton, R.A., Somoza, C., Kim, H.B., Liang, F., Biediger, R.J., Boatman, P.D., Shindo, M., Smith, C.C., Kim, S., 1994a. First total synthesis of taxol. 1. Functionalization of the B ring. Am. Chem. Soc. 116, 1597–1598.

Holton, R.A., Kim, H.B., Somoza, C., Liang, F., Biediger, R.J., Boatman, P.D., Shindo, M., Smith, C.C., Kim, S., 1994b. First total synthesis of taxol. 2. Completion of the C and D rings. Am. Chem. Soc. 116, 1599–1600.

Howat, S., Park, B., Oh, I.S., Jin, Y.W., Lee, E.K., Loake, G.J., 2014. Paclitaxel: biosynthesis, production and future prospects. New Biotechnol. 31 (3), 242–245.

Huang, Q., Roessner, C.A., Croteau, R., Scott, A.I., 2001. Engineering *Escherichia coli* for the synthesis of taxadiene, a key intermediate in the biosynthesis of taxol. Bioorg. Med. Chem. 9, 2237–2242.

Jennewein, S., Park, H., DeJong, J.M., Long, R.M., Bollon, A.P., Croteau, R.B., 2005. Coexpression in yeast of *Taxus* cytochrome P450 reductase with cytochrome P450 oxygenases involved in taxol biosynthesis. Biotechnol. Bioeng. 89, 588–598.

Jia, M., Chen, L., Xin, H.L., Zheng, C.J., Rahman, K., Han, T., Qin, L.P., 2016. A friendly relationship between endophytic fungi and medicinal plants: a systematic review. Front. Microbiol. 7, 906.

Khosroushahi, A.Y., Valizadeh, M., Ghasempour, A., Khosrowshahli, M., Naghdibadi, H., Dadpour, M.R., Omidi, Y., 2006. Improved taxol production by combination of inducing factors in suspension cell culture of *Taxus baccata*. Cell Biol. Int. 30 (3), 262–269.

Kim, B.J., Gibson, D.M., Shuler, M.L., 2004. Effect of subculture and elicitation on instability of taxol production in *Taxus* sp. suspension cultures. Biotechnol. Prog. 20, 1666–1673.

Kovacs, K., Zhang, L., Linforth, R.T., Whittaker, B., Hayes, C., Fray, R., 2007. Redirection of carotenoid metabolism for the efficient production of taxadiene [taxa-4(5),11(12)-diene] in transgenic tomato fruit. Transgenic Res. 16, 121–126.

Kundu, S., Jha, S., Ghosh, B., 2017. Metabolic engineering for improving production of taxol. In: Jha, S. (Ed.), Transgenesis and Secondary Metabolism. Reference Series in Phytochemistry, Springer, Cham, pp. 463–484.

Kusari, S., Hertweck, C., Spiteller, M., 2012. Chemical ecology of endophytic fungi: origins of secondary metabolites. Chem. Biol. 19 (7), 792–798.

Kusari, S., Singh, S., Jayabaskaran, C., 2014. Rethinking production of Taxol®(paclitaxel) using endophyte biotechnology. Trends Biotechnol. 32 (6), 304–311.

Larkin, P.J., Scowcroft, W.R., 1981. Somaclonal variation—a novel source of variability from cell cultures for plant improvement. Theor. Appl. Genet. 60 (4), 197–214.

Lenka, S.K., Nims, N.E., Vongpaseuth, K., Boshar, R.A., Roberts, S.C., Walker, E.L., 2015. Jasmonate-responsive expression of paclitaxel biosynthesis genes in *Taxus cuspidata* cultured cells is negatively regulated by the bHLH transcription factors TcJAMYC1, TcJAMYC2, and TcJAMYC4. Front. Plant Sci. 6, 115.

Li, Y.C., Tao, W.Y., 2009. Interactions of taxol-producing endophytic fungus with its host (*Taxus* spp.) during taxol accumulation. Cell Biol. Int. 33 (1), 106–112.

Li, Y.C., Tao, W.Y., Cheng, L., 2009. Paclitaxel production using co-culture of *Taxus* suspension cells and paclitaxel-producing endophytic fungi in a co-bioreactor. Appl. Microbiol. Biotechnol. 83 (2), 233.

Li, L.Q., Li, X.L., Fu, C.H., Zhao, C.F., Yu, L.J., 2013. Sustainable use of *Taxus media* cell cultures through minimal growth conservation and manipulation of genome methylation. Process Biochem. 48 (3), 525–531.

Linden, J.C., Haigh, J.R., Mirjalili, N., Phisaphalong, M., 2001. Gas concentration effects on secondary metabolite production by plant cell cultures. In: Zhong, J.J., et al. (Eds.), Plant Cells. Advances in Biochemical Engineering/Biotechnology. vol. 72. Springer, Berlin, Heidelberg, pp. 27–62.

Malik, S., Cusidó, R.M., Mirjalili, M.H., Moyano, E., Palazón, J., Bonfill, M., 2011. Production of the anticancer drug taxol in *Taxus baccata* suspension cultures: a review. Process Biochem. 46 (1), 23–34.

Malik, S., Mirjalili, M.H., Fett-Neto, A.G., Mazzafera, P., Bonfill, M., 2013. Living between two worlds: two-phase culture systems for producing plant secondary metabolites. Crit. Rev. Biotechnol. 33 (1), 1–22.

McClenahan, S.R., Rogers, W.M., 2018. Investigation of DNA Methylation and Paclitaxel Production Levels in *Taxus* Cell Cultures. Worcester Polytechnic Institute.

Ming, Q., Su, C., Zheng, C., Jia, M., Zhang, Q., Zhang, H., Qin, L., 2013. Elicitors from the endophytic fungus *Trichoderma atroviride* promote *Salvia miltiorrhiza* hairy root growth and tanshinone biosynthesis. J. Exp. Bot. 64 (18), 5687–5694.

Mirjalili, N., Linden, J.C., 1996. Methyl jasmonate induced production of taxol in suspension cultures of *Taxus cuspidata*: ethylene interaction and induction models. Biotechnol. Prog. 12 (1), 110–118.

Moon, W.J., Yoo, B.S., Kim, D.I., Byun, S.Y., 1998. Elicitation kinetics of taxane production in suspension cultures of *Taxus baccata* Pendula. Biotechnol. Tech. 12 (1), 79–81.

Murthy, H.N., Lee, E.J., Paek, K.Y., 2014. Production of secondary metabolites from cell and organ cultures: strategies and approaches for biomass improvement and metabolite accumulation. Plant Cell Tissue Organ Cult. 118 (1), 1–16.

Nadeem, M., Rikhari, H.C., Kumar, A., Palni, L.M.S., Nandi, S.K., 2002. Taxol content in the bark of Himalayan Yew in relation to tree age and sex. Phytochemistry 60 (6), 627–631.

Navia-Osorio, A., Garden, H., Cusidó, R.M., Palazón, J., Alfermann, A.W., Piñol, M.T., 2002. Taxol® and baccatin III production in suspension cultures of *Taxus baccata* and *Taxus wallichiana* in an airlift bioreactor. J. Plant Physiol. 159 (1), 97–102.

Ngernprasirtsiri, J., Akazawa, T., 1990. Modulation of DNA methylation and gene expression in cultured sycamore cells treated by hypomethylating base analog. Eur. J. Biochem. 194 (2), 513–520.

Nicolaou, K.C., Yang, Z., Liu, J.J., Ueno, H., Nantermet, P.G., Guy, R.K., Claiborne, C.F., Renaud, J., Couladouros, E.A., Paulvannan, K., Sorensen, E.J., 1994. Total synthesis of taxol. Nature 367, 630–634.

Onrubia, M., Moyano, E., Bonfill, M., Cusidó, R.M., Goossens, A., Palazón, J., 2013. Coronatine, a more powerful elicitor for inducing taxane biosynthesis in *Taxus media* cell cultures than methyl jasmonate. J. Plant Physiol. 170 (2), 211–219.

Ottaggio, L., Bestoso, F., Armirotti, A., Balbi, A., Damonte, G., Mazzei, M., Sancandi, M., Miele, M., 2008. Taxanes from shells and leaves of *Corylus avellana*. J. Nat. Prod. 71 (1), 58–60.

Palazón, J., Cusidó, R.M., Bonfill, M., Morales, C., Piñol, M.T., 2003. Inhibition of paclitaxel and baccatin III accumulation by mevinolin and fosmidomycin in suspension cultures of *Taxus baccata*. J. Biotechnol. 101 (2), 157–163.

Rahpeyma, S.A., Moieni, A., Jalali Javaran, M., 2015. Paclitaxel production is enhanced in suspension-cultured hazel (*Corylus avellana* L.) cells by using a combination of sugar, precursor, and elicitor. Eng. Life Sci. 15 (2), 234–242.

Ramirez-Estrada, K., Osuna, L., Moyano, E., Bonfill, M., Tapia, N., Cusido, R.M., Palazon, J., 2015. Changes in gene transcription and taxane production in elicited cell cultures of *Taxus× media* and *Taxus globosa*. Phytochemistry 117, 174–184.

Rateb, M.E., Hallyburton, I., Houssen, W.E., Bull, A.T., Goodfellow, M., Santhanam, R., Jaspars, M., Ebel, R., 2013. Induction of diverse secondary metabolites in *Aspergillus fumigatus* by microbial co-culture. RSC Adv. 3 (34), 14444–14450.

Rodriguez, R.J., White Jr., J.F., Arnold, A.E., Redman, A.R.A., 2009. Fungal endophytes: diversity and functional roles. New Phytol. 182 (2), 314–330.

Sabater-Jara, A.B., Tudela, L.R., López-Pérez, A.J., 2010. In vitro culture of *Taxus* sp.: strategies to increase cell growth and taxoid production. Phytochem. Rev. 9 (2), 343–356.

Sabater-Jara, A.B., Onrubia, M., Moyano, E., Bonfill, M., Palazón, J., Pedreño, M.A., Cusidó, R.M., 2014. Synergistic effect of cyclodextrins and methyl jasmonate on taxane production in *Taxus× media* cell cultures. Plant Biotechnol. J. 12 (8), 1075–1084.

Salehi, M., Moieni, A., Safaie, N., 2017. A novel medium for enhancing callus growth of hazel (*Corylus avellana* L.). Sci. Rep. 7, 15598.

Salehi, M., Karimzadeh, G., Naghavi, M.R., Badi, H.N., Monfared, S.R., 2018a. Expression of artemisinin biosynthesis and trichome formation genes in five Artemisia species. Ind. Crop. Prod. 112, 130–140.

Salehi, M., Karimzadeh, G., Naghavi, M.R., Badi, H.N., Monfared, S.R., 2018b. Expression of key genes affecting artemisinin content in five Artemisia species. Sci. Rep. 8 (1), 1–11.

Salehi, M., Moieni, A., Safaie, N., 2018c. Elicitors derived from hazel (*Corylus avellana* L.) cell suspension culture enhance growth and paclitaxel production of *Epicoccum nigrum*. Sci. Rep. 8, 12053.

Salehi, M., Karimzadeh, G., Naghavi, M.R., 2019a. Synergistic effect of coronatine and sorbitol on artemisinin production in cell suspension culture of Artemisia annua L. cv. Anamed. Plant Cell Tissue Organ Cult. 137 (3), 587–597.

Salehi, M., Moieni, A., Safaie, N., Farhadi, S., 2019b. Elicitors derived from endophytic fungi *Chaetomium globosum* and *Paraconiothyrium brasiliense* enhance paclitaxel production in *Corylus avellana* cell suspension culture. Plant Cell Tissue Organ Cult. 136 (1), 161–171.

Salehi, M., Moieni, A., Safaie, N., Farhadi, S., 2019c. New synergistic coculture of *Corylus avellana* cells and *Epicoccum nigrum* for paclitaxel production. J. Ind. Microbiol. Biotechnol. 46 (5), 613–623.

Salehi, M., Naghavi, M.R., Bahmankar, M., 2019d. A review of *Ferula* species: biochemical characteristics, pharmaceutical and industrial applications, and suggestions for biotechnologists. Ind. Crop. Prod. 139, 111511.

Salehi, M., Farhadi, S., Moieni, A., Safaie, N., Ahmadi, H., 2020a. Mathematical modeling of growth and paclitaxel biosynthesis in *Corylus avellana* cell culture responding to fungal elicitors using multilayer perceptron-genetic algorithm. Front. Plant Sci. 11, 1148.

Salehi, M., Moieni, A., Safaie, N., Farhadi, S., 2020b. Whole fungal elicitors boost paclitaxel biosynthesis induction in *Corylus avellana* cell culture. PLoS One 15 (7), e0236191.

Salehi, M., Farhadi, S., Moieni, A., Safaie, N., 2021. A hybrid model based on general regression neural network and fruit fly optimization algorithm for forecasting and optimizing paclitaxel biosynthesis in *Corylus avellana* cell culture. Plant Methods 17, 13.

Sanchez-Muñoz, R., Bonfill, M., Cusidó, R.M., Palazón, J., Moyano, E., 2018. Advances in the regulation of *in vitro* paclitaxel production: methylation of a Y-patch promoter region alters BAPT gene expression in *Taxus* cell cultures. Plant Cell Physiol. 59 (11), 2255–2267.

Schif, P.B., Fant, J., Horwitz, S.B., 1979. Promotion of microtubule assembly *in vitro* by taxol. Nature 277 (5698), 665–667.

Shinwari, Z.K., Qaiser, M., 2011. Efforts on conservation and sustainable use of medicinal plants of Pakistan. Pak. J. Bot. 43 (1), 5–10.

Soliman, S.S.M., Raizada, M.N., 2013. Interactions between co-habitating fungi elicit synthesis of taxol from an endophytic fungus in host *Taxus* plants. Front. Microbiol. 4, 3.

Soliman, S.S.M., Trobacher, C.P., Tsao, R., Greenwood, J.S., Raizada, M.N., 2013. A fungal endophyte induces transcription of genes encoding a redundant fungicide pathway in its host plant. BMC Plant Biol. 13, 93.

Somjaipeng, S., Medina, A., Magan, N., 2016. Environmental stress and elicitors enhance taxol production by endophytic strains of *Paraconiothyrium variabile* and *Epicoccum nigrum*. Enzyme Microb. Technol. 90, 69–75.

Stierle, A., Strobel, G., Stierle, D., 1993. Taxol and taxane production by *Taxomyces andreanae*, an endophytic fungus of Pacific yew. Science 260 (5105), 214–216.

Syklowska-Baranek, K., Furmanowa, M., 2005. Taxane production in suspension culture of *Taxus× media* var. Hicksii carried out in flasks and bioreactor. Biotechnol. Lett. 27 (17), 1301–1304.

Syklowska-Baranek, K., Szala, K., Pilarek, M., Orzechowski, R., Pietrosiuk, A., 2018. A cellulase-supported two-phase in situ system for enhanced biosynthesis of paclitaxel in *Taxus× media* hairy roots. Acta Physiol. Plant. 40 (11), 201.

Tabata, H., 2004. Paclitaxel production by plant-cell-culture technology. Biomanufacturing 87, 1–23.

Wang, W., Zhong, J.J., 2002. Manipulation of ginsenoside heterogeneity in cell cultures of *Panax notoginseng* by addition of jasmonates. J. Biosci. Bioeng. 93 (1), 48–53.

Wang, H.Q., Yu, J.T., Zhong, J.J., 1999. Significant improvement of taxane production in suspension cultures of *Taxus chinensis* by sucrose feeding strategy. Process Biochem. 35 (5), 479–483.

Wang, Y., Dai, C., Zhao, Y., Peng, Y., 2011. Fungal endophyte-induced volatile oil accumulation in *Atractylodes lancea* plantlets is mediated by nitric oxide, salicylic acid and hydrogen peroxide. Process Biochem. 46 (3), 730–735.

Wani, M.C., Taylor, H.L., Wall, M.E., Coggon, P., Mc Phail, A.T., 1971. Plant antitumor agents. VI. The isolation and structure of taxol. A novel antileukemic and antitumor agent from *Taxus brevifolia*. J. Am. Chem. Soc. 93 (9), 2325–2327.

Wheeler, N.C., Jech, K., Masters, S., Brobst, S.W., Alvarado, A.B., Hoover, A.J., Snader, K.M., 1992. Effects of genetic, epigenetic, and environmental factors on taxol content in *Taxus brevifolia* and related species. J. Nat. Prod. 55 (4), 432–440.

Wickremesinhe, E.R., Arteea, R.N., 1993. *Taxus* callus cultures: initiation, growth optimization, characterization and taxol production. Plant Cell Tissue Organ Cult. 35 (2), 181–193.

Wilson, S.A., Roberts, S.C., 2012. Recent advances towards development and commercialization of plant cell culture processes for the synthesis of biomolecules. Plant Biotechnol. J. 10 (3), 249–268.

Yang, Y., Zhao, H., Barrero, R.A., Zhang, B., Sun, G., Wilson, I.W., Xie, F., Walker, K.D., Parks, J.W., Bruce, R., Guo, G., Chen, L., Zhang, Y., Huang, X., Tang, Q., Liu, H., Bellgard, M.I., Qiu, D., Lai, J., Hoffman, A., 2014. Genome sequencing and analysis of the paclitaxel-producing endophytic fungus *Penicillium aurantiogriseum* NRRL 62431. BMC Genomics 15 (1), 69.

Yuan, Y.J., Li, C., Hu, Z.D., Wu, J.C., Zeng, A.P., 2002. Fungal elicitor-induced cell apoptosis in suspension cultures of *Taxus chinensis* var. mairei for taxol production. Process Biochem. 38 (2), 193–198.

Zeng, F., Li, X., Qie, R., Li, L., Ma, M., Zhan, Y., 2019. Triterpenoid content and expression of triterpenoid biosynthetic genes in birch (*Betula platyphylla* Suk) treated with 5-azacytidine. J. For. Res. 31 (5), 1843–1850.

Zhai, X., Jia, M., Chen, L., Zheng, C.J., Rahman, K., Han, T., Qin, L.P., 2017. The regulatory mechanism of fungal elicitor-induced secondary metabolite biosynthesis in medical plants. Crit. Rev. Microbiol. 43 (2), 238–261.

Zhang, C.H., Wu, J.Y., 2003. Ethylene inhibitors enhance elicitor-induced paclitaxel production in suspension cultures of *Taxus* spp. cells. Enzyme Microb. Technol. 32 (1), 71–77.

Zhang, X., Yazaki, J., Sundaresan, A., Cokus, S., Chan, S.W.L., Chen, H., Henderson, I.R., Shinn, P., Pellegrin, M., Jacobsen, S.E., Ecker, J.R., 2006. Genome-wide high-resolution mapping and functional analysis of DNA methylation in Arabidopsis. Cell 126 (6), 1189–1201.

Zhang, H., Lang, Z., Zhu, J.K., 2018. Dynamics and function of DNA methylation in plants. Nat. Rev. Mol. Cell Biol. 19 (8), 489–506.

Zhao, J., Davis, L.C., Verpoorte, R., 2005. Elicitor signal transduction leading to production of plant secondary metabolites. Biotechnol. Adv. 23 (4), 283–333.

Zhao, K., Ping, W., Li, Q., Hao, S., Zhao, L., Gao, T., Zhou, D., 2009. *Aspergillus niger* var. *taxi*, a new species variant of taxol-producing fungus isolated from *Taxus cuspidata* in China. J. Appl. Microbiol. 107 (4), 1202–1207.

Zhong, J.J., 2002. Plant cell culture for production of paclitaxel and other taxanes. J. Biosci. Bioeng. 94 (6), 591–599.

6

Botany of paclitaxel producing plants

S. Karuppusamy[a] and T. Pullaiah[b]

[a]Department of Botany, The Madura College, Madurai, India, [b]Department of Botany, Sri Krishnadevaraya University, Anantapur, India

6.1 History of taxol

In the early 1960s, the collection of plants was organized by the National Cancer Institute (NCI), United States to identify anticancer drugs sources. The plant collection program was undertaken by botanists working for the United States Department of Agriculture (USDA), namely Arthur Barclay, Robert Perdue, and James Duke. They have collected many plant samples to explore new drugs for cancer treatment in agreement with the NCI in 1962. In this program, 35,000 plant samples were screened for exploring anticancer phytocompounds in the NCI. A part of Barclay's plant samples were shipped to the natural product chemistry laboratory for testing the potency of anticancer properties, and this work was carried out by Monroe Wall, Morris Kupchan, Jack Cole, Norman Farnsworth, and Robert Pettit at Research Triangle Institute of North Carolina. The results of 2 years of research by this team discovered an extract from the barks of the Pacific Yew tree (*Taxus brevifolia*) samples, possessing significant cytotoxicity activity. Barclay's team has collected the Pacific yew sample at a site 7 miles north of Packwood, Washington, in the Gifford Pinchot National Forest on August 21, 1962. The solvent extract was confirmed with antiproliferative activity in an animal test system, and by September of 1964, *T. brevifolia* bark samples were collected in bulk with the aim of purifying and identifying the most active principle in the bark extract. After several years of continuous research efforts in this regard has yielded the compound, named as paclitaxel, which is marketed under the trade name of taxol. The chemical structural elucidation of taxol was carried out by X-ray crystallographic studies (Wani et al., 1971). It provided the new insight for development of anticancer drug. After the successful purification and characterization, studies on taxol production confirmed that *Taxus* species as the best and greatest taxol

Paclitaxel. https://doi.org/10.1016/B978-0-323-90951-8.00008-4
Copyright © 2022 Elsevier Inc. All rights reserved.

yielding species. Further, the effect of genotype and environment to yield was also made in *Taxus* species. All these explorations led to the conclusion that *T. brevifolia* as best source of the anticancer alkaloid, taxol (Isah, 2015).

The generic name *Taxus* is reflected in the name of the poisonous taxanes found in the tree. Some botanists did not consider Yew to be a true conifer, since it does not bear its seeds in a cone. However, proper consideration of its evolutionary relationships now places the yew family (Taxaceae) firmly within the conifers (http://www.plantsofthe-worldonline.org/taxon/urn:lsid:ipni.org:names:306036-2).

Due to difficulties in the harvest of Pacific Yew bark for taxol extraction and complexities involved in its synthesis and purification, developments to use it as a clinical drug was very slow. The average yield of taxol in *T. brevifolia* was 100 mg/kg bark, but NCI clinical trial required about 27,000 kg of the bark (Wall and Wani, 1995). A typical 100-year-old *T. brevifolia* tree yields 3 kg of the bark, which means the trial studies required about 9000 trees (Hartzell, 1991). The continuous collection of Yew tree barks for taxol extraction eradicated the trees, and the population of *T. brevifolia* became threatened with extinction (Chase, 1991). Further, the discovery of 10-deacetylbaccatin III from the needles of *Taxus* species in greater quantity provided another renewable source for the semisynthesis of taxol to meet demand for the drug production, and assured a continuous supply without threat to the species' population (Jean-Noel and Greene, 1988; Guenard et al., 1993). It stimulated the search for alternative sources of taxol producing plants (Choi et al., 1995). The search also extended to microbes, and the first report of taxol production from endophytes isolated from the Pacific Yew tree recorded (Collins et al., 2003). Since then, several endophytes from *Taxus* and other plants are reported to produce taxol.

6.2   Botany of *Taxus*

The origin of the genus name, *Taxus* is uncertain, and it may be derived from the Latin word, *texere* (to weave), due to the arrangement of the distichous (2-ranked) leaves, while others suggest that the name, *Taxus* is derived from the Greek *toxus* or *toxon*, meaning a bow or from the Greek *toxikon (toxicon, toxicum)*, meaning poison. Whatever its origin, the genus name, *Taxus* was firstly proposed by Tournefort in 1717, and was subsequently adopted by Linnaeus in his *Species Planatarum* in 1753 (Chadwick and Keen, 1974). The modern delimitation of *Taxus* species has been difficult, due to lack of adherence in applying universal taxonomic standards, the morphological similarities among species, and the vast number of varieties within species (Patel, 1998; Spjut, 2003). *Taxus* species are widespread and native to moist, temperate forests of the world, particularly the Pacific and Atlantic coasts, the mid-Atlantic and Great Lakes regions of North America, western, northern and southern Europe, Algeria, southeastern Russia, eastern China, Nepal, Myanmar, Laos, Thailand, Vietnam, Iran, and as far south as Sumatra and Celebes (Voliotis, 1986; Hartzell, 1991; Patel, 1998). In the genus *Taxus*, 12 accepted species are listed, and its native range is Temperate Eurasia to Malesia, North America to Guatemala (POWO, 2020). These species are *T. baccata* L., *T. brevifolia* Nutt., *T. calcicola* L.M. Gao & Mich. Möller, *T. canadensis* Marshall, *T. chinensis* (Pilg.) Rehder, *T. contorta* Griff., *T. cuspidata* Siebold & Zucc., *T. floridana* Nutt. ex Chapm., *T. florinii* Spjut, *T. globosa* Schltdl., *T. mairei* (Lemée & H.Lév.) S.Y.Hu and *T. wallichiana* Zucc. More recently the horticulture potential of *Taxus* species was reviewed for their botanical identity and anticancer compound production of Taxol (DeLong and Prange, 2019).

6.2.1 Classification of *Taxus*

Taxus is one of the most difficult conifer genera, i.e., their morphology looks very similar. They were all treated as subspecies of *T. baccata* (Pilger, 1903). Other taxonomists identified 25 species with over 50 varieties, and they have established new species on the basis of even minute morphological differences (Spjut, 2007). Overall, efforts to subdivide the diversity of the genus have generally reached disparate conclusions except in those cases, where taxa are clearly discrete on the basis of geography. Only in the 21st century molecular investigations finally started to reveal with objective of clarity, the heterogeneity of the genus, and it has been concluded with 12 accepted species, worldwide (POWO, 2020).

Certain *Taxus* species have long been generally recognized on the basis of their geographic distribution (https://www.conifers.org/ta/Taxus.php). Among the known species, eight species are indigenous to the northern hemisphere, these include: *T. brevifolia* (Pacific Yew), *T. baccata* (English Yew), *T. canadensis* (Canadian Yew, ground hemlock), *T. cuspidata* (Japanese Yew), *T. chinensis* (Chinese Yew), *T. floridana* (Florida Yew), *T. wallichiana* (Himalayan Yew), and *T. globosa* (Mesoamerican Yew) (Bailey and Bailey, 1976; Spjut, 2003). Although the *Taxus* genus is distinguished by its cone (aril) and leaf morphology (Spjut, 2003), differentiating *Taxus* species during cultivation can be very difficult; however morphological features, such as growth and branching habit, leaf shape, color, and arrangement may be helpful (Cope, 1998; Spjut, 2003, 2010).

The distribution of *Taxus* species include *T. baccata* in Europe and far western Asia, *T. canadensis* in the northeast part of United States and adjacent Canada, *T. floridana* in Florida, United States, *T. brevifolia* in the far western United States and adjacent Canada, *T. globosa* from Mexico to El Salvador, and *T. cuspidata* in Japan and adjacent mainland of Asian countries. The remaining species are distributed in the Himalayas, China and through South East Asia into Malaysia. Though having a more complicated history in the taxonomic literature, few of the species are now generally assigned to *T. contorta* in the western Himalaya, *T. sumatrana* in Malaysia and tropical South East Asia, and a number of taxa that occur in the eastern Himalaya, subtropical South East Asia, and China that have frequently overlapping distributions, and are morphologically similar. *T. chinensis, T. mairei,* and *T. wallichiana* are recently been elucidated as a separate species using a combination of morphological and molecular techniques (https://www.conifers.org/ta/Taxus.php).

In the north and south America, four native species are recognized by their geographic location: (1) the Pacific Yew (*T. brevifolia*), distributed in western Canada and the United States; (2) the Mexican Yew (*T. globosa*), native to Mexico, El Salvador, Honduras, and Guatemala; (3) the Florida Yew (*T. floridana*), native to northern Florida; and (4) the Canadian Yew (*T. canadensis*), which grows from Newfoundland to Manitoba and southward to Iowa and north Carolina. The English Yew (*T. baccata*) flourishes in Europe and North Africa (Patel, 1998). The Japanese Yew (*T. cuspidata*) is endemic to eastern Asia (China, Japan, Korea, and Russia), while the Himalayan Yew (*T. wallichiana*) grows from the eastern Afghanistan to Tibet and China (Patel, 1998). *Taxus* spp. can also be differentiated on the basis of the morphological attributes of the leaves, resulting in three main groupings: the Baccata group including *T. canadensis* and some of the cultivated species, e.g., *T. baccata* and *T. cuspidata*; the Wallichiana group—including *T. brevifolia, T. floridana*, and *T. globosa*; and the Sumatrana group. The Baccata and Wallichiana groups are native to the North America, while the Sumatrana group is largely represented by the south-east Asian species (Spjut, 2003).

Attempts have been made to classify *Taxus* spp. on the basis of the levels of anticancer taxane compounds (e.g., baccatin III, 10-deacetylbaccatin III; paclitaxel, in particular) present in the needles (van Rozendaal et al., 2000; Parc et al., 2002). Although cultivars of a species or hybrid (e.g., *Taxus × media*) have differing leaf taxane content, presently it is not possible to create consistent high-to-low taxane categories, due to the inherent variation encountered, when needle taxane levels are measured (Poupat et al., 2000; van Rozendaal et al., 2000). Interestingly, species like *T. canadensis* show unique taxane profiles that may be potentially exploited for the pharmaceutical industry (Parmar et al., 1999; Zamir et al., 1992; Nikolakakis et al., 2004). Thus, segregating *Taxus* species, hybrids, or cultivars on the basis of commonly shared taxane compounds may be more useful than delineation based on leaf taxane content or leaf morphology (van Rozendaal et al., 1999). Also, genome analysis would aid in understanding the genetic relationships among many species and cultivars of the *Taxus* genus.

6.2.2 Modern classification of *Taxus* species

Hao et al. (2008) considered multiple cpDNA markers and one nuclear (ITS) marker in a suite, representing the most widely-recognized taxa in the Taxaceae, in an analysis that confirmed the monophyly of *Taxus* and placed it sister to *Pseudotaxus* in a clade sister to *Austrotaxus*. This result was consistent across the various markers, and was largely consistent with a family-wide morphological analysis performed by Ghmire and Heo (2014) that differed in finding *Austrotaxus* sister to all the rest of the family. Within the genus, the analysis of Hao et al. (2008) was ambiguous between various molecular markers, but the weight of evidence placed *T. baccata, T. brevifolia,* and *T. globosa* in a common clade (repeating a similar finding by Li et al., 2001), and *T. canadensis* in a clade with *T. cuspidata* (plausible; the two species have a natural hybrid, *T. x hunnewelliana* (Collins et al., 2003). Results for the remaining east Asian/Malaysian taxa were more complicated and unclear, partly due to the inclusion of a number of hybrid taxa in the analysis, but one interesting point was inclusion of *T. contorta* sister to the *baccata-brevifolia-globosa* clade, a biogeographically interesting finding that corroborates findings by Hao et al. (2008) and suggested a common ancestor for *T. baccata* and *T. contorta*, which are now separated by the deserts and mountains of northern Iran and Afghanistan.

Möller et al. (2013) used combined morphological and molecular (cpDNA) evidence to differentiate *T. chinensis, T. mairei,* and *T. wallichiana* in the eastern Himalaya, China, and Indochina; they were also divided into two clades of *T. florinii* from *T. chinensis* and *T. calcicola* from *T. wallichiana*. These latter species are of limited distribution (*T. florinii* in SW Sichuan and NW Yunnan; *T. calcicola* on limestone karst near the Gulf of Tonkin). These two taxa can't be segregated pending further information on the distribution and description. The analysis of Möller et al. (2013) found seven discrete haplotype lineages within this group of taxa, but the haplotypes are generally not exclusive to species, instead varying in frequency across taxa. Conversely, principal components analysis of the morphological data clearly distinguished five distinct groups; four of these groups corresponded to previously-described taxa, while the authors erected *T. calcicola* to accommodate the fifth. They also found a number of "atypical" specimens, primarily from the areas around the Sichuan basin, that they interpret as providing evidence of past hybridization events between

T. chinensis, T. mairei, and *T. wallichiana* (https://www.conifers.org/ta/Taxus.php).

6.3 Enumeration of taxol producing plant species

Originally, taxol was isolated from the bark of *T. brevifolia* (Wani et al., 1971). It was confirmed that the taxol production is confined to a narrow taxonomic group of Gymnosperms, belonging to the family Taxaceae, specifically the genus, *Taxus*. Among the other genera in the family Taxaceae, namely *Amentotaxus, Austrotaxus, Cepahalotaxus, Pseudotaxus* and *Torreya*, only *Austrotaxus* (Gueritte-Voegelein et al., 1987) and *Pseudotaxus* have been reported with some simpler taxanes. Some other related coniferous species, such as *Cephalotaxus* (Cephalotaxaceae) and *Podocarpus glacilor* (Podocarpaceae) are also noted for the taxol production (Stahlhut et al., 1999). The nine species of taxol yielding *Taxus* recorded include *T. wallichiana, T. globosa, T. floridana, T. chinensis, T. canadensis, T. baccata, T. cuspidata, T. fauna* and *T. sumatrana*. The hybrid species *T. media* (cross between *T. baccata* and *T. cuspidata*) and *T. hunnewelliana* (cross between *T. cuspidata* and *T. canadensis*) are also recognized as species to possess taxol. At a point, the extant species were considered subspecies of *T. baccata*, but molecular evidences showed the existence of many distinct and closely related species, and majorly nine species are recognized to have a high content of taxol. After a remarkable discovery of antitumor properties of the bark extract of *T. brevifolia*, screening for taxol in other related taxa was made with resultant isolation of several taxoids. The classification of taxol producing plants can be categorized broadly into Gymnosperms and Angiosperms. Among the Gymnosperms, 10 species of the *Taxus* genus are reported to have taxol and taxol related chemical derivatives in their plant parts, and their details are detailed later.

6.3.1 *Austrotaxus spicata* Campton (New Claedona Yew)

It is an evergreen coniferous species of shrub or small tree, about 5–20 m tall with reddish flaked bark. Leaves are lanceolate in outline, about 8–12 cm in length, and arranged spirally on the stem. Male cones measure 10–15 mm in length with a slender shape. Female cones (measuring 20–25 mm length) are drupe-like with a fleshy aril covered over the single seed (Fig. 6.1A). This species is endemic to New Caledonia, where it grows in the central and northern parts of the island on serpentine soils at elevations of 300–1350 m above sea level.

Chapter 6 Botany of paclitaxel producing plants **161**

Fig. 6.1 (A) *Austrotaxus spicata*; (B) *Pseudotaxus chienii*; (C) *Taxus baccata*;
(D) *Taxus brevifolia*; (E) *Taxus canadensis*; (F) *Taxus chinensis*. (A) httpsen.
wikipedia.orgwikiAustrotaxus#mediaFileAustrotaxus_spicata;
(B) httpsconifersociety.orgconiferspseudotaxus-chienii; (C) httpswww.flickr.
comphotosbeto_frota6186100281; (D) httpsen.wikipedia.orgwikiTaxus;
(E) httpswww.minnesotawildflowers.infoshrubcanada-yew; (F) httpswww.
hweitugroup.comindex.phpmain_page = product_info&products_id = 509027.

6.3.2 *Pseudotaxus chienii* (W.C. Cheng) W.C. Cheng (White Berry Yew)

Synonyms: *T. chienii* W.C. Cheng, *Nothotaxus chienii* (W.C. Cheng) Florin, *Pseudotaxus liiana* Silba, *Pseudotaxus chienii* subsp. *liana* Silba. Description: It is shrub to 4 m tall with gray-brown, peeling barks. Leaves are linear, borne at 40–45 degrees to axis when young but at 50–90 degrees on mature trees, straight or slightly falcate, 1–2.6 cm × 2–4.5 mm, not leathery, seeds are arillated with 5–7 mm diameter, single ovoid, measuring 5–8 × 4–5 mm (Fig. 6.1B). This species is distributed in China: N Guangdong, N Guangxi, NW and S Hunan, SW Jiangxi, S Zhejiang; cultivated as an ornamental in Zhejiang. Naturally occurs in the cool, moist temperate montane forest dominated by evergreen or deciduous angiosperms.

6.3.3 *Taxus baccata* L. (European, English, or Common Yew)

The hybrid *T. baccata* × *T. cuspidata* is *T. × media*. Description: It is an evergreen tree to 40 m tall with thin, scaly bark. Leaves are flat, arranged spirally but appearing 2-ranked, 10–40 × 2–3 mm, dark green. Seed consists of soft, bright red aril to 8–15 mm long (Figs. 6.1C, 6.2, and 6.3). This tree is native to Britain and sometimes referred to as the English yew, and it is also found across much of Europe, western Asia, and North Africa.

6.3.4 *Taxus brevifolia* Nuttall

Syn: *Taxus baccata* L. subsp. *brevifolia* (Nuttall) Pilger; *T. baccata* var. *brevifolia* (Nuttall) Koehne; *T. baccata* var. *canadensis* Bentham; *T. bourcieri* Carrière; *T. lindleyana* A. Murray. Description: It is a dioecious shrub or small tree to 25 m tall scaly, reddish purple bark. Leaves are green, linear, acute, mucronate, 8–35 mm long × 1–3 mm wide, often falcate, whorled but appearing 2-ranked, with a 5–8 mm long decurrent leaf base. Seed is ovoid, 2–4-angled, 5–6.5 mm, maturing late summer-fall, enclosed in a red aril, ca. 10 mm diameter (Fig. 6.1D). It is distributed in SE Alaska, Montana, Idaho, Oregon, Washington and California N of Sonora Pass; Canada: British Columbia, Alberta.

6.3.5 *Taxus canadensis* Marshall (Canada Yew)

Syn.: *Taxus baccata* L. subsp. *canadensis* (Marshall) Pilger; *T. baccata* L. var. *minor* Michaux; *T. minor* (Michaux) Britton ex Small et Vail; *T. baccata* L. var. *procumbens* Loudon. Description: This species is shrub

Chapter 6 Botany of paclitaxel producing plants **163**

Fig. 6.2 *Taxus baccata* in Kashmir valley, India. Image by Zubair Ahmad Rather with permission.

Fig. 6.3 *Taxus baccata* male cone. Image by Zubair Ahmad Rather with permission.

to 2 m, usually monoecious, straggling, spreading to prostrate with thin reddish bark. Leaves are linear, 1–2.5 cm × 1–2.4 cm wide, pale green. Seed is somewhat flattened, 4–5 mm (Fig. 6.1E). This taxon is widely distributed in Canada: extreme SE Manitoba, Ontario, Québec, Prince Edward Island, New Brunswick, Nova Scotia, Newfoundland; France: St. Pierre and Miquelon; United States: Connecticut, Illinois, Indiana, Iowa, Kentucky, Maine, Massachusetts, Michigan, Minnesota, New Hampshire, New York, Ohio, Pennsylvania, Rhode Island, Tennessee, Vermont, Virginia, Wisconsin, and West Virginia.

6.3.6 *Taxus chinensis* (Pilg.) Rehd (Chinese Yew)

Syn.: *Taxus baccata* L. subsp. *cuspidata* (Siebold & Zucc.) Pilg. var. *chinensis* Pilg. ex Engler, *Taxus baccata* L. var. *sinensis* A. Henry, *Taxus cuspidata* Siebold & Zucc. var. *chinensis* (Pilg.) C.K. Schneid., *Taxus wallichiana* Zucc. var. *chinensis* (Pilg.) Florin. Description: It grows shrub or tree to 20 m tall with a rounded or pyramidal crown and thin, red-purplish bark. Leaves are distichous, flat, short-lanceolate to oblong, 15–20 × 2–3.2 mm, twisted at the short-petiolate base with an obtuse to mucronate apex. Seed is ovoid, mostly solitary, sessile, with a green aril maturing red or orange, 10–13 × 7–10 mm (Fig. 6.1F). It is distributed in Vietnam; China: Anhui, Chongqing, Fujian, S Gansu, N Guangxi, Guizhou, W Hubei, NE Hunan, S Shaanxi, Sichuan, E Yunnan, Zhejiang. This species is commonly associated with *Pseudotsuga sinensis, Pinus fenzeliana, Chamaecyparis hodginsii, Tsuga chinensis, Podocarpus pilgeri, Nageia fleuryi* and occasionally *Cupressus vietnamensis.*

6.3.7 *Taxus cuspidata* Siebold et Zucc.

Syn.: *T. baccata* L. var. *cuspidata* (Siebold & Zucc.) Carr., *T. baccata* L. subsp. *cuspidata* (Siebold et Zucc.) Pilg., *T. baccata* L. var. *microcarpa* Trautv., *T. cuspidata* Siebold et Zucc. var. *microcarpa* (Trautv.) Kolesn., *T. cuspidata* Siebold et Zucc. var. *umbraculifera* (Siebold ex Endl.) Makino, *Cephalotaxus umbraculifera* Siebold ex Endl., *T. cuspidata* Siebold et Zucc. var. *latifolia* (Pilg.) Nakai, *T. cuspidata* var. *nana* Hort. ex Rehder. Description: It is an erect shrub or small tree, to 16 m tall with thin scaly bark. Leaves are spirally arranged, appearing 2-ranked, linear, 15–25 × 2–3 mm, abruptly narrowed to a short spinescent apex. Cones are axillary, seeds with aril, red when ripe (Fig. 6.4A). It is distributed in N Korea; China: Heilonjiang, Jilin, Nei Mongol, Liaoning, Hebei, Shaanxi, and Shanxi; Russia: E and S from the Amur River basin; Japan: Hokkaido, Honshu, Shikoku, and Kyushu.

Chapter 6 Botany of paclitaxel producing plants **165**

Fig. 6.4 (A) *Taxus cuspidata*; (B) *Taxus floridana*; (C) *Taxus globosa*; (D) *Taxus meirei*; (E) *Taxus sumatrana*; (F) *Taxus wallichiana*. (A) httpssi.gardenexplorer. orgtaxon-6129.aspx; (B) httpswww.conifers.orgtaTaxus_floridana.php; (C) httpswww.inaturalist.orgtaxa135516-Taxus-globosa; (D) httpwww. plantsoftheworldonline.orgtaxonurnlsidipni.orgnames828365; (E) httpscommons. wikimedia.orgwikiFileTaxus_sumatrana_kz2.jpg; (F) httpstreesandshrubsonline. orgarticlestaxustaxus-wallichianapreview = 8721#8721.

6.3.8 *Taxus floridana* Nuttall ex Chapman (Florida Yew)

Syn.: *T. canadensis* L. var. *floridana* (Nuttall ex Chapman) Pilger. Description: This species is dioecious shrub or small tree to 2–5 m tall with smooth purple-brown bark. Leaves are pliable, dark green, 1–2.6 cm × 1–2 mm, slightly falcate. Seed cones are underside of shoots, seeds ellipsoid with green fleshy aril maturing orange to red (Fig. 6.4B). This species is distributed in Florida, United States. Endemic to a small area of bluffs and ravines on the east bank of the Appalachicola River, typically in moist, shaded ravines in hardwood forests at 15–30 m elevation.

6.3.9 *Taxus globosa* Schlectendahl (Mexican Yew)

Syn.: *Taxus baccata* subsp. *globosa* Pilger. Description: It grows to 15 m tall tree with spreading or ascending branches with purple-brown bark. Leaves are distichous, mostly linear, slightly curved, 20–30 × 2–2.5 mm, mucronate to cuspidate tip. Seed is solitary, ovoid, aril at first green and covering lower half of seed, later orange or red (Fig. 6.4C). This species is narrowly distributed in Mexico: Nuevo León, Tamaulipas, San Luis Potosí, Veracruz, Querétaro, Hidalgo, Tlaxcala, Puebla, Oaxaca; Guatemala; Honduras; El Salvador at 2000–3000 m elevation. It is associated with other conifers including *Abies guatemalensis* and *Cupressus lusitanica*; in Mexico, cloud forest covers a broader elevation band and broadleaf trees enter the mix, including species of *Quercus, Magnolia, Liquidambar, Platanus, Ulmus, Carpinus, Fagus,* and *Acer* as well as the conifers *Pinus chiapensis* and *Podocarpus matudae.*

6.3.10 *Taxus meirei* (Lemee & Lev.) S.Y. Hu ex T.S. Liu (Meires Yew)

Syn.: *Tsuga mairei* Lemee & Lev., *Taxus chinensis* (Pilg.) Rehd. var. *mairei* (Lemee & Lev.) W.C. Cheng & L.K. Fu, *Taxus wallichiana* Zucc. var. *mairei* (Lemee & Lev.) L.K. Fu & Nan Li, *Taxus speciosa* Florin, *Taxus mairei* (Lemee & Lev.) S.Y. Hu **ex** T.S. Liu var. *speciosa* (Florin) Spjut, *Taxus kingstonii* Spjut. Description: It is shrub, or trees to 30 m tall with a single trunk and a pyramidal crown with thin red, purple, or brown flaking bark. Leaves are 2-ranked, linear, 15–35 × 2–4 mm, usually falcate, thick, coriaceous, with revolute margins, with a cuspidate to mucronate apex. Seed is solitary or in pairs with aril at first green and covering lower half of seed, swelling to orange or red, 10–13 × 7–10 mm (Fig. 6.4D). It is distributed in Taiwan and China at elevations of 1000–3000 m.

6.3.11 *Taxus sumatrana* (Miquel) de Laubenfels (Tampinur batu)

Syn.: *Cephalotaxus sumatrana* Miq., *Podocarpus celebicus* Hemsl., *Cephalotaxus celebica* Warb., *Taxus celebica* (O. Warburg) H.L. Li. Description: It is an evergreen shrub or tree to 14 m tall with thin scaly bark. Leaves are linear-lanceolate, falcate, spirally arranged, spreading in two ranks, 1.2–2.7 cm in length and 2–2.5 mm broader, abruptly pointed at the apex. Seeds are drupe-like, the fleshy arillate coat reddish at maturity (Fig. 6.4E). This species is distributed in Sumatra, Philippines, Celebes at 1400–2300 m in moist subtropical forests, tropical highland ridges, and moss forests in the subcanopy, locally dominant.

6.3.12 *Taxus wallichiana* Zucc. (Himalayan Yew)

It is a large evergreen shrub or small tree that grows to mature heights of 30 m tall with pyramidal shape with dense, erect, spreading branches and reddish to purplish peeling bark. Leaves are linear to lanceolate shaped and lustrous dark green in color, measuring 15–35 mm in length and 2–4 mm broader, and gradually taper to an acuminate or cuspidate tip. Seeds are solitary, ovoid, broad with a red or orange aril (Fig. 6.4F). This species is native to southwestern China, southern Anhui, Fujian, southern Gansu, northern Guangdong, northern Guangxi, Guizhou, western Henan, western Hubei, Hunan, Jiangxi, southern Shaanxi, Sichuan, Taiwan, southeastern Xizang, Yunnan, and Zhejiang provinces; as well as in Bhutan, northern India, Laos, Myanmar, and Vietnam at elevations of 100–3500 m above sea level.

6.4 Taxol from angiosperms

A continuous search for taxol producing plant sources was extended to angiosperms where the first report available from the hazelnut tree (*Corylus avellana*). The taxanes were reported from defatted leaves, green and brown shell extract of the tree and the yield in different tissues of various concentrations comparable with *Taxus* sp. (Hoffman et al., 1998). The species could offer an alternative supply of taxol when compared to *Taxus* sp. due its fast growth (Gallego et al., 2017).

6.4.1 *Corylus avellana* L. (common hazel or European hazel)

It is typically a shrub reaching 4–8 m tall, occasionally more than 10 m, and the stem is usually branched. The bark is gray with white and large spots. The leaves are deciduous, rounded, 6–12 cm long, with a double serrate margin and hairy on both sides. The flowers appear in early spring, before the leaves, and are monoecious with single-sex wind-pollinated catkins. Male catkins are usually grouped together (—two to four flowers) and are yellowish-brown and up to 10 cm long, while female catkins are very small. During flowering, the inflorescence becomes slender and doubles in length. The fruit is a nut, grouped in clusters of one to four together. Each nut is held in a short leafy involucre (husk) which encloses about half of the nut. The nut is roughly spherical, up to 2 cm long, yellow-brown with a pale scar at the base. It is usually an understory shrub, very common in naturally regenerated mixed-hardwood stands. It can be found throughout Europe, from Norway to the Iberian Peninsula and east as far as the Urals. This species is very appreciated for its nuts, for which it is cultivated worldwide, especially in European countries such as Turkey, Italy, and Spain, and further afield in the United States and Canada (Enesca et al., 2016).

6.5 Conclusions and future direction

Taxol is initially obtained from *T. brevifolia*, then continuous scientific screening yielded a number of *Taxus* species possesses to have production of taxol in various concentration. Most of the taxol producing plants are distributed in the narrow geographical range with specific strata of climate which is alarming to continuous of extraction of value added or commercially important medicinal compounds from the natural population. The identification of alternate natural sources and precise evaluation of potentially target compound yielding plant species is prime importance in medicinal plant research. Overall, the present chapter has attempted to enumerate the botanical details of taxol producing plant species and their distribution ranges. It would help with the precise identification of taxol producing plants for further research and improvement in taxol production. It also helps to assess the natural diversity of taxol producing plants and their conservation. However, the lower yield of taxol from these available plant sources is a major limitation. Hence, more scientific explorations should be encouraged in the direction of bioprospecting novel plant species with the potentiality of greater taxol yield.

References

Bailey, L.H., Bailey, E.Z., 1976. Hortus Third. A Concise Dictionary of Plants Cultivated in the United States and Canada. vol. 2, L–Z, Barnes and Noble Books, New York.
Chadwick, L.C., Keen, R.A., 1974. A study of the genus *Taxus*. Ohio Agric. Res. Dev. Cent. Res. Bul. 1086.

Chase, M., 1991. Clashing priorities: a new cancer drug may extend lives – at cost of rare trees – that angers conservationists, who say taxol extraction endangers the prized yew – pressure on Bristol-Myers too. The Wall Street Journal, New York. 9 April 1991.

Choi, M.S., Kwak, S.S., Liu, J.R., Park, Y.G., Lee, M.K., An, N.H., 1995. Taxol and related compounds in Korean native Yews (*T. cuspidata*). Planta Med. 61, 264–266.

Collins, D., Mill, R.R., Moller, M., 2003. Species separation of *Taxus baccata, T. canadensis* and *T. cuspidata* (Taxaceae) and origins of their reputed hybrids inferred from RAPD and cpDNA data. Am. J. Bot. 90, 175–182.

Cope, E.A., 1998. Taxaceae: the genera and cultivated species. Bot. Rev. 64, 291–319.

DeLong, J.M., Prange, R.K., 2019. *Taxus* spp.: botany, horticulture, and source of anticancer compound. Hort. Rev. 32, 299–327.

Enesca, C.M., Durrant, T.H., de Rigo, D., Caudullo, G., 2016. *Corylus avellana* in Europe: distribution, habitat, usage and threats. In: European Atlas of Tree Species. https://www.researchgate.net/publication/299468333_Corylus_avellana_in_Europe_distribution_habitat_usage_and_threats/citations.

Gallego, A., Malik, S., Yousefzadi, M., Makhzoum, A., Tremaillaux-Gviller, J., Bonfill, M., 2017. Taxol from *Coryllus avellana*: paving the way for a new source of this anticancer drug. Plant Cell Tissue Organ Cult. 129, 1–16.

Ghmire, B., Heo, K., 2014. Cladistic analysis of Taxaceae s.l. Plant Syst. Evol. 300, 217–223. https://doi.org/10.1007/s00606-013-0874-y.

Guenard, D., Gueritte-Voegelein, F., Poiter, P., 1993. Taxol and Taxotere: discovery, chemistry and structure activity relationship. Acc. Chem. Res. 26, 160–167.

Gueritte-Voegelein, F., Guénard, D., Potier, P., 1987. Taxol and derivatives: a biogenetic hypothesis. J. Nat. Prod. 50, 9–18.

Hao, D.C., Xiao, P.G., Huang, B.-L., Ge, G.B., Yang, L., 2008. Interspecific relationships and origins of Taxaceae and Cephalotaxaceae revealed by partitioned Bayesian analyses of chloroplast and nuclear DNA sequences. Plant Syst. Evol. 276, 89–104. https://doi.org/10.1007/s00606-008-0069-0.

Hartzell, H.J., 1991. The Yew Tree: A Thousand Whispers. Hulgosi Books.

Hoffman, A., Khan, W., Worapong, J., Griffin, D., Arbogast, B., Zheng, P., 1998. Bioprospecting: using high performance liquid chromatography-thermospray mass spectrometry to identify useful metabolites in plant extracts. Spectroscopy 13, 22–32.

Isah, T., 2015. Natural sources of taxol. Br. J. Pharm. Res. 6, 214–227.

Jean-Noel, D., Greene, A.E., 1988. A highly efficient, practical approach to natural taxol. J. Am. Chem. Soc. 110, 5917–5919.

Li, J., Davis, C.C., Del Tredici, P., Donoghue, M.J., 2001. Phylogeny and biogeography of *Taxus* (Taxaceae) inferred from sequences of the internal transcribed spacer region of nuclear ribosomal DNA. Harv. Pap. Bot. 6, 267–274.

Möller, M., Gao, L.-M., Mill, R.R., Liu, J., Zhang, D.-Q., Poudel, R.C., Li, D.-Z., 2013. A multidisciplinary approach reveals hidden taxonomic diversity in the morphologically challenging *Taxus wallichiana* complex. Taxon 62 (6), 1161–1177.

Nikolakakis, A., Haidara, K., Sauriol, F., Mamer, O., Zamir, L.O., 2004. Semi-synthesis of an *O*-glycosylated docetaxel analogue. Bioorg. Med. Chem. 11, 1551–1556.

Parc, G., Canaguier, A., Landré, P., Hocquemiller, R., Chriqui, D., Meyer, M., 2002. Production of taxoids with biological activity by plants and callus culture from selected *Taxus* genotypes. Phytochemistry 59, 725–730.

Parmar, V.S., Jha, A., Bisht, K.S., Taneja, P., Singh, S.K., Kumar, A., Jain, P.R., Olsen, C.E., 1999. Constituents of the yew trees. Phytochemistry 50, 1267–1304.

Patel, R.N., 1998. Tour de paclitaxel: biocatalysis for semisynthesis. Annu. Rev. Microbiol. 52, 361–395.

Pilger, R., 1903. Taxaceae-Taxoideae—Taxeae. *Taxus*. In: Engler, Das Pflanzenreich IV, pp. 110–116.

Poupat, C., Hook, I., Gueritte, F., Ahond, A., Guenard, D., Adeline, M.T., Wang, X.P., Dempsey, D., Breuillet, B., Potier, P., 2000. Neutral and basic taxoid contents in the needles of *Taxus* species. Planta Med. 66, 580–584.

POWO, 2020. http://www.plantsoftheworldonline.org/?q=Taxus. Searched on 28/12/2020.

Spjut, R.W., 2003. Introduction of Taxus: Methodology, Taxonomic Relationships, Leaf and Seed Characters, Phytogeographic Relationships, Cultivation, and Chemistry. www.worldbotanical.com/Introduction.htm.

Spjut, R.W., 2007. Taxonomy and nomenclature of *Taxus*. J. Bot. Res. Inst. Texas 23, 203–289.

Spjut, R.W., 2010. Overview of the Genus *Taxus* (Taxaceae): The Species, Their Classification, and Female Reproductive Morphology. http://www.worldbotanical. com/TAXNA.HTM. (Accessed 27 January 2018).

Stahlhut, R., Park, G., Petersen, R., Ma, W., Hylands, P., 1999. The occurrence of the anticancer diterpene Taxol in *Podocarpus gracilior* Pilger (Podocarpaceae). Biochem. System. Ecol. 27 (6), 613–622.

van Rozendaal, E.L., Kurstjens, S.J.L., van Beek, T., van den Berg, R.G., 1999. Chemotaxonomy of *Taxus*. Phytochemistry 52, 427–433.

van Rozendaal, E.L., Lelyveld, G.P., van Beek, T.A., 2000. Screening of the needles of different yew species and cultivars for paclitaxel and related taxoids. Phytochemistry 53, 383–389.

Voliotis, D., 1986. Historical and environmental significance of the yew (*Taxus baccata* L.). Israel J. Bot. 35, 47–52.

Wall, M.E., Wani, M.C., 1995. Cain Memorial Award Lecture: Camptothecin and taxol: discovery to clinic—thirteenth Bruce. Cancer Res. 55, 753–760.

Wani, M.C., Taylor, H.L., Wall, M.E., Coggon, P., McPhail, A.T., 1971. Plant antitumor agents VI. The isolation and structure of Taxol, a novel antileukemic and antitumor agent from *Taxus brevifolia*. J. Am. Chem. Soc. 93, 2325–2327.

Zamir, L.O., Nedea, M.E., Garneau, F.X., 1992. Biosynthetic building blocks of *Taxus canadensis* taxanes. Tetrahedron Lett. 333, 5235–5236.

Propagation of paclitaxel biosynthesizing plants

T. Pullaiah[a], S. Karuppusamy[b], and Mallappa Kumara Swamy[c]

[a]Department of Botany, Sri Krishnadevaraya University, Anantapur, India, [b]Department of Botany, The Madura College, Madurai, India, [c]Department of Biotechnology, East West First Grade College, Bengaluru, Karnataka, India

7.1 Introduction

Paclitaxel, an anticancer drug is extracted from the stem bark and needles of *Taxus* species. About 10,000 kg of bark of *Taxus* species has to be processed to obtain 1 kg of paclitaxel. The present requirement of paclitaxel for medicinal use is 250 kg per year, and meet this yield, nearly 750,000 trees of *Taxus* species have to be sacrificed. Despite high cost, the demand for this natural drug is increasing at an annual rate of 20% every year. The global demand of paclitaxel is also steadily increasing, i.e., 800–1000 kg, annually. Paclitaxel is usually accumulated in the bark of *Taxus* species, but the tree dies, when stripped of its bark for the drug extraction. *Taxus* species are facing the threat of extinction, because of growing demand for both wood and paclitaxel production and difficulty in propagation. Most of the fruits are eaten by birds, and are difficult to collect. Propagation of this economically important plant species is one of the key determinants in its utilization.

The seed germination efficiency in *T. brevifolia* ranges from 0% to 16% with an average of 5% (Anderson and Owens, 2001). Abortive ovules and post fertilization anomalies are the common facts for lower rate of reproduction in *Taxus* species. The possible reasons for reproductive defects include poor fruit set, pollination failure, insect damage, limitation of light along with other physico-environmental factors.

Taxus species are becoming endangered in its natural habitat, due to the merciless felling of the trees or unauthorized stripping of the bark for paclitaxel extraction, leading to slow death of the trees. The natural population of *T. baccata* and other species of *Taxus* are lost greatly in their population, due to their overexploitation for paclitaxel extraction. The overexploitation coupled with poor regeneration process, slow growth rate and prolonged seed dormancy are leading the great loss of natural diversity of *Taxus* species. Therefore, vegetative

Paclitaxel. https://doi.org/10.1016/B978-0-323-90951-8.00011-4
Copyright © 2022 Elsevier Inc. All rights reserved.

multiplication by stem cuttings may be the only practical solution for its large-scale propagation. Since *Taxus* trees are slow growing, it takes many years for reaching the seed-bearing stage. New stocks of plants are required to replenish the natural resources. Micropropagation is an alternative method for obtaining large number of microshoots and plantlets within a short span of time. Even though the juvenile shoots produced in vitro are not desirable for planting, they provide an opportunity in the production of taxane class of compounds. Micropropagation by using mature explants is desirable as the advantages far outweigh the disadvantages faced, when using mature explants in vitro.

In this chapter, methods to overcome dormancy by embryo culture, micropropagation using embryo explants and stem explants are described. The application of different media, plant growth regulators and their combinations and concentration required for shoot induction, shoot elongation, root induction are given. In addition, methods of acclimatization of in vitro regenerated plants are highlighted.

7.2 Propagation

T. baccata can be propagated through stem cuttings, artificial seed germination, grafting and layering (Hartzell, 1991; Steinfield, 1992; Mitter and Sharma, 1999). Most of the *Taxus* species can also be propagated by seed sowing approach. Propagation from seed will take 3 years while those from cuttings will take two (Steinfield, 1992).Túri-Farkas and Kovács (2016) gave an account on the propagation of *Taxus* by seed. According to them, the seeds need to be layered for a year before sowing. These species can be propagated by striking in unheated plant growing facilities from August to September, and in heated greenhouses in the months of January to March. The most frequently used plant growth media are as follows:

(a) Sand: The best sand is dredged from the bottom of large rivers, i.e., washed river sand.
(b) Washed and graded pearl pebble: Planting in pearl pebbles is recommended only if we can ensure high humidity by regular overhead sprinkling either automatically or manually.
(c) Sand, or the 1:1 mixture of sandy peat and perlite or peat and fine gravel are recommended.
(d) Expanded perlite: Perlite is a rock of volcanic origin, which is expanded at a temperature of over 1000°C after extraction and grinding. The high temperature perfectly sterilizes the medium at the same time. The large grain (at least 2–5 mm in diameter), i.e., horticultural perlite is advised for striking. Being extremely light is a definite advantage (it is easy to replace) and at the same time it is perfectly sterile, airy and has a good water-retaining capacity.

(e) Fibrous peat moss: Light brown fibrous peat moss is used for striking. Peat is a slowly renewing resource when measured on a human scale. It has an outstandingly good water-retaining capacity once soaked, while it remains loose and airy even when saturated with water. It is strongly acidic (pH 3.5–4.5). Acidity hinders the spread of fungal diseases and the humic acids in the peat stimulate cuttings. In Western-Europe, tree nurseries almost entirely use peat as rooting medium; pure in exceptional cases but the more frequently mixed with coarse sand. The mixing ratio differs from nursery to nursery as well as from plant to plant, but ranges between peat mass and coarse sand with a ratio of 1:3 or 3:1. The lime content of sand is generally high enough to set the required pH value, while its weight and density improves the physical properties of the peat. We can safely strike cuttings in this mixture of peat and sand with all the possible humidifying methods.

7.2.1 Vegetative propagation of *Taxus* spp.

Vegetatively propagated elite *Taxus* clones can serve as a potential and economic tissue source for enhancing paclitaxel production (Ho et al., 1998). The rooting of stem cuttings is a challenging task from old trees of Gymnosperms. If old tree cuttings are selected, it takes more than 1 year to root. Also, poor initial growth and survival performance of rooted cuttings is observed in the traditional vegetative propagation methods. *T. baccata* is currently propagated by grafting in Western European ornamental tree nurseries, which is however a costly and slow propagation method. Young seedling stocks of *Taxus* species can be increased by stem cutting experiments (Ho et al., 1998; Mitter and Sharma, 1999), but it needs years together required to mass propagate these clones.

Stem cutting experiments in a study has shown that about 50% of various ages of cuttings developed roots, but the total clonal variance was high up to 71.2%, however, it was found to depend on the age of cutting stock (Thomas and Polwart, 2003). There were no clonal differences found among the rooting ability between female and male clones (Schneck, 1996), but Nandi et al. (1996) reported that auxin treatment of stem cuttings induced higher rooting performance in male trees (55%) when compared to female trees (15%). The plastic tree shelters are not recommended for *T. baccata*, because the cuttings fail to grow due to overheating.

The collection of erect shoots for cuttings is very difficult, since the trees are very large, increased height of crowns, and also coppice sprouts are very rare. All stecklings were obtained from the lateral branches, and they may grow plagiotropically. The stecklings derived from orthotropic growth could not grow in pots or in field for more than 6 years. However, bud explants from the lateral branches of 1.5-years-old trees can be grown in vitro and successfully developed healthy shoots for about 6 months (Chang et al., 1998). Hence, in vitro propagation might be a very useful technique to mass clonal propagation of elite superior yew trees and the production of high-quality plantlets for paclitaxel extraction.

Normally, vegetative propagation is achieved through stem cuttings. In case of *Taxus* spp., the percentage of rooting varied between cultivars and species depending on the conditions of temperature, misting and use of plant growth regulator, indole butyric acid (IBA). About 63%–100% rooting was observed in many *Taxus* species and cultivars (Eccher, 1988). Chee (1995a) reported that stem cuttings of *T. cuspidata* treated with an aqueous solution of IBA (0.2%) and naphthalene acetic acid (NAA) (0.1%) and thiamine (0.08%) induced 73.5% rooting frequency. But, the success rate was not very high for *T. wallichiana* (20%–30%) and the cuttings took at least 3 months to root. The success of grafting and layering was also not very encouraging (Chatterjee and Dey, 1997). A variety of reasons have been attributed for this low rate of success in rooting (Mitchell, 1997), and they are mentioned below:

- Great age of parent trees (more than 100 years) and the associated low growth vigor.
- Variation in rooting among parent trees from different geographical locations due to variations in a number of ecological and environmental factors.
- Variation in rooting percentage between different species and cultivars as well as between male and female trees.
- The yews tend to grow in stressful environments (too cold, too exposed) which might reduce the shoot vigor, finally adversely affecting rooting success.

The roots of mature Yew trees are associated with ectomycorrhizal mats (ECM) and vesicular arbuscular mycorrhiza (VAM) in their natural habitats. ECM and VAM might play important roles in rooting and establishment of cuttings derived from mature trees by suppressing root pathogens and increasing the available nutrients (Griffiths et al., 1995).

The stem cutting technique was used to preserve *T. wallichiana* at Da Lat Institute of Biology, Vietnam, during 1994–96. The rooting performance rate after 90 days was found to be only 38% (Fig. 7.1A; Nhut et al., 2007). The branches of *T. wallichiana* recorded vigorous root formation after 3 months with an average root length of 6–8 cm (Fig. 7.1B; Nhut et al., 2007). Kaul (2008) investigated the optimal conditions required for propagating *T. wallichiana* vegetatively during spring season. This study suggested that IBA (0.5 mM) treatment as the most suitable plant growth regulator enhance adventitious rooting response in 1 year old long and dwarf shoots of both male and female trees. Researchers have emphasized to apply auxins, especially

Fig. 7.1 Cutting technique applied for *Taxus wallichiana*. (A) Rooting of branches. (B) Rooted cutting after 3 months. Reproduced with permission from Nhut, D.T., Hien, N.T.T., Don, N.T., Khiem, D.V., 2007. In vitro shoot development of Taxus wallichiana Zucc., a valuable medicinal plant. In: Jain, S.M., Häggman, H. (Eds.), Protocols for Micropropagation of Woody Trees and Fruits. Springer, Dordrecht, pp. 107–116, with permission.

IBA exogenously to promote adventitious root formation from stem cuttings of *Taxus* species (Aslam et al., 2007). Saumitro and Jha (2014) also witnessed that IBA as the best plant growth regulator, as its treatment to cuttings exhibited overall superior rooting response in *T. wallichiana*. In another study, application of IBA at 500 ppm level resulted the higher number of roots, rooting percentage, and length of roots in the juvenile shoot cuttings of *T. wallichiana* (Aslam et al., 2017).

Anjum et al. (2011) have studied the effect of different auxins on direct rooting response of *T. baccata* shoot cuttings. In their experiment, stem cuttings were treated with indole acetic acid (IAA), IBA and NAA in various concentrations. Based on the results, IBA was found to be the most suitable hormone for direct rooting, followed by NAA and IAA respectively. The study revealed that IBA at 500 ppm was the best concentration for rooting of the stem cuttings of *T. baccata*, and also for inducing higher rooting percentage, number of roots per shoots and root length.

Khali and Sharma (2003) and Singh (2006) reported that among all the phytohormones tried, IBA is the most effective hormone for the induction of higher rooting percentage in the juvenile cuttings of *Taxus baccata*. Application of IBA at 10,000 ppm concentration is most suitable for the induction for a better rooting percentage, higher number, and greater length of primary roots in the *Taxus baccata* cuttings.

Many factors influence the stem cutting propagation of trees, for example physiological age of cuttings, the propagation environment and the concentration and types of auxins. IBA is reported to have a better rooting response, because of its long retention time at the site of application, and also is effective in promoting rooting response of a many numbers of tree species. Das and Jha (2018) investigated the effect of genotype, physiological age of stem cuttings and IBA on rooting of *T. baccata* shoot cuttings. Among the six candidate trees (CTs) of *T. baccata* that are collected from various places of Eastern Himalaya and performed the rooting response of stem cuttings, CT 2 (collected from BSI, Shillong) had given the highest rooting response (46.28%). The study noted the juvenile cuttings with higher rooting efficiency; however, the basal callusing was more prominent in mature stem cutting. The effect of IBA treatment was also most significant for rooting response, where 1000 ppm was effective for stimulating rooting juvenile stem cuttings and 2000 ppm induced the rooting in mature cuttings (Fig. 7.2; Das and Jha, 2018). *T. baccacta* stem cuttings treated with various concentration of IBA showed the best results at 500 IBA, and also registered higher callusing percentage, rooting performance, number of roots and root length in the juvenile shoot cuttings (Aslam and Rather, 2008).

Nicholson and Munn (2003) described methods for the propagation of *T. globosa*. The basal ends of cuttings from current season growth (7–15 cm) were treated with 50% ethyl alcohol solution and 10,000 ppm of IBA. The cuttings were put into a medium of coarse sand perlite (1:1) under a clear plastic propagation tent with bottom heat (24°C) supplied by heating cables. After 5 months, cuttings were removed from the medium, and recorded for rooting frequency. The plants produced an overall rooting percentage of 89.5% with an average of 11.37 roots per cutting.

Nazir et al. (2018) tested the consequence of different plant growth regulators, such as IBA, IAA and NAA on adventitious rooting of *T. wallichiana* yew shoot cuttings in 4 different seasons (spring, summer, autumn, and winter) under nursery conditions. Shoot cuttings were treated with diverse concentrations of IBA, IAA, and NAA. However, to stimulate root length, root numbers and rooting percentage, IBA at 1000 ppm applied during the spring season (March-May) showed the best response. In a similar study on *T. baccata*, Sobha and Chauhan (2018) reported that IBA at 500 ppm as the most optimum plant growth regulator for inducing higher frequency of rooting, fresh weight, and dry weight of rooted stem cuttings.

Susilowati et al. (2019) investigated shoot cuttings technique to propagate paclitaxel producing *T. sumatrana*. They experimented 43-years old trees for rooting performance. The use of soil: rice husk (1:1 v/v) media with auxin treatment produced the significant rooting ability (30%). In general, auxin and media composition affected induction of primary rooting of shoot cuttings. Das and Jha (2014) found that hardwood cuttings of *Taxus wallichiana* took a considerably longer time than softwood to root.

Three-months-old rooted cuttings of *T. mairei* clones T1, Nj11, K3, and Wx were grown hydroponically by Ho et al. (2007) in 1/10 and 1/20 strengths of Johnson's nutrient solution (JS) for 6, 9, and 11 months at different times (September, December, and February). They were

Fig. 7.2 Propagation of *Taxus baccata* L. by branch cutting (A) A candidate tree of *Taxus baccata*; (B) Callused formation in mature cutting from CT 3; (C) Vigorous rooting of juvenile cuttings obtained from CT 2; (D) Vigorous rooting of mature cuttings obtained from CT 2. (E) Rooting of IBA treated juvenile cuttings of *Taxus baccata*; (F) Rooting of IBA treated mature cuttings of *Taxus baccata*. Reproduced with permission from Das, S., Jha, L.K., 2018. Effect of physiological age of stem and IBA treatment on rooting of branch cuttings of *Taxus baccata* L. Curr. Bot. 9, 1–7. http://creativecommons.org/licenses/by/4.0/.

sampled and analyzed in March 2000 for yielding of paclitaxel and its derivatives. Both dry weights and the yields of 10-deacetyl baccatin III (10-DAB), paclitaxel, and total taxanes per plant at the three harvesting times did not significantly differ between the two strengths of JS; only the 11-month-old plants produced greater paclitaxel in 1/20 JS. Great variations of these two parameters in three organs (leaves, stems and roots) of plants among the four clones were found. Wx grew poorly and produced only few taxanes. Nj11 grew fast with the greatest amount of taxanes. T1 had a growth rate similar to K3, but with greater taxanes. Nj11 was found to be a 10-DAB-rich clone, and after 11-months of its growing, it produced 5.83 mg per plant. T1 was the taxol-rich clone, and it produced 1.16 mg per plant. 10-DAB was usually rich in needles, while taxol was rich in the roots. However, T1's needles contained greater taxol than its roots. The concentrations of both 10-DAB and taxol were highest in December compared to the other two harvesting months. Although 11-month-old plants had the greatest biomass, the taxanes had only increased a little as compared to 9-month-old ones. Since most potted plants died and grew poorly during the culture period, Ho et al. (2007) suggests that hydroponics is a good tool to rapidly evaluate the clonal performance of Taiwan yew.

The rooting efficiency of *Taxus × media* var. *densiformis* stem cuttings was evaluated with the treatment of two different AMF, namely *Gigaspora gigantea* and *Glomus intraradices* in a greenhouse conditions up to 9–15 months duration (Gemma et al., 1998). The effect of AMF treatments was found to be significant, and leaves of treated plants showed significantly more chlorophyll (about 1.3–4.1 times higher) than untreated plants. Also, results showed that mycorrhizal fungi treated plants had significantly larger (1.3–1.4 times) and longer (1.7–2.1 times) roots than non-mycorrhizal fungi treated plants. Further, inducing of higher rate of branched roots (1.3–2.9 times) was observed in mycorrhizal fungi treated plants. But, there were no differences in diameter, height of the shoot and dry weight at the end of the experiments. They also reported that number of buds was significantly higher in the cuttings inoculated with *G. intraradices* after 15 months (Gemma et al., 1998).

7.3 Micropropagation

Even though success with vegetative propagation has not been significant in *Taxus* spp., not much work has been done with respect to micropropagation studies in *Taxus* spp. Precocious germination from seeds of *Taxus* spp., de novo adventitious bud development and somatic embryogenesis are some of the studies carried out in *Taxus* spp. so far, and are presented in Tables 7.1 and 7.2. In addition, much of the information has been in the form of patents, thus restricting its use and availability in the public domain. Jaziri et al. (1996) have extensively reviewed tissue and organ culture up to the year 1996.

Table 7.1 In vitro emrbryo culture and seed germination studies in *Taxus* species.

Name of the species	Explant/s	Media	Plant growth regulators	Observations	References
T. baccata	Mature embryos	Heller	–	Leaching of mature embryos, role of ABA implied in seed dormancy	Le Page-Degivry (1970, 1973a,b) and Lepage-Degivry (1973)
T. × media	Immature embryos	White/MS	–	70% germination, 30% seedlings after 2 weeks	Flores and Sgrignoli (1991)
T. brevifolia	Embryos	White/MS/B5	–	60%–70% germination, 30% seedlings after 2 weeks	Flores and Sgrignoli (1991) and Flores et al. (1993)
Taxus sp.	Immature and mature embryos	B5	GA_3	63% seedlings after 2 weeks	Hu et al. (1992)
T. baccata	Mature embryos	Modified MS/ Heller	–	Leaching of whole seeds for 1 week. 100% seedlings after 2 weeks	Zhiri et al. (1994)
T. brevifolia T. stricta T. baccata T. cuspidata	Mature embryos	B5	–	36% seedlings 63% 16% 14%	Chee (1994)
T. mairei T. baccata T. canadensis T. media T. cuspidata	Mature and immature embryos	½ MS + PVP	GA_3	15% germination from immature embryos and 90% germination from mature embryos	Chang and Yang (1996)
T. chinensis	Mature embryos	B5, new medium with AC, 20% sugar	–	97% germination	Chien et al. (1998)
T. chinensis T. brevifolia	Mature embryos			98% after 30 days	Li (2001)
T. wallichiana	Mature and immature embryos	½ WPM	1% AC	43% embryonic growth, 32% seedling development	Datta and Jha (2004)
T. media	Mature embryos	½ MS	BAP, 2,4-D	62% germination and seedling development	Liao et al. (2006)
T. baccata	Embryos	MS	5 g/L AC	Healthy seedlings	Zarek (2007)
T. baccata	Mature embryos	½ MS + 2 g/L AC	–	Best embryo development	Davarpanah et al. (2014)

Table 7.2 Micropropagation studies in different species of *Taxus*.

Taxus species	Explant/s	Medium	Plant growth regulators	Observations	References
Taxus sp.	Shoot tips	WPM	BAP	2–4 shoots per explant	Amos and McCown (1981)
T. floridana	Shoot tips	WPM	BAP	2–4 shoots per explant	Barnes (1983)
T. floridana	Needles	MS	2,4-D	Glutamine induces embryogenic structures on callus	Salandy et al. (1993)
T. × media	Shoot tips	Hogland	Kn	–	Cerdeira et al. (1994)
T. baccata *T. brevifolia* *T. × media*	Embryos	BL	2,4-D/BAP	Two stage culture	Wann and Goldner (1994)
T. brevifolia	Embryos	½ B5	BAP	Multiple shoots, 58% of rooting, plantlets	Chee (1995b)
T. baccata *T. brevifolia* *T. cuspidata*	Male flowers, immature embryos	Modified MS (DCR) medium	Kn, BAP, 2,4-D for callus induction, Kn, IBA for embryogenesis	Cotyledons formed but no root formation	Ewald et al. (1995)
T. cuspidata	Immature embryos	Modified B5, MS, WPM	NAA, 2,4-D, Kn	Embryogenesis after transfer to hormone free medium	Lee and Son (1995)
T. brevifolia	Immature embryos	WPM	2,4-D, BAP, NAA for callus growth, 2,4-D, NAA, BAP for embryogenesis	5% emblings produced on hormone free ½ WPM	Chee (1995a,b, 1996)
Taxus sp.	Juvenile needles from in vitro shoot cultures	Modified WPM	NAA, 2-iP	Nodules	Ellis et al. (1996)
T. mairei	Shoot tips and stem from 6 to 18 months old seedlings	MS	BAP	Both orthotropic and plagiotropic shoots obtained. 60% rooting	Chang et al. (1998)
T. chinensis	Mature embryos			Callus formation and somatic embryogenesis	Qiu et al. (1998)

Species	Explant	Medium	Plant growth regulators	Remarks	Reference
T. baccata	Seeds and in vitro seedlings	WPM	BAP	Shoots developed in 70 days	Majada et al. (2000)
T. mairei	Bud explants	MS	BAP	Shoots from axillary buds	Chang et al. (2001)
T. baccata	Seeds; Apical buds	MS	BAP + IBA	Dissected embryos germinated; Shoot proliferation and elongation	Taiebeh and Shahrzad (2003)
T. wallichiana	Embryos	½ WPMSH	BAP + 2,4-D or NAA for callus induction and BAP for shoot induction	Two stage (callus induction and shoot induction) culture	Datta et al. (2006)
T. baccata	Closed buds, shoot tips	WPM	Zeatin	1.2–1.8 buds per month, shoots from axillary buds	Ewald (2007)
T. wallichiana	Buds from young branches	MS	BAP 1 mg/L	Two buds per explant	Nhut et al. (2007)
T. baccata	Bud explants, embryos	MS	BAP 2 mg/L + 0.3 mg/L NAA	Callus formation and 8 shoots per embryo	Abbasin et al. (2010)
T. wallichiana	Stem bit explants	WPM	TDZ, BAP	Shoots from axillary buds	Kulkarni (2010)
T. wallichiana	Stem, leaves, shoot tip	MS	2,4-D for callus induction BAP for shoot elongation	Only callus formation from stem and leaves Shoots from shoot tip	Hussain et al. (2013)
T. wallichiana	Leaves (needles), internode	MS	2,4-D + Kn for callus induction and BAP + IAA for shoot induction	Two stage (callus induction and shoot induction) culture	Gul et al. (2020)

The first in vitro cultures of *Taxus* gametophyte and pollen were investigated by Larue (1953) and Tuleke (1959). Tuleke was interested in the organogenetic potential of *Taxus* pollen, but that time paclitaxel was not known. Since 1970, many numbers of papers have been published on in vitro germination of *Taxus* embryo, and all these investigations focused on breaking of seed dormancy involving the use of abscisic acid (Le Page-Degivry, 1970, 1973a,b; Le Page-Degivry and Garello, 1973). The first report on in vitro *Taxus* callus proliferation was achieved in 1973 (Rohr, 1973a,b), followed by David's group, and successfully cultured calli of *Taxus* using different explant sources (David and Plastira, 1974, 1976; David, 1977). They evaluated various types of explants, media with mineral and hormonal compositions for enhancing callus proliferation. They also observed that *Taxus* callus accumulated tannins and phenolics, which are the major reasons for the slow growth of callus. Later, micropropagation protocol of *Taxus* species was schemed out by Amos and McCown (1981) and Barnes (1983, 1985) using shoot apex and intermodal explants.

The embryo culture of *T. baccata* was first described by Le Page-Degivry (1970). In this study, abscisic acid was successfully removed through a leaching process, before culturing of embryos. Twenty years later, Flores and Sgrignoli (1991) investigated the use of immature embryos of *T. brevifolia* and *T. × media* for in vitro germination studies. They observed the development of the complete seedlings from approximately 30% of the precociously germinated embryos. Likewise, the similar germination efficiency was reported in other *Taxus* sp. (Chee, 1994). At the same time, simple and efficient in vitro protocol for breaking the seed dormancy of *T. baccata* was developed by Zhiri et al. (1994). The experiment resulted in the highest rate of germination of mature embryos (100%), when the seeds were washed with tap water for 7 days and then cultured on modified MS or Heller media. Organogenetic response of *T. brevifolia* culture was reported by Chee (1995b). Zygotic embryos were cultured on half strength B5 medium supplemented with 10 μM of benzyladenine (BA) for 14 days for induction of adventitious bud primordia. The in vitro adventitious shoots formed elongated roots and plantlets with vigorous growth within 3 months of culture with approximately 58% of response. According to a research group, microcutting cultures of *T. × media* can be obtained from the stem cuttings with direct in vitro rooting (Cerdeira et al., 1994). They evaluated different salt media for the promotion of bud breaking. Shoots were induced in Hoagland medium supplemented with 1 mg/L Kn and 20 g/L sucrose. Reduced salt concentrations with low pH induced root formation of microcuttings.

7.3.1 Embryo and seed culture to overcome seed dormancy

The seed viability of *T. baccata* can be high (up to 100%), but germination rates may be usually normal with about 50%–70% (USDA, 1974). The artificial seed germination of *T. baccata* enhanced germination efficiency up to 4 years by providing damp sand or peat at low temperatures (USDA, 1974). The seeds can also be stored for 5–6 years by drying at room temperature with 10% moisture content, and then storing at 1–2°C. By employing this protocol, about 95% germination rate can be achieved over a period of 3-years (Heit, 1968, 1969). Even at lower temperature of − 3°C, nearly 50% seed germination can be achieved (Suszka, 1985).

Taxus seeds undergo a dormancy period for at least two long freezing seasons, before germination. Chilling treatment at low temperatures can help to overcome this dormancy (Hartzell, 1991). The *Taxus* embryo is still immature, when the seeds are fully developed, and after the ripening period only the complete development of the embryo occurs, naturally (Wareing and Phillips, 1981). The embryos from dormant seeds require either leaching and treatment with gibberellic acid (GA_3) or chilling treatment for germination, and it has been reported that the seed germination after the treatment with GA_3 can record about 28% of germination (Le Page-Degivry and Garello, 1973). Later, a protocol was developed for germinating Yew seeds without any pretreatment procedures (Majada et al., 2000). According to them, in vitro culture maintained embryos in darkness for 2–4 weeks and their subsequent transfer to fresh medium under 16 h photoperiod was effective for a successful germination. The seed dormancy of *T. baccata* can be broken by culturing excised embryos from mature seeds. Through the in vitro seed germination, seedlings can be obtained within 8 weeks, which otherwise requires up to 2 years for stratified seeds.

Excised Yew seed embryos could be induced to germinate rapidly by in vitro culturing for 15 weeks in Heller liquid medium. However, the germination was found to be inhibited, and is attributed to the presence of a dormancy compound, i.e., similar to abscisic acid (Le Page-Degivry, 1969). Consequently, the leaching and temperature treatment followed by the isolation of the embryos are linked to the breakage of dormancy, and may also be associated with the suppression of an endogenous inhibitor compounds in *Taxus* seeds.

The germination under ex vitro conditions has been reported to require alternating temperatures and cold stratification for 8.5 months to achieve 50% germination (Chang and Yang, 1996). Embryo dormancy is quite natural in *Taxus* species, and it requires cold stratification for a year to release it. This dormancy can be overcome by removal of the seed coat and leaching for about a week in running tap water (Abbasin et al., 2010). The percentage of germination capacity after stratification treatment was 47 to 68% for *T. baccata*, *T. cuspidata* and *T. mairei* (Chien et al., 1994; Rudolf, 1974). In *T. mairei* seed dormancy can be broken by a treatment involving warm stratification followed by cold stratification (Chien et al., 1998). Embryo culture technique has also been used for micropropagation of *Taxus chinensis* by Li et al. (2000).

Zarek (2007) developed in vitro protocol for overcoming the dormancy of *T. baccata* from isolated embryos. But in vitro seed germination fails to germinate directly, when mature seeds are used. However, the seeds soaked in distilled water at 4°C for at least about 48 h, followed by inoculation of isolated embryos on the MS medium supplemented with 5 g/L activated charcoal can encourage a higher germination efficiency.

Kulkarni (2000) could successfully induce germination in vitro with 95%–100% success rate in excised embryos of *T. baccata* ssp. *wallichiana* on DCR basal medium lacking plant growth regulators (Fig. 7.1). Unfortunately, the seedlings could not be acclimatized under ex vitro conditions.

The in vitro embryo culture for 30 days can shorten the dormancy period with early germination and germination percentage in *T. chinensis* and *T. brevifolia* (above 98%) (Li, 2001). The survival rate of the seedlings can be increased (from 49% to 70%), when the seeds were stored at low temperature, and treated for 4 months. The seedling's growth can be improved, if the main inorganic element content of the medium is reduced. In vitro seed and embryo germination studies in *Taxus* species have been listed in Table 7.1.

7.3.2 Somatic embryogenesis

Previously, Wann et al. (1999) have given a detailed review on somatic embryogenesis in *Taxus* species. More recent studies on micropropagation and somatic embryogenesis are listed in Table 7.2. In most of the cases, the embryogenesis was achieved from surface sterilized seeds with dissected embryos, and the choice of explants was preferably freshly harvested seeds (Ewald et al., 1995). Usually, embryo explants were cultured on the media either incubated in dark or under a16 h photoperiod at 22–26°C.

An embryogenic callus was developed from *T. floridana* on the media fortified with 2,4-dichloro phenoxyacetic acid (2,4-D) (1 mg/L) and glutamine (1 g/L) (Salandy et al., 1993). The enhancement of somatic embryogenesis in two *Taxus* sp. has also been patented (Wann and Goldner, 1994; Lee and Son, 1995). Somatic embryos were induced from immature embryo explants on media added with a combination of 2,4-D and BA, but the germination efficiency of these in vitro derived somatic embryos was not studied.

Ewald et al. (1995) obtained the embryo like structures in tissue culture of *T. baccata*, *T. brevifolia* and *T. cuspidata*. They cultured male flowers on modified MS medium supplemented with various concentrations and combinations of Kn and IBA. The best callus proliferation was obtained with the addition of 0.108 mg/L Kn and 1.016 mg/L IBA to Murashige and Skoog (1962) media. Fresh immature embryo explants were cultured on modified MS medium fortified with various concentrations and combinations of Kn, NAA, BAP, and 2,4-D. However, the immature embryo culture failed to give a typical embryo suspensor masses, even after 60 days of culture period. The surviving calli of all species were transferred after 90 days on fresh media for supporting further subculture. A year-old callus culture formed dense of globular embryo-like structure, but only from a few hormonal combinations. Some of these globular structures developed further and formed cotyledon-like organs. The continuous culture of these embryos on further subculture failed to form roots (Ewald et al., 1995).

The zygotic embryos were cultured on Woody Plant Medium supplemented with various concentrations and combinations of 2,4-D + 5 µM BAP + 5 µM NAA by Chee (1996). The study showed the formation of a yellowish-white friable callus after 4–5 weeks of culture. The calli were further transferred to secondary culture medium. After 8 weeks, callus culture from secondary medium was transferred to the medium containing 160 µM 2,4-D + 5 µM BAP and 5 µM NAA to induce embryogenesis. Almost 20% of the subcultured calli proliferated and formed somatic embryos. Zygotic embryo explants developed the somatic embryos. In rare cases, only a single somatic embryo was produced from the explants, while in other cases 5–10 embryogenic structures formed from different areas of the explants. When embryogenic calli were transferred to mononuclear cell medium (MCM) supplied with 40 µM ABA and 1% activated charcoal (AC), somatic embryos ceased proliferation and began to mature. The putative somatic embryos were isolated from nonembryogenic calli and incubated on maturation medium. After 1 month of culturing on maturation medium (MCM added with 40 µM ABA and 1% AC), they developed 90% somatic embryos with a defined and elongated bipolar structure. However, only 5% of these somatic embryos developed into plantlets with shoots and roots upon transfer to a conversion medium. All shoots grown 2 cm length within a month of culture on the medium. The remaining 95% somatic embryos either produced only shoots or roots, or sometimes ceased their growth after repeated transfer.

Qiu et al. (1998) reported plant regeneration via somatic embryogenesis. Calli were successfully induced by culturing the mature embryos of *T. chinensis*, *T. chinensis* var. *mairei* (Fujian) and *T. chinensis* var. *mairei* (Sichuan). The highest calli-inducing frequency was found to be 95.7%, 98.4%, and 91.7%, respectively. The cellular and morphological observation revealed that somatic embryos at different growth stages could be produced from the calli, which were cultured for some time under appropriate culture condition. For the above 3 taxa

186 Chapter 7 Propagation of paclitaxel biosynthesizing plants

explants, the largest number of somatic embryos per 100 calli were recorded to be 127, 91, and 33, respectively.

Nicholson and Munn (2003) described a method of propagating *T. globosa*. The fleshy red aril surrounding the seed was removed and the seeds were divided into equal lots of 50 seeds. Three pregermination treatments were tried prior to sowing: (a) Lot no. 1: 3 months of cold moist chilling (stratification) at 2°C prior to sowing; (b) Lot no. 2: insertion into a moist medium for 5 months at 23°C followed by 3 months at 2°C prior to sowing; (c) Lot no. 3: seeds received a 1 h soak in a 1000 ppm GA_3 followed by 3 months of moist chilling at 2°C prior to sowing. Germination occurred only with seeds receiving pretreatment with a germination rate of 34%.

Taiebeh and Shahrzad (2003) described a method for the micropropagation of *T. baccata* using mature seeds as explants. The mature seeds were washed with running tap water for 15 days, and surface sterilized with sodium hypochlorite 3% (w/v). The embryos were aseptically dissected and transferred on modified MS medium. The radicle emergence up to 77% after 4 weeks was noticed, and about 65% of germinating embryos developed into seedlings. In vitro seedlings after acclimation were transferred on to soil pots under greenhouse condition with a good survival rate.

The mature seeds were surface sterilized with concentrated HCl for 10 min and rinsed 3 times with sterile distilled water (Datta et al., 2006). In another set of mature seed explants were surface sterilized in 0.1% $HgCl_2$ (w/v) for 30 min and rinsed five times in sterile distilled water. After surface sterilization, the seeds and the embryos were excised aseptically, and then rinsed in sterile distilled water to shake off the endosperm cells. These excised embryos were cultured on modified MS media. The results showed the proliferation of callus from the excided embryos. From this investigation, it was established that the surface sterilants can affect germination and callus proliferation of embryo explants of *Taxus* species.

Among the three media tested (White, B_5 and MS) for germination of embryos of *T. brevifolia* and *T. media*, the highest germination efficiency was recorded on White's and MS media (Flores and Sgrignoli, 1991). The excised embryos from immature green seeds with underdeveloped arils showed the highest germination rates. Several investigations reported that the lengthy seed dormancy of *Taxus* species can be overcome by culturing excised embryos from relatively mature seeds (Chee, 1994). The best culture condition was established to be B_5 basal salt medium in continuous darkness for 4 weeks of culture, and subsequently transfer of the developing embryos on fresh medium under a 16 h photoperiod. Seedlings can be obtained from excised embryos within 8 to 10 weeks as compared to 1 to 2 years for stratified seeds.

The half-strength medium is more effective than full strength medium for improving bud and root formation in *Taxus* species (Chee, 1995b; Chang and Yang, 1996; Abbasin et al., 2010; Davarpanah, 2003; Davarpanah et al., 2014). The sugar concentration has an important role in root formation, and woody plants especially gymnosperms prefer low salt concentration for root formation (Davarpanah et al., 2014). Therefore, high osmotic potential should be considered as an inhibitor of root formation and penetration in the medium can be avoided by using the low salt medium, for example half strength MS with decreased sugar concentration. Chee (1995b) reported that adventitious bud primordia were induced by culturing zygotic embryos on half-strength B_5 medium fortified with 10 µM BAP for 14 days. Further, vegetative buds were produced following subculture to half-strength Woody Plants Medium (WPM) (Lloyd and McCown, 1981) containing 1% AC. Culturing of immature zygotic embryos on WPM supplemented with 160 µM 2,4-D and 5 µM BAP and 5 5 µM NAA for 4 weeks. Following the transfer of cultures to WPM supplemented with 4 µM BA, 5 µM Kn, and 1 µM NAA for 6 to 8 weeks obtained embryogenic callus and putative embryoids (Chee, 1996).

Lee and Son (1995) used mature embryo as explants for somatic embryogenesis in *T. cuspidata*. They cultured the embryos on MS and modified B_5 medium supplemented with 11 µM NAA, 2 µM 2,4-D, and 5 µM Kn for callus induction and 14 µM zeatin and 14 µM 2,4-D for somatic embryogenesis.

Chang and Yang (1996) reported that half strength MS medium with Polyvinyl pyrrolidone supported the best embryonic growth and seedling development in *T. mairei, T. baccata, T. canadensis, T. cuspidata* and *T. media*. The survival rate of in vitro plantlets grown in the greenhouse was about 90%. Another study investigated the effect of different media, such as MS, ½ MS, WPM, and ½ WPM on embryonic growth, and found that ½ MS as the better medium as it produced fewer abnormal forms. According to Liao et al. (2006), the removal of endosperm from mature seed explants is necessary for the germination of embryos. Half strength MS medium supplemented with 1 mg/L BAP and 0.1 mg/L 2,4-D was found to be the best hormonal combination for germinating embryos. About 62% of the embryos could germinate and develop into seedlings on this optimized media. Davarpanah et al. (2014) found that addition of 2 g/L AC was the best for embryo germination, and seedlings developed well and showed less abnormalities like lack of shoot growth or root, twisting or thickening the roots. Likewise, Datta and Jha (2004) reported the addition of AC to ½ WPM supports better embryonic growth (43%) and seedling development (32%). Further, the authors noticed that the addition of TDZ, BAP and GA_3 had no effect on the germination of embryos.

Datta et al. (2006) described two-step method for obtaining microshoots in *T. wallichiana*. Zygotic embryos were inoculated on ½ WPM medium supplied with Schenk and Hildebrandt vitamin (1/2WPMSH) + 0.5 mg/L BAP in combination with 1–2 mg/L 2,4-D or

NAA. The results of the culture produced two morphologically varied types of calli, i.e., compact green callus (CG) and compact yellow (CY) callus after 4 weeks of culture. When this CG calli were cultured on ½ WPMSH fortified with 2.5 mg/L BA optimum frequency (63%) of adventitious shoot bud induction was achieved (3 shoot buds per gram of callus). The addition of 1% AC to ½ WPMSH (shoot elongation medium) led to maximum shoot elongation (2.15 cm).

In a study by Abbasin et al. (2010), culturing of isolated embryos during the first 4 weeks resulted in callus growth, and later developed into shoots. Embryos cultured on half-strength MS medium supplemented with 2 mg/L BAP and 0.3 mg/L NAA resulted in increased callus proliferation and the number of regenerated shoots increased up to 8 per embryo.

7.3.3 Explant sources for micropropagation

The juvenile tissue explants have been used for micropropagation of *T. media* and *T. mairei* (Cerdeira et al., 1994; Chang et al., 1998), while in vitro propagation of mature *Taxus* tree explants has been demonstrated by Chang et al. (2001). Stem cuttings of selected adult trees harvested in the month of September were directly rooted by using rooting paste containing 2 g/L IBA under high pressure fog in a greenhouse (Ewald, 2007). These in vitro rooted plants were used as donor plants for establishing tissue culture explants, such as closed buds or shoot tips after pretreatment with a fungicide for 24 h. Bud explants selected from 1000-year-old field grown yew trees and 1-year-old stecklings raised from rooted cuttings have also been used for micropropagation of *T. mairei* (Chang et al., 2001). Steckling-derived explants collected in different seasons were evaluated for survival rates and good growth performance in vitro. The investigation proved that the steckling-derived explants cultures performed better than mature tree-derived explants in terms of shoot multiplication and rooting ability. Whereas, the survival of mature tree-derived explants was highly season dependent with significant survival only in the early spring. Both in vitro derived shoots exhibited plagiotropic growth, but steckling-derived shoots had a smaller plagiotropic angle (near vertical) than mature tree-derived shoots which shows parallel angles. The plagiotropic angles were decreased after these shoots subcultured on fresh medium under greenhouse conditions for half a year in both cases. Orthotropic growth was restored in 25% of the steckling-derived shoots. However, only 10% of mature tree-derived shoots recorded orthotropic growth (Chang et al., 2001).

Taiebeh and Shahrzad (2003) described a method for micropropagation using apical buds from the selected mature yew trees. After various sterilizing agents' treatments, the explants were cultured on different medium, including MS, B5 and MCM medium. Among the media tested, MS medium containing 0.5 mg/L BAP + 0.1 mg L IBA was the best medium for shoot multiplication and elongation. Basal ends of the microshoots were dipped in 50 mg/L IBA solution for 24 h, and then they were transplanted on auxin-free medium with 1% AC. This protocol enhanced the rooting of plantlets. After acclimatization, all the rooted plantlets were transferred to soil under greenhouse condition.

A study by Nhut et al. (2007) used the dormant bud explants from juvenile branches, green vigorous growing sprouts and without infection for micropropagation. Young sprouting branches were cut into 5 cm long segments (Fig. 7.3) and washed several times with distilled

Fig. 7.3 Plantlets for explant collection. (A) Five-year-old *Taxus wallichiana* yew cultivated in Da Lat Institute of Biology. (B) A young branch bearing dormant buds. (C) Buds (D) Young shoots. Reproduced with permission from Nhut, D.T., Hien, N.T.T., Don, N.T., Khiem, D.V., 2007. In vitro shoot development of Taxus wallichiana Zucc., a valuable medicinal plant. In: Jain, S.M., Häggman, H. (Eds.), Protocols for Micropropagation of Woody Trees and Fruits. Springer, Dordrecht, pp. 107–116, with permission.

water. Surface sterilization was done with 70% ethanol for 30 s, and 1% mercuric chloride was added with 2 to 3 drops of 0.01% Tween 80 detergent solution for 10–12 min. Then the explants segments were rinsed 4 times with sterile distilled water. The sterilized segments were cut into 2–2.5 cm long explants and inoculated them vertically on to the culture media.

Abbasin et al. (2010) used juvenile branches from mature trees as explants. The young sprouting twigs were rinsed for 2–4 h in running tap water, and then surface sterilized with 70% ethyl alcohol for 1 min followed by 5% sodium hypochlorite for 15 min, and then rinsed four times in sterile water. The bud explants were separated from mature shoot tips of about 1–1.5 cm in length, and the needle leaves were removed from their bases after sterilization. Gul et al. (2020) have advised to sterilize the leaf and internode explants with 0.2% $MgCl_2$, and then wash three times with distilled water. While, Ewald (2007) used 0.2% mercuric chloride followed by Tween-80 for sterilization of the stem explants.

Kulkarni (2000) used stem cuttings with axillary buds of *T. wallichiana* as explant source. The author had given a liquid soap treatment of 1% (v/v) Teepol with constant agitation for 5 min. After thorough washing off soap solution, one-minute long treatment with 95% (v/v) ethyl alcohol followed by 10 min duration treatment with 0.1% (w/v) MgCl2 under vacuum was given. Again, $HgCl_2$ was removed off with 6 times washes with a sterile distilled water inside the laminar airflow.

Chang et al. (2001) have used explants from stecklings of rooted cuttings. After 2 months of culture, new shoots of 2 months old in vitro raised steckling source were separated into shoot tips (about 0.5 cm long) without needles were cultured in a shoot induction medium. The results showed a better growth response of shoot tips with increased proliferation rate and multiple shoots were induced.

The major problem associated with micropropagation of *Taxus* species is browning of explants, due to accumulation of phenols (Jiaru et al., 1999; Chang et al., 2001). This problem generally could be overcome by the addition of activated charcoal to the medium (Chang and Yang, 1996; Nhut et al., 2007). By the successive subculturing of *Taxus* explants in vitro, it can be circumvented by supplementing the basal medium with 1 g/L AC and 100 mg/L silver nitrate (Chang et al., 2001). Nhut et al. (2007) have observed the addition of 2 g/L AC to reduce the release of phenolic compounds into the media, and to promote shoot elongation. However, the addition of PVP is the most effective antioxidant component for embryo explants of yew trees, as it can encourage more than 80% seedlings development.

A number of basal media, including MS medium (Murashige and Skoog, 1962), Woody plants medium (WPM) (Lloyd and McCown, 1981) and B5 medium of Gamborg et al. (1968) either at full strength or at half strength (with major salts reduced to half strengths) were used for an axillary bud break in *Taxus* species. Among them MS medium and WPM was found to be a better response for regeneration in many investigations. For callus induction auxins like 2,4-D and NAA have been used while for bud break and shoot regeneration cytokinins like BAP, Kn and 2-iP have been used. In some experiments phenylurea TDZ has been used for shoot induction.

WPM medium supplemented with 20 g/L sucrose was used to propagate and elongate shoots in *T. baccata* (Ewald, 2007). Addition of zeatin at 1.5 g/L concentration to WPM was used for shoot bud induction. During *Taxus* in vitro propagation, the major problem is the occurrence of endophytic bacteria, which can be overcome by the addition of 500 mg/L ticarcillin in the medium. Lateral stem elongation was stimulated by a subculture of in vitro derived show with free of phytohormones in the medium, subsequently subculture with low concentration of thidiazuran (TDZ) (0.01 mg/L).

Shoot induction was achieved in *T. wallichiana* MS medium supplemented with 20 g/L sucrose and 1 mg/L IBA (Nhut et al., 2007) using different explants. Shoot-tip cultures were produced only one elongated shoot per explant, while stem with nodes recorded the elongation of axillary buds. The higher concentration of sucrose inhibited the development of apical primordial, and induced callus formation. Whereas, explants cultured in MS medium fortified with BAP 1 mg/L after 6 weeks induced 2 buds. Subsequently, 12 weeks of culture produced 80.7% of adventitious buds. Nearly 17.4% of explants produced 2 buds, and 1.8%, 0.9%, and 0.9% of explants formed 3, 4, and 6 buds, respectively.

According to Kulkarni (2000), ½ WPM supplemented with lower concentrations of BAP (0.05 mg/L) gave higher sprouting percentage (83.33%) as compared to 1/2 WPM supplemented with lower concentrations of Kn (0.05 mg/L) (75.77%). It was established that BAP as a better cytokinin to achieve high sprouting. However, shoot quality and elongation capacity induced by BAP was not good, as it failed to elongate shoots. The shoot quality and elongation ability were better achieved in ½ WPM supplemented with Kn. Kulkarni (2000) also reported the maximum initiation of primary sprouting when higher concentrations of TDZ (80% at 0.5 mg/L) and BAP (58.3% at 5 mg/L) was added to ½ WPM. However, further growth and development of primary shoots was hampered and the shoots failed to elongate even up to a length of 1 cm in subsequent subcultures. On the contrary, primary sprouts developed on lower concentrations of BAP (0.1 and 2.0 mg/L) and TDZ (0.01 and 0.05 mg/L) could elongate to a length of 2 cm, but released phenolic exudates in limited quantities into the nutrient medium. This led to browning of shoots and yellowing of leaves, and the shoots were not able to survive beyond 2 months.

7.3.4 Callus proliferation and organogenesis

The callus proliferation was optimally achieved on MS medium supplemented with 2,4-D (2.0 mg/L) and Kn (0.5 mg/L) from zygotic embryo explants, and with 2,4-D (2.0 mg/L), activated charcoal (1.5 mg/L) from stem explants of *T. baccata* (Datta et al., 2006; Hussain et al., 2013). Researchers have suggested that callus production can be achieved better by using 2, 4-D and Kn supplied in B_5 medium (Jha et al., 1998). Recently, Gul et al. (2020) used two stage cultures for the regeneration of plantlets from *Taxus* species. They noticed callus from leaf and intermodal explants after 15 days of culture on MS medium supplemented with 2,4-D and Kn. Among the different combinations of auxins and cytokinins, 1.7 mg/L 2,4-D and 1 mg/L Kn was found to be the most significant for callus induction from leaf explants. Regeneration of shoots from callus was better on MS medium supplemented with 2 mg/L BAP and 2.5 mg/L IAA. WPM added with 2,4-D and BAP also showed *T. wallichiana* regeneration (Fig. 7.4; Datta et al., 2006). However, calli induced from stem or leaf explants in *T. wallichiana* have failed to show shoot organogenesis (Wickremesinhe and Arteea, 1993; Datta et al., 2006). Similarly, Hussain et al. (2013) reported that callus culture from stem and needle explant is not suitable for organogenesis in *T. wallichiana*. Calli developed from zygotic explant materials as described by Datta et al. (2006) is very efficient in the formation of shoot bud primordial on ½ WPM basal medium.

Abbasin et al. (2010) reported that different ecotypes of explants sources responded to the culture differently. Whereas, Ecotype J, Eco-A, Eco-C and Eco-I showed a varied response with 8 shoots per explant with 4 cm length, while other ecotypes showed only 2 shoots per explant.

Among different cytokinins tested, BAP was found to be more significant for shoot induction compared to Kn, 2-iP and thidiazuron (TDZ). BAP was also found to be effective in inducing shoots from stem cultures of *T. mairei* seedlings (Chang et al., 1998). Chee (1995b) reported that shoot regeneration from embryo cultures of *T. brevifolia*, which was observed better in BAP than TDZ or Kn supplied media. Conifer explants are sensitive to cytokinins, as pointed out by Amos and McCown (1981), where they recommended less than 1 mg/L BA in the medium for inducing shoots.

In another study, the highest number of vigorous shoots were produced from isolated embryos in MS medium supplemented with BAP at 2 mg/L in *Taxus* species. Also, increasing the concentration of BAP from 2 to 4 mg/L failed to affect the shoot multiplication and growth. Overall, other cytokinins, i.e., Kn and Zeatin exhibited relatively less shoot morphogenesis, and produced only two shoots per explants (Abbasin et al., 2010). Likewise, Chang et al. (2001) have reported that

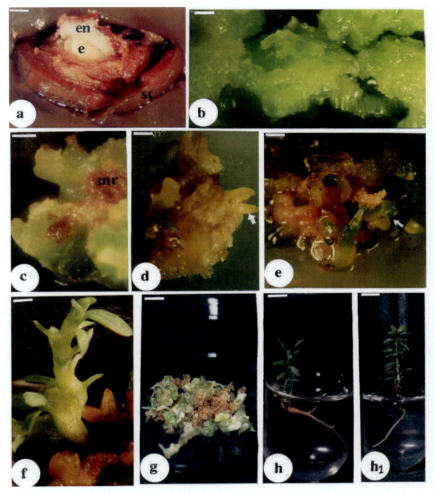

Fig. 7.4 (A–H₁) Organogenesis (shoot formation) in callus cultures derived from zygotic embryos of *Taxus wallichiana*. (A) Longitudinally halved mature seed (Stage II, 6 mm long) showing embryo (e) surrounded by endosperm (en) and seed coat (sc). *Bar:* 2 mm. (B) Compact, *green* (CG) callus induced after 4 weeks on 1/2 WPMSH basal medium supplemented with 2,4-D (2.0 mg/L) and BA (0.5 mg/L). Bar: 5 mm. (C) Meristemoids (mr) induced on CG callus cultured on 1/2 WPMSH basal medium supplemented with BA (2.5 mg/L). Bar: 5 mm. (D) Shoot bud primordial *(arrow)* induced on CG callus cultured on 1/2 WPMSH basal medium supplemented with BA (2.5 mg/L). Bar: 3 mm. (E) Clusters of shoot buds *(arrow)* produced on CG callus cultured on 1/2 WPMSH basal medium supplemented with BA (2.5 mg/L). Bar: 3 mm. (F) Elongated microshoot in vitro. Bar: 5 mm. (G) Long term (2 year old) CG callus cultures. Bar: 8 mm. (H, H₁) Initiation of roots from the base of the microshoots cultured on MS-N basal medium Bar: 10 mm. Reproduced with permission from Datta, M.M., Majumder, A., Jha, S., 2006. Organogenesis and plant regeneration in *Taxuswallichiana* (Zucc). Plant Cell Rep. 25, 11–18.

BAP induces higher shoot induction rate as compared to Kn. They have stated that 2.5 mg/L BAP as the optimum concentration for shoot production. The elongation of shoots is better on media containing lower concentrations of plant growth regulators (PGRs), hence BAP-induced multiple shoots need to be transferred into PGR-free medium to promote elongation (Fig. 7.5; Chang et al., 2001).

7.3.5 Shoot topophysis

The growth potential of apical or lateral shoots as influenced by the position on the morphogenesis of explants is known as topophysis. It greatly influences on the micropropagation of conifers (Fig. 7.5; Chang et al., 2001). The lateral shoot-derived coniferous explants to grow at an angle are known as plagiotropic growth, in contrast with the vertical or orthotropic growth of erect shoot-derived explants. Chang et al. (2001) have noticed these phenomena in stem cuttings and in vitro cultures of *T. mairei*. When shoot tips and stem explants derived from erect shoots of field-grown coppices or stecklings were cultured in vitro, erect growth (orthotropic) was maintained by newly produced shoots. If plagiotropic explants cultured on media, most of the new shoots grew at an angle, and fewer than 25% of them grew vertically, even after sub culturing for 6 months (Fig. 7.5; Chang et al., 2001). The explants derived from plagiotropic shoots were capable of producing smaller plagiotropic angles than explants derived from orthotropic tree branches. In *T. mairei* explants develop needles in a spiral arrangement from orthotropic explants, whereas needles in the branches showed the opposite pattern in plagiotropic shoots. For instance, needles leaf arrangements of erect sprouts from plagiotropic explants became spiral in prolonged subculture. A similar response was noticed from the plagiotropic shoot explants derived from 1.5-year-old *T. mairei* seedlings (Chang et al., 1998). The explants derived from juvenile seedlings of 1.5 years old that can produce more orthotropic shoots (buds with 31.7% and shoots with 24.4%, respectively) than explants derived from coppices, stecklings, or trees (Fig. 7.5; Chang et al., 1998, 2001). The mature plagiotropic explants from 15-year-old *T. baccata* produced plagiotropic angles between 14 and 28 degrees (Goo et al., 1990).

7.3.6 Rooting and acclimatization

A study has shown that in vitro regenerated *T. baccata* shoots cultured on ½ MS medium containing 8 mg/L IBA produced 80% rooting of shoots with an average root length of 6 cm (Abbasin et al., 2010). The rooting efficiency of IBA, NAA and IAA was evaluated in *T. brevifolia* by Mitchell (1997) and *T. mairei* by Chang et al. (2001). The low concentration of IBA promotes the rooting efficiency in *T. brevifolia*

Fig. 7.5 (A–I) In vitro cultures of *Taxus mairei*. (A, B) Shoots induced from a steckling-derived large stem (A) and a tree-derived bud explant (B). (C) Multiple shoots initiated from a tree derived bud culture. (D) Shoot cultures turned brown after a 3-month subculture, (E) their growth recovery on media containing antibrowning agents. (F) Root induction of a tree derived shoot in a medium containing 2.5 mg/L IBA. (G–I) Plagiotropic angles of in vitro propagated shoots: 160 degrees (G), 90 degrees (H), and 10 degrees (I). In plagiotropic shoots the needles are arranged opposite to each other (G, H) but became spiral in a nearly vertical shoot (I). Bars: 1 cm. Reproduced with permission from Chang, S.-H., Ho, C.-K., Chen, Z.-Z., Tsay, J.-Y., 2001. Micropropagation of Taxus mairei mature trees. Plant Cell Rep. 20, 496–502.

up to 72%. According to them, the optimum concentration of 2.5 mg/L IBA for steckling and tree-derived microshoots showed 55% and 35% of rooting efficiency, respectively (Fig. 7.5F; Chang et al., 2001). Nhut et al. (2007) have evaluated the rooting response in both full strength and half strength MS medium, and found that ½ MS induces better rooting response. Among the various auxins with different concentration tested (IBA, IAA and NAA), IBA at 1 mg/L was the most effective one for root induction in conifers.

The elongated shoot tips of *T. brevifolia* were cultured on full strength and half strength MS media supplemented with different concentration of IBA (Hussain et al., 2013). The results showed that the significant rooting response observed on ½ MS media fortified with 8 mg/L IBA and full-strength MS media supplemented with 3.5 mg/L IBA after 60–80 days of culturing. Similar in vitro rooting response was reported in *T. brevifolia* (Chee, 1996) and *T. wallichiana* (Das et al., 2008).

In vitro microshoots cultured on shoot elongation medium (1/2 WPMSH) containing 1% AC obtained rooting spontaneously after 6 months of subculture at 4 weeks interval period. But, the role of AC in rooting is unclear, though it might have adsorbed growth inhibitors in the culture media. The higher rooting frequency (40%) was produced on modified MS basal medium, when the concentration of nitrates was reduced to one-fifth in normal concentration after 4 months of culture. The regenerated microshoots of *T. brevifolia* could be rooted on the application of rooting powder in the culture media (Chee, 1995b), but in *T. wallichiana*, the microshoots treated with rooting powder failed to produce any roots (Datta et al., 2006).

Kulkarni (2000) failed to succeed in vitro roots formation in *T. wallichiana*, when microshoots were cultured on ½ WPMSH medium fortified with different types of auxins, either singly or in combinations. Ewald (2007) has successfully induced roots from shoot tip explants on L9 nutrient medium containing 2 mg/L IBA.

Taxus species have symbiotic relationship with AMF, which significantly improved the roots in soil (Gemma et al., 1998). Abbasin et al. (2010) reported that rooting (94%) and root length of 8 cm was better in ½ strength MS medium in combination with the forest soluble soil, which suggests that there were some growth-promoting components in the soil.

The effect of various strains of *Agrobacterium rhizogenes* on root induction was reported (Ewald, 2007). *A. rhizogenes* induced more rooting efficiency and survival rate than IBA treatment of microshoots of *Taxus* species. Basal callus formation with decay of shoot bases was also reduced during the long-lasting process of rooting by *A. rhizogenes,* and especially of the *A. rhizogenes* Marburg strain. This variant strain of *Agrobacterium* was the best in relation to the general

Chapter 7 Propagation of paclitaxel biosynthesizing plants **197**

Table 7.3 In vitro rooting response of various *Taxus* species on culture media.

Taxus species	Media	Plant growth regulators	Observations	References
T. mairei	MS	–	10% rooting	Ho et al. (1998)
T. brevifolia	½ MS	–	Successfully rooted	Mitchell (1997)
T. mairei	½ MS	IBA 2.5 mg/L	55% rooting	Chang et al. (2001)
T. wallichiana	MS-N	No PGR	40% rooting	Datta et al. (2006)
T. baccata	LS	IBA 2 mg/L	Rooting in 2 weeks	Ewald (2007)
T. wallichiana	½ MS	IBA	1 to 2 roots per explants	Nhut et al. (2007)
T. baccata	½ MS	IBA 8 mg/L	94% rooting, 1–2 roots, 2–3 cm long by 3–7 weeks	Abbasin et al. (2010)
T. wallichiana	MS	IBA 3.5 mg/L	Successfully rooted	Hussain et al. (2013)

behavior of shoots during rooting. The rooted plants were transferred into containers (4.5×4.5 cm, 20 cm high, each for 40 plantlets) with rooting compost and placed under high pressure fog (95% air humidity) (Ewald, 2007). Most of the plantlets survived with high success rate and showed normal growth.

Rooted microshoots were acclimatized on a sterilized soil mixture consisting of peat moss, sand, and soil in 1:1:1 ratio (v/v) (Abbasin et al., 2010). The successive rate of rooted plantlets achieved almost 93% in greenhouse conditions that are acclimatized ex vitro and potted plants. Plants grew well without changes of any morphological characters after 2 months of transplanting (Abbasin et al., 2010). Various investigation on in vitro rooting in different species of *Taxus* have been listed in Table 7.3.

7.4 Conclusions

The preliminary attention of research on medicinal plants till date has been focused on the area of phytochemistry, pharmacology and horticulture. The increasing demand for such natural products in commercial purposes has sparked the need to find strategies to enhance their production without disturbing their natural population. For such instance, the evaluation and standardization of rapid propagation techniques are essential for commercially valued plant species. Usually, paclitaxel yielding Yew trees are recalcitrant to conventional and in vitro culture. Applications of biotechnological approaches can be very useful for *Taxus* species mass propagation and conservation, thereby helping

198 Chapter 7 Propagation of paclitaxel biosynthesizing plants

to replenish their population in nature. The natural population of Yew species in large number can be produced by making use of these modern propagation techniques. It will protect the Yew trees from rarity and endanger in wild condition. Microcuttings of stem explants and embryoculture of Yew trees can be improve the characters and quality of the plants for commercial production of target medicinal compounds.

References

Abbasin, Z., Zamani, S., Movahedi, S., Khaksar, G., Sayed Tabatabaei, B.E., 2010. In vitro micropropagation of Yew (*Taxus baccata*) and production of plantlets. Biotechnology 9 (1), 48–54.

Amos, R.R., McCown, B.H., 1981. Micropropagation of members of the coniferae. Hortic. Sci. 16, 453.

Anderson, E.D., Owens, J.N., 2001. Embryo development, mega-gametophyte storage product accumulation and seed efficiency in *Taxus brevifolia*. Can. J. For. Res. 31, 1046–1056.

Anjum, Q., Sharma, L.K., Ganie, S.A., Rather, M.M., Rather, H.A., 2011. Effect of auxins on macropropagation of *Taxus baccata* Linn. through stem cuttings. Indian Forester 137 (12), 1382–1385.

Aslam, M., Rather, M.S., 2008. Macropropagation of *Taxus baccata* Linn: a novel method for conserving acritically endangered medicinal plant. Indian Forester 134 (8), 1058–1063.

Aslam, M., Arshid, S., Rather, M.S., Salathia, H.S., Seth, C.M., 2007. Auxin induced rooting in *Taxus baccata* Linn. stem cuttings. Indian J. For. 30, 221–226.

Aslam, M., Raina, P.A., Rafiq, R., Siddiqi, T.O., Reshi, Z.A., 2017. Adventitious root formation in branch cuttings of *Taxus wallichiana* Zucc. (Himalayan yew): a clonal approach to conserve the scarce resource. Curr. Bot. 8, 127–135.

Barnes, L., 1983. Micropropagation of endangered native conifers, *Torreya taxifolia* and *Taxus floridana*. Hortic. Sci. 18, 617.

Barnes, L.R., 1985. Clonal propagation of endangered native plants *Rhododendron championii* Gray, *Taxus floridana* Nutt. and *Torreya taxifolia* Arn. (Ph.D. thesis). University of Florida.

Cerdeira, R.M., McChesney, J.D., Burandt Jr., C., 1994. Microcuttings of *Taxus x media* cv. Hicksii. In Vitro Cell. Dev. Biol. 30, 77.

Chang, S.-H., Yang, J.-C., 1996. Enhancement of plant formation from embryo cultures of *Taxus mairei* using suitable culture medium and PVP. Bot. Bull. Acad. Sin. 37, 35–40.

Chang, S.-H., Ho, C.-K., Tsay, J.-Y., 1998. Micropropagation of *Taxus mairei* seedlings at different ages and recoverability of their plagiotropic shoots. Taiwan J. For. Sci. 13, 29–39.

Chang, S.-H., Ho, C.-K., Chen, Z.-Z., Tsay, J.-Y., 2001. Micropropagation of *Taxus mairei* mature trees. Plant Cell Rep. 20, 496–502.

Chatterjee, S., Dey, S., 1997. A preliminary survey of the status of *Taxus baccata* in Tawang district of Arunachal Pradesh. Indian Forester 8, 746–754.

Chee, P.P., 1994. In vitro culture of zygotic embryos of *Taxus* species. Hortic. Sci. 29, 695–697.

Chee, P., 1995a. Stimulation of adventitious rooting of *Taxus* species by thiamine. Plant Cell Rep. 14, 753–757.

Chee, P.P., 1995b. Organogenesis in *Taxus brevifolia* tissue cultures. Plant Cell Rep. 14, 560–565.

Chee, P.P., 1996. Plant regeneration from somatic embryos of *Taxus brevifolia*. Plant Cell Rep. 16, 184–187.

Chien, C.T., Yang, J.J., Chung, Y.I., Lin, T.P., 1994. Germination promotion of *Tauxs mairei* seed. Taiwan Forestry J. 20 (10), 46–47.

Chien, C.-T., Kuo-Huang, L.-L., Lin, T.-P., 1998. Changes in ultrastructure and abscisic acid level, and response to applied gibberellins in *Taxus mairei* seeds treated with warm and cold stratification. Ann. Bot. 81, 41–47.

Das, S., Jha, L.K., 2014. Effect of wounding and plant growth regulators (IBA and NAA) on root proliferation of *Taxus wallichiana* shoot cuttings. Res. J. Agric. Forestry Sci. 2 (12), 8–14.

Das, S., Jha, L.K., 2018. Effect of physiological age of stem and IBA treatment on rooting of branch cuttings of *Taxus baccata* L. Curr. Bot. 9, 1–7.

Das, K., Dang, R., Ghanshala, N., Rajasekharan, P.E., 2008. *In vitro* establishment and maintenance of callus of *Taxus wallichiana* Zucc. For the production of secondary metabolites. Nat. Prod. Rad. 7 (2), 150–153.

Datta, M.M., Jha, S., 2004. Embryo culture of *Taxus wallichiana* (Zucc.). J. Plant Biotechnol. 6 (4), 213–219.

Datta, M.M., Majumder, A., Jha, S., 2006. Organogenesis and plant regeneration in *Taxuswallichiana* (Zucc). Plant Cell Rep. 25, 11–18.

Davarpanah, S.J., 2003. Study of biotechnology of paclitaxel in *Taxus* sp. tissue culture and some of its endophytic fungi (M.Sc. dissertation). Ferdowsi University of Mashhad, Iran.

Davarpanah, S.J., Lahouti, M., Karimian, R., 2014. Micropropagation of common yew using embryo culture. J. Appl. Biotech. Rep. 1 (2), 81–84.

David, A., 1977. La culture des tissus de deux gymnospermes *Pinus pinaster* Sol. et *Taxus baccata* L. In: Gautheret, G. (Ed.), La Culturedes Tissus et des Cellules des Vegetaux. Masson, Paris, New York. pp 9: 3–99.

David, A., Plastira, V., 1974. Histophysiologie vegetale. Realisation de cultures de tissus d'Ilf *(Taxus baccata* L.) a partir de fragments de pousses ages d'unean; obtentiun d'une culture indefinie. C. R. Acad. Sci. Pairs 279, 1757–1759.

David, A., Plastira, V., 1976. Histophysiologie vegetale. Comportementen culture *in vitro* de ceUules isolées de deux gymnospermes: *Taxus baccata* L. et *Pinus pinaster* Sol. C. R. Acad. Sci. Pairs 282, 1159–1162.

Eccher, T., 1988. Response of cuttings of 16 *Taxus* cultivars to rooting treatments. Acta Hortic. 227, 251–253.

Ellis, D.D., Zeldin, E.L., Brodhagen, M., Russin, W.A., McCown, B.H., 1996. Taxol production in nodule cultures of *Taxus*. J. Nat. Prod. 59 (3), 246–250.

Ewald, D., 2007. Micropropagation of Yew (*Taxus baccata* L.). In: Jain, S.M., Häggman, H. (Eds.), Protocolsfor Micropropagation of Woody Trees and Fruits. Springer, Dordrecht, pp. 117–123.

Ewald, D., Weckwerth, W., Naujoks, G., Zocher, R., 1995. Formation of embryo-like structures in tissue cultures of different yew species. J. Plant Physiol. 147, 139–143.

Flores, H.E., Sgrignoli, P.J., 1991. In vitro culture and precocious germination of *Taxus* embryos. In Vitro Cell Dev. Biol. 27, 139–142.

Flores, T., Wagner, L.J., Flores, H.E., 1993. Embryo culture and taxane production in *Taxus* spp. In Vitro Cell Dev. Biol. 29, 160–165.

Gamborg, O.L., Miller, R.A., Ojima, K., 1968. Nutrient requirements of suspension cultures of soybean root cells. Exp. Cell Res. 50, 151–158.

Gemma, J.N., Koske, R.E., Roberts, E.M., Hester, S., 1998. Response of *Taxus x media* var. *densiformis* inoculation with arbuscular mycorrhizal fungi. Can. J. For. Res. 28 (1), 150–153.

Goo, G.H., Lee, K.Y., Youn, K.S., Kwon, Y.H., 1990. Effect of ramet age and types of cuttings on rooting, cyclophysis and topophysis of rooted cuttings in *Taxus cuspidata* S. et Z. J. Korean For. Soc. 79, 359–366.

Griffiths, R.P., Chadwick, A.C., Robatzek, M., Schauer, K., Schaffroth, K.A., 1995. Association of ectomycorrhizal mats with Pacific yew and other understory trees in coniferous forests. Plant Soil 173, 343–347.

Gul, N., Baig, S., Ahmed, R., Shahzadi, I., Zaman, I., Shah, M.M., Baig, A., 2020. Conservation of an endangered medicinal tree species *Taxus wallichiana* through callus induction and shoot regeneration. Plant Tissue Cult. Biotechnol. 30 (1), 161–166.

Hartzell Jr., H., 1991. The Yew Tree: A Thousand Whispers. Biography of a Species. Hulogosi, Eugene, Oregon.

Heit, C.E., 1968. Thirty-five years' testing of tree and shrub seed. J. For. 66, 632–633.

Heit, C.E., 1969. Propagation from seed. Part 18: testing and growing seeds of popular *Taxus* forms. Am. Nurserym. 129, 10–11. 118–128.

Ho, C.-K., Chang, S.-H., Tsai, J.-Y., 1998. Selection breeding, propagation and cultivation of *Taxus mairei* in Taiwan. Taiwan For. Res. Inst. 88, 65–82.

Ho, C.-K., Chang, S.-H., Chen, K.-P., Tsai, J.-Y., 2007. Clonal test of *Taxus mairei* using water culture. Taiwan J. For. Sci 22 (2), 113–123.

Hu, C.-Y., Wang, L., Wu, B., 1992. *In vitro* culture of immature *Taxus* embryos. Hortic. Sci. 27, 698.

Hussain, A., Qarshi, I.A., Nazir, H., Ullah, I., Rashi, M., Shinwari, Z.K., 2013. *In vitro* callogenesis and organogenesis in *Taxus wallichiana Zucc.*, the Himalayan Yew. Pak. J. Bot. 45, 1755–1759.

Jaziri, M., Zhiri, A., Guo, Y.W., Dupont, J.P., Shimomura, K., Hamada, H., Vanhaelen, M., Homes, J., 1996. *Taxus* sp. cell, tissue and organ cultures as alternative sources for taxoid production: a literature survey. Plant Cell Tissue Organ Cult. 46, 59–75.

Jha, S., Sanyal, D., Ghosh, B., Jha, T.B., 1998. Improved taxol yield in cell suspension culture of *Taxus wallichiana* (Himalayan yew). Planta Med. 64, 270–272.

Jiaru, L., Manxi, L., Huirong, C.H., Zhenbin, W., Junjian, W., 1999. Callus initiation and subculture of *Taxus chinensis*. J. For. Res. 10, 11–14.

Kaul, K., 2008. Variation in rooting behavior of stem cuttings in relation to their origin in *Taxus wallichiana* Zucc. New For. 36, 217–224.

Khali, R.P., Sharma, A.K., 2003. Effect of phytohormones on propagation of Himalayan yew (*Taxus baccata* L.) through stem cuttings. Indian Forester 129, 289–294.

Kulkarni, A.A., 2000. Micropropagation and secondary metabolite studies in *Taxus* spp. and *Withania somnifera* (L.) Dunal (Ph.D. thesis). National Chemical Laboratory, Pune.

Larue, C.D., 1953. Studies on growth and regeneration in gametophytes and sporophytes of gynmosperms. In: Abnormal and pathological plant growth. Report of symposium held August 3–5, 1953. Brookhaven National Laboratory, Upton, New York, pp. 187–208.

Le Page-Degivry, M.T., 1969. Mise en evidence de l'acide (+) abscissiquechez une Gymnosperme. C. R. Acad. Sci. 269, 2534–2536.

Le Page-Degivry, M.T., 1970. Physiologie Végétale. Acide abscissique et dormance chez les embryons de *Taxus baccata* L. C. R. Acad. Sci. Paris 271, 482–484.

Le Page-Degivry, M.T., 1973a. Influence de l'acid abscissique sur le développement des embryons de *Taxus baccata* L. cultivés *in vitro*. Z. Pflanzenphysiol. 70, 406–413.

Le Page-Degivry, M.T., 1973b. Physiologie Végétale. Intervention d'un inhibiteur liédans la dormance embryonnaire de Taxus baccata L. C. R. Acad. Sci. Paris 77, 177–180.

Le Page-Degivry, M.T., Garello, G., 1973. La dormance embryonnaire chez *Taxus baccata* Inflence de la composition du milieu liquide sur l'induction de la germination. Physiol. Plant. 29, 204–207.

Lee, B.S., Son, S.H., 1995. A Method for Producing Taxol and Taxanes From Embryo Cultures of *Taxus* spp. WO Patent No. 95/02063.

Lepage-Degivry, M.T., 1973. Etude en culture in vitro de la dormanceembryonnaire chez *Taxus baccata* L. Biol. Plant. 15, 264–269.

Li, Z.-L., 2001. Embryo culture of *Taxus chinensis* (Pilger) Rehd. and *Taxus brevifolia* Nutt. in vitro. J. Plant Resour. Environ. 10, 62–63 (in Chinese).

Li, X.K., Huang, Y.Q., Deng, Z.B., Su, Z.M., Li, J.F., 2000. The propagation and growth dynamics of seedlings of *Taxus chinensis* (Pilger) Rend. J. Plant Resour. Environ. 9, 48–50.

Liao, Z., Gong, Y., Pi, Y., Chen, M., Tan, Q., Tan, F., Sun, X., Tang, K., 2006. Rapid and efficient in vitro germination of embryos from *Taxus media* Rehder. Asian J. Plant Sci. 5 (1), 139–141.

Lloyd, G., McCown, B.H., 1981. Commercially feasible micropropagation of mountain laurel *Kalmia latifolia* by use of shoot tip culture. Proc. Inter. Plant Prop. Soc. 30, 421–437.

Majada, J.P., Sierra, M.L., Sanches-Tames, R., 2000. One step more towards taxane production through enhanced *Taxus* propagation. Plant Cell Rep. 19, 825–830.

Mitchell, A.K., 1997. Propagation and growth of Pacific Yew (*Taxus brevifolia* Nutt.) cuttings. Northwest Sci. 71, 56–62.

Mitter, H., Sharma, A., 1999. Propagation of *T. baccata* Linn. by stem cuttings. Indian Forester 125, 159–162.

Murashige, T., Skoog, F., 1962. A revised medium for rapid growth and bioassays with tobacco tissue cultures. Physiol. Plant. 15, 473–497.

Nandi, S.K., Palni, L.M.S., Kosi-Katarmal, A., 1996. Chemical induction of adventitious root formation in *Taxus baccata* cuttings. Plant Growth Regul. 19, 117–122.

Nazir, N., Kamili, A.N., Shah, D., Zargar, M.Y., 2018. Adventitious rooting in shoot cuttings of *Taxus wallichiana* Zucc., an endangered medicinally important conifer of Kashmir Himalaya. For. Res. 7, 221. https://doi.org/10.4172/2168-9776.1000221.

Nhut, D.T., Hien, N.T.T., Don, N.T., Khiem, D.V., 2007. In vitro shoot development of *Taxus wallichiana* Zucc., a valuable medicinal plant. In: Jain, S.M., Häggman, H. (Eds.), Protocols for Micropropagation of Woody Trees and Fruits. Springer, Dordrecht, pp. 107–116.

Nicholson, R., Munn, D.X., 2003. Observation on the propagation of *Taxus globosa* Schltdl. Bol. Soc. Bot. México 72, 129–130.

Qiu, D., Li, R., Li, L., 1998. Studies on the somatic embryogenesis of *Taxus chinensis* and *Taxus chinensis* var. *mairei*. Sci. Silvae Sin. 34, 50–54.

Rohr, R., 1973a. Production de cals par les gametophytes males de *Taxus baccata* L. cultives sur un milieu artificiel. Etude en microscopiephotonique et electronique. Caryologia 25, 177–189.

Rohr, R., 1973b. Cytologie vegetale. Ultrastructure des spermatozo'fdesde *Taxus baccata* L. obtenus a partir de cultures aseptiquesde microspores sur un milieu artificiel. C. R. Acad. Sci. Paris 277, 1869–1871.

Rudolf, P.O., 1974. Taxus L. Yew. In: Schopmeyer, C.S. (Ed.), Seeds of Woody Plants in the United States. Agriculture Handbook No 450. Forest Service, U.S. Dept. of Agriculture, Washington, DC, pp. 799–802.

Salandy, A., Grafton, L., Uddin, M.R., Shafi, M.I., 1993. Establishing an embryogenic cell suspension culture system in Florida Yew (*Taxus floridana*). In Vitro Cell Dev. Biol. 29, 75A.

Saumitro, D., Jha, L.K., 2014. Effect of wounding and plant growth regulators (IBA and NAA) on root proliferation of *Taxus wallichiana* shoot cuttings. Res. J. Agric. For. Sci. 2 (12), 8–14.

Schneck, V., 1996. Studies on influence of clone on rooting ability and rooting quality in the propagation of cuttings from 40-to 350-year-old *Taxus baccata* L. ortets. Silvae Genet. 45, 246–249.

Singh, S.P., 2006. Effect of phytohormones on propagation of Himalayan Yew (*Taxus baccata* L.) through stem cuttings. Bull. Arunachal For. Res. 22, 64–67.

Sobha, Chauhan, J.S., 2018. Regeneration of *Taxus baccata* through vegetative propagation using different concentrations of IBA and NAA. J. Pharmacogn. Phytochem. 7 (4), 1853–1857.

Steinfield, D., 1992. Early lessons from propagating Pacific yew. Rocky Mountain Forest and range. Exp. Sta. Tech. Res. Rept., 221.

Susilowati, A., Rachmat, H.H., Kholibrina, C.R., Hartini, K.S., Rambe, H.A., 2019. Propagation for conserving endangered taxol producing tree *Taxus sumatrana* through shoot cuttings technique. J. Phys. Conf. Ser. 1282, 012110. https://doi.org/10.1088/1742-6596/1282/1/012110.

Suszka, B., 1985. Conditions for after-ripening and germination of seeds and for seedling emergence of the English yew (*Taxus baccata* L.). Arbor. Kórnickie 30, 285–338.

Taiebeh, S.N., Shahrzad, S., 2003. In vitro propagation of *Taxus baccata* and *Sequoia sempervirens*. Research Institute of Forest and Rangeland, p. 64.

Thomas, P.A., Polwart, A., 2003. *Taxus baccata* L. J. Ecol. 91, 489–524.

Tuleke, W., 1959. The pollen cultures of C.D. Larue: a tissue from the pollen of *Taxus*. Bull. Torrey Bot. Club 86, 283–289.

Túri-Farkas, Z., Kovács, D., 2016. Propagation of *Taxus baccata* 'green diamond' by cutting. Rev. Agric. Rural Dev. 5 (1–2), 71–76.

USDA, 1974. Seeds of woody plants in the United States. In: Agricultural Handbook 450. United States Department of Agriculture, Forest Service, Washington, DC, USA.

Wann, S.R., Goldner, W.R., 1994. Induction of Somatic Embryogenesis in *Taxus*, and the Production of Production of Taxane-Ring Containing Alkaloids Therefrom. WO Patent No. 93/19585.

Wann, S.R., Kahphammer, J., Veazey, R.L., 1999. Somatic embryogeneis in *Taxus*. In: Jain, S.M., Gupta, P.K., Newton, R.J. (Eds.), Somatic Embryogenesis in Woody Plants. Forestry Science, vol. 55. Springer, Dordrecht.

Wareing, P.F., Phillips, I.D.J., 1981. Growth and Differentiation in Plants, third ed. Pergamon Press, Oxford.

Wickremesinhe, E.R., Arteea, R.N., 1993. Taxus callus cultures: initiation, growth optimization, characterization and taxol production. Plant Cell Tissue Organ Cult. 35 (2), 181–193.

Zarek, M.A., 2007. Practical method for overcoming the dormancy of *Taxus baccata* isolated embryos under *in vitro* conditions. In Vitro Cell. Dev. Biol. Plant 43, 623–630.

Zhiri, A., Jaziri, M., Homès, J., Vanhaelen, M., Shimomura, K., 1994. Factors affecting the *in vitro* rapid germination of *Taxus* embryos and the evaluation of taxol content in the plantlets. Plant Cell Tissue Organ Cult. 39, 261–263.

8

Endophytes for the production of anticancer drug, paclitaxel

Mallappa Kumara Swamy[a], Tuyelee Das[b], Samapika Nandy[b], Anuradha Mukherjee[c], Devendra Kumar Pandey[d], and Abhijit Dey[b]

[a]Department of Biotechnology, East West First Grade College, Bengaluru, Karnataka, India, [b]Department of Life Sciences, Presidency University, Kolkata, India, [c]MMHS, Joynagar, India, [d]Department of Biotechnology, Lovely Faculty of Technology and Sciences, Lovely Professional University, Phagwara, India

8.1 Introduction

The incidence of different forms of tumors is a persistent menace to human health, and has a major influence on humankind across the world. Cancer is considered as one of the major causes of death globally. A probable 1,806,590 new cases of cancer and 606,520 cancer-associated deaths are estimated to occur in the United States in 2020. The cancer death rate is 158.3 per 100,000 men and women per year, which is based on the total number of deaths between the years, 2013 and 2017 (https://www.cancer.gov/about-cancer/understanding/statistics). Between 2010 and 2020, a rise in the number of new cases of cancer of around 24% in menfolk to above 1 million cases per year, and by around 21% in females above 900,000 cases per year is estimated (https://www.cdc.gov/cancer/dcpc/research/articles/cancer_2020.htm). The annual death rate due to cancer is growing steadily, worldwide. Hence, a number of research groups have been enthusiastically involved in developing new anticancer medicines all over the globe. As stated earlier, the total budget required to develop a new medication is approximately US$ 2.6 billion. The worldwide yearly spending on anticancer drugs is about US$ 100 billion, and is expected to rise to US$ 150 billion by the end of 2020 (Prasad et al., 2017; Uzma et al., 2018). However, the currently available anticancer drugs are less effective against different cancer types. Moreover, they display adverse side effects and cause toxicity to multiplying normal cells. Thus, there is a

Paclitaxel. https://doi.org/10.1016/B978-0-323-90951-8.00012-6
Copyright © 2022 Elsevier Inc. All rights reserved.

need to discover novel bioactive compounds from natural resources, including plant, animal, and microbes (Remesh, 2017; Uzma et al., 2018).

Natural resources have the potential for discovering novel drug molecules as they possess diverse chemostructural classes with distinctive biological activities for cancer remedy. Among microbial sources, endophytic fungi offer a rich source of biologically active compounds having chemotherapeutic properties, and also these bioactive compounds can be modified structurally for cancer therapy. Endophytic fungi may exist in all plant tissue types via symbiotic associations without showing any external signs. Endophytes may stimulate the plant growth and development, and provide resistance to plant diseases, pests, drought, cold, and other extreme conditions (Rai et al., 2014; Jia et al., 2016; Swamy et al., 2016; Alurappa et al., 2018; Strobel, 2018). Endophytes produce significant metabolites of pharmacological and commercial importance, such as unique anticancer agents, antimicrobial compounds, immunosuppressive compounds, antibiotics, antioxidants, etc. Moreover, endophytic association with plants may moderate secondary metabolites production in the host plant (Jalgaonwala et al., 2011; Strobel, 2014; Alurappa et al., 2018; Strobel, 2018).

One of the most effective antitumor drugs is the Taxol (paclitaxel), a diterpenoid compound initially sequestered from the *Taxus brevifolia* (Pacific yew tree) bark extract in the middle of 1960s (Mamadalieva and Mamedov, 2020). With distinctive mode of action, paclitaxel is effective against several types of cancers, and hence there is an increasing demand for its worldwide supply. It is difficult to meet the present market requisite using existing industrial production approaches for the drug, i.e., total chemical synthesis and semi-synthesis by extracting paclitaxel precursors from *T. brevifolia*. Therefore, other methods have been meticulously explored, and one such encouraging strategy is the use of endophytic fungi (Hao et al., 2013). In this chapter, sources of endophytic fungi, their association with host plants and their influence in producing the anticancer compound, paclitaxel from their host plants are discussed. Also, paclitaxel-producing endophytic fungi and their possible scientific exploration are explained in detail.

8.2 Paclitaxel sources in nature

Paclitaxel has been reported to be isolated from different plants of yew family Taxaceae which include *Taxus brevifolia* Nutt., *Taxus baccata* L., *Taxus cuspidata* Siebold and Zucc., *Taxus wallichiana* Zucc., *Taxus chinensis* (Pilg.) Rehder, *Taxus floridana* Nutt. ex Chapm., *Taxus canadensis* Marshall, *Taxus globosa* Schltdl., and *T. calcicola* L.M. Gao and Mich. Möller, (BoLin et al., 1995; Bala et al., 1999; Chakravarthi et al., 2008; Choi et al., 1999; Fang et al., 1993; Hirasuna et al., 1996; Phisalaphong and Linden, 1999; Rao et al., 1996; Ruiz-Sanchez et al., 2010; Stierle et al., 1993; Witherup et al., 1990).

In addition to species of *Taxus*, paclitaxel is also present in *Corylus avellana* L. (hazel plant). Bestoso et al. (2006) reeported that *Corylus avellana* can produce paclitaxel without the association of endophytic fungi. In the hazel plant paclitaxel, production possesses by the metabolic pathway. They compared Hazel with *Taxus* species and demonstrated that in vitro cultivation of *Corylus avellana* is more advantageous than *Taxus* in vitro culture, hence, hazel could be used as an alternative source of paclitaxel. They added methyl jasmonate and chitosan as elicitor and observed enhancement of paclitaxel production in cell suspension cultures (Bestoso et al., 2006). Another study resulted in *Corylus arellana* produced small amounts of paclitaxel from brown hard shells, green shell cover, and leaves. Paclitaxel was detected by HPLC and electro-spray mass spectrometry (Hoffman and Shahidi, 2009). Paclitaxel contents were found in leaves and stem of hazelnut as 23.65 and 4.09 µg/mL, respectively (Luo et al., 2011). One of the other hazelnut species (*Corylus mandshurica* Maxim.) was also detected in paclitaxel synthesis (Ma et al., 2013). *Podocarpus gracilior* Pilger is an African pine fern from where a small amount of paclitaxel (70 µg) was first time isolated other than Taxaceae family (Stahlhut et al., 1999).

8.3 Available approaches for paclitaxel production

Since the earlier days, wild species of *Taxus* are the main source for extracting paclitaxel. Due to depletion of the wild species in nature, many leading drug firms rapidly implemented extensive farming of *Taxus* species. Even today, some part of the paclitaxel demand is met by the raw tree materials. The prevalent impediment encountered during the early years of paclitaxel production is the requisite of *Taxus brevifolia* (Pacific yew) trees of 100 years old (Demain and Vaishnav, 2011; Uzma et al., 2018). Undesirably, the yield of paclitaxel from wild *Taxus* tree species occurs in very low quantity. For instance, the yield is around 0.01% of a dry weight (DW) of phloem in *T. brevifolia* (Pacific yew tree) (Hao et al., 2013). The limitations in the obtainability, sequestration, and synthesis of paclitaxel encouraged the investigators to consider alternative resources for its production.

Further, the taxane contents differed among and within *Taxus* species. The Paclitaxel content is considerably influenced by trees' age, altitude, geographical locality, and the season of harvesting. Even in

Hicks yew (*Taxus × media* "Hicksii"), a hybrid variety of Pacific yew yields paclitaxel compound of only about 109 to 112 mg/kg of the dried needles (Wheeler et al., 1992; Ne'meth-Kiss et al., 1996; Poupat et al., 2000; Nadeem et al., 2002; Mukherjee et al., 2002; Hao et al., 2013). Owing to sluggish growth of *Taxus* tree species and the truncated harvesting, the global demand for paclitaxel using this conventional approach cannot be met. Hence, alternative approaches were explored, including biotechnological applications.

The total chemical synthesis has been firstly accomplished autonomously by the Holton's and Nicolaou's group during the year, 1994 (Holton et al., 1994a, b; Nicolaou et al., 1994; Mukaiyama et al., 1999). Later, another synthetic path was achieved by Danishefsky's group in 1996 (Danishefsky et al., 1996). For chemists, the total synthesis of paclitaxel is a complicated process, due to the fact that the compound comprises of four complex (A, B, C and the oxetane) rings. Further, the molecule also possesses 11 chiral centers. Hence, the synthetic process involves in excess of 20 steps. The Holton's group produced only 0.07% and the Nicolaou's group produced up to 2.7% of paclitaxel through total synthetic approach. Later, the asymmetric total synthesis of paclitaxel was also proposed by another group (Mukaiyama et al., 1999). Likewise, intermediate of paclitaxel production method was developed by Doi et al. (2006) using an automatic synthesizer involving 36 steps of chemical synthesis series. Owing to the complicated structure of paclitaxel, involvement of costly chemical reagents and the stringent requisite for reactions, comprising several-steps to synthesize paclitaxel, the total synthesis approach becomes expensive and impracticable for commercial applications to manufacture paclitaxel.

Alternatively, 10-deacetylbaccatin III, a precursor molecule of taxol biosynthesis has been identified in *Taxus* species. A number of coupling approaches of a phenylisoserine part with protected with 10-deacetylbaccatin III have been described by a number of researchers (Denis et al., 1986; Ojima et al., 1991, 1992; Holton et al., 1995). 10-deacetylbaccatin III can be sequestered from the needles of *T. baccata* and the hybrid species, *T. media* with moderately greater yield (Denis et al., 1988; Witherup et al., 1990; ElSohly et al., 1995; Van Rozendaal et al., 2000). An innovation to produce paclitaxel semisynthetically using baccatin III and C-13 taxol side chain has significantly boosted the pharma industries to produce it in a scale (Denis et al., 1986; Ojima et al., 1992; Holton et al., 1995; Patel et al., 2000; Hao et al., 2013). This process is widely used even today also, and has made the accessibility of the drug to patients with low cost. Nevertheless, dependency on the resource of yew plant materials is a major concern for the usefulness of this commonly used process.

At present, few plant cell and tissue culture-based methods have been commercialized for producing plant bioactive compounds that are used in applications by pharmaceutical, food, and cosmetic industries (Rency et al., 2019; Satish et al., 2020). Manufacture of plant metabolites through in vitro cell cultures is renewable, economically feasible and environmentally friendly. The use of *Taxus* spp. cell cultures is measured as a quick approach to achieve adequate quantity of tree biomass (Frense, 2007; Majumder and Jha, 2009; Hao et al., 2013). Different in vitro approaches have been explored extensively to upsurge the paclitaxel content in *Taxus* cell cultures. Some of them include selection of high-paclitaxel-yielding genotypes (Parc et al., 2002), application of nutrients and plant growth regulators, and the employment of elicitation technique, i.e., using chemical elicitors (silver thiosulfate, methyl jasmonate, etc), the heat shock treatment, providing mechanical stimulus, the use of two-phase cultures, and many others (Ketchum and Gibson, 1996; Roberts and Shuler, 1997; Wu and Lin, 2003; Tabata, 2004; Khosroushahi et al., 2006; Zhang and Fevereiro, 2007; Hao et al., 2013). These approaches have significantly improved the production of paclitaxel. However, truncated and unstable yield of paclitaxel, high manufacturing budget, and impurity due to byproduct are some of the key bottlenecks for viable commercial utilization of in vitro cell culture approaches. A German and Canadian biotechnology firm, Phyton Biotech is one of the leading suppliers of paclitaxel in the world, which commercially produces or provides starting material for paclitaxel and docetaxel API (Active Pharmaceutical Ingredient) using their "green" Plant Cell Fermentation (PCF) technology facility (https://phytonbiotech.com/our-company/).

8.4 Endophytes producing paclitaxel from different host plant species

Paclitaxel as an anticancer agent already proven its efficiency, however, many challenges are still present due to the slow growth of the *Taxus* plant, which is the primary source of paclitaxel. Cancer cases overall are rising and the demand for paclitaxel as well increasing, but due to the slow growth of the *Taxus* plant and small yield of paclitaxel, it cannot co-up with the high demand for the treatment of cancer patients. It has been reported that for the treatment of 500 patients, 1 kg paclitaxel is required and for that approximately 300 trees harvested (Zhou et al., 2010). Hence, alternative strategies for paclitaxel extraction are in demand. In the last 40 years, many effective biotechnological approaches, such as callus culture, chemical synthesis, cell suspension, hairy root culture, field cultivation, and tissue culture, for paclitaxel production have been developed (Croteau et al., 2006;

Debbab et al., 2009; Stierle et al., 1993). However, these approaches are not impactful due to a large number of reaction steps, long incubation time, and low yields that limits its practicality (Nicolaou et al., 1994; Patel, 1998; Yukimune et al., 1996).

Endophytes are given tremendous promises for the production of paclitaxel in association with the host plant. Endophytes make associations with higher plants by mutualistic or commensalism or pathogenic interaction and are capable of producing associated plant secondary metabolites compounds by modulating genes that are responsible for secondary metabolites biosynthesis pathways (Kusari and Spiteller, 2011; Wastewater and Rodr, 2016). Endophytes are organisms, which colonize to plant tissue intercellularly or intracellularly and present into the healthy host plant tissues in asymptomatic conditions (Sturz and Nowak, 2000; Wilson, 1995). In 1866, Bary for the first time introduced the term "endophytes", and defined endophytes (Bary, 1866) as "Fungi and bacteria which, for all or part of their life cycle, invade the tissues of living plants and cause inapparent, asymptomatic infections entirely within the plant tissues but cause no symptoms of disease" (Wilson, 1995; Nisa et al., 2015). Paclitaxel is a high-value drug for cancer, also known commercially as taxol, that first time isolated from *Taxus brevifolia* plant endophytic fungus *Taxomyces andreanae*, used successfully in the treatment of a wide variety of cancers (Stierle et al., 1993). Endophytes can be isolated from diverse host plant species that belong to different families under various ecological and geographical conditions. Endophytes are identified by both morphological (phenotypic characteristics) and molecular techniques. Different extraction methods are used for the isolation of fungal-derived paclitaxel. Mostly encountered endophytes are from fungi, however, endophytic bacteria are also researched and documented. Endophytic fungi association with plants is over 1 million and have been found in association with angiosperm, gymnosperm, bryophytes, pteridophytes, and lichens (Ibrahim et al., 2017). Endophytic fungi have been known as a novel and alternative resource of naturally derived bioactive products. Due to diverse environment ecology geographic distribution, endophytes produced diverse types of bioactive compounds, i.e., secondary metabolites, that the help plant to co-up in both biotic and abiotic stresses (Rodriguez et al., 2009).

So far, researchers have found more than 35 species of endophytic fungi, including *Phyllosticta tabernaemontanae, Pestalotiopsis terminaliae, Chaetomella raphigera, Fusarium oxysporum, Pestalotiopsis microspora, Alternaria alternata, Colletotrichum gloesporioides, Glomerella cingulata, Nigrospora sphaerica, Lasiodiplodia theobromae, Cladosporium oxysporum*, etc., that can produce paclitaxel. The range of hosts of paclitaxel-producing fungi describes broadly in Table 8.1. Paclitaxel production not only restricted to the endophytes of Taxaceae family

Host	Family	Endophytic fungi	Isolated from	Paclitaxel produced by the test fungus in medium	Paclitaxel yield	Solvent used for Paclitaxel extraction	Reported genes	Detection methods	Reference/s
Taxus sp.	Taxaceae	*Aspergillus fumigatus* KU-837249	Bark, stem, needle; Shimla, Himachal Pradesh India	S7	1.60 g/L	Ethyl acetate	*dbat*	HPLC, TLC,UV, MS, FTIR, NMR	Kumar et al. (2019)
Taxus baccata L.	Taxaceae	*Alternaria* sp.	Wood; central-northern Italy	M1D	10–20 ng/L	Dichloromethane	–	HPLC, LC-MS, EIA	Caruso et al. (2000)
	Taxaceae	*Beauveria* sp.	Wood; central-northern Italy	BAMB; SFB	10–20 ng/L	Dichloromethane	–	HPLC, LC-MS, EIA	Caruso et al. (2000)
	Taxaceae	*Epicoccum* sp.	Leaves; central-northern Italy	BAM; CYB; MIDB	20–50 ng/L	Dichloromethane	–	HPLC, LC-MS, EIA	Caruso et al. (2000)
	Taxaceae	*Fusarium* sp.	Wood; central-northern Italy	CYB	20–50 ng/L	Dichloromethane	–	HPLC, LC-MS, EIA	Caruso et al. (2000)
	Taxaceae	*Mycelia sterilia*	Wood; central-northern Italy	BAM	50–100 ng/L	Dichloromethane	–	HPLC, LC-MS, EIA	Caruso et al. (2000)
	Taxaceae	*Mycelia sterilia*	Leaves; central-northern Italy	BAM	10–20 ng/L	Dichloromethane	–	HPLC, LC-MS, EIA	Caruso et al. (2000)
	Taxaceae	*Stemphylium sedicola* SBU-16	Inner bark; Tehran, Iran	PDB	6.9 µg/L	Methanol	ts	HPLC	Mirjalili et al. (2012)
	Taxaceae	*Paraconiothyrium variabile* SS1	Twig; Bedfordshire, U.K	M1D medium	1.75 µg/L	Dichloromethane	–	HPLC, LC-MS/MS	Somjaipeng et al. (2015)
	Taxaceae	*Epicoccum nigrum* SS2	Twig; Bedfordshire, U.K	Twig; Bedfordshire, U.K	1.32 µg/L	Dichloromethane	–	HPLC, LC-MS/MS	Somjaipeng et al. (2015)
	Taxaceae	*Cladosporium* sp. F1 and F3	Northern forests, Iran	PDA	129 mg/kg dw	Yeast extract: 1% (w/v), peptone:2% (*w/v*), glucose: 2% (w/v)	ts	HPLC (anticancer	Kasaei et al. (2017) Zaiyou et al. (2017)
Taxus baccata L. subsp. *wallichiana* (Zucc.) Pilger	Taxaceae	*Fusarium redolens* TBPJ-B	Bark; Doda, Jammu and Kashmir, India	S-7	66 µg/L	Methanol	bapt	HPLC-MS	Garyali et al. (2013)
Taxus brevifolia Nutt.	Taxaceae	*Taxomyces andreanae*	Inner bark; northern Montana	S-7	24–50 ng/L	Dichloromethane	–	TLC, HPLC, MS	Stierle et al. (1993)
Taxus celebica (*Warb.*) H.L. Li	Taxaceae	*Fusarium solani*	Stem cutting; Bedgeburg, National Pinetum, UK	PDA or MID	1.6 µg/L	Methylene chloride	–	TLC, HPLC, LC-ESI-MS	Chakravarthi et al. (2008)

Continued

Table 8.1 List of endophytes producing paclitaxel from different host plant species—cont'd

Host	Family	Endophytic fungi	Isolated from	Paclitaxel produced by the test fungus in medium	Paclitaxel yield	Solvent used for Paclitaxel extraction	Reported genes	Detection methods	Reference/s
Taxus chinensis var. mairei (Lemée and H. Lév.) W.C. Cheng and L.K. Fu	Taxaceae	BT2	Twig, old inner bark; Jin Yun Mountain, China	PDA	4–7 µg/L	Methylene chloride	–	HPLC, LCMS, CIEIA	Guo et al. (2006)
	Taxaceae	Aspergillus aculeatinus Tax-6	Bark; Anhui Province, China	PDA	334.92–1337.56 µg/L	Dichloromethane	–	HPLC, ESI-MS	Qiao et al. (2017)
	Taxaceae	Didymostilbe sp. (DF110)	Bark; Sichuan province, Southwest China	PDA	8–15 µg/L	Methylene chloride	–	LC-MS, EIA	Wang and Tang (2011)
Taxus chinensis Roxb.	Taxaceae	Fusarium solani, Tax-3	Bark; Qinba mountains, China	PDB	163.35 µg/L	Ethyl acetate	–	HPLC	Deng et al. (2009)
	Taxaceae	Metarhizium anisopliae H-27	Bark; Qinba, China	PDB	846.1 ng/L	Ethyl acetate	–	HPLC, LC-MS, ESI-MS	Liu et al. (2009)
	Taxaceae	Mucor rouxianus DA10	Bark; Jing Gang Mountain, China	PDA	30 µg/L	Dichloromethane	dbat, bac-catin III	LC-MS, ELISA	Miao et al. (2009)
Taxus cuspidata Sieb. & Zucc	Taxaceae	Aspergillus niger HD86–9	Inner bark; China	MEA	273.46 µg/L	Methanol, ethyl acetate	–	LC-MS	Zhao et al. (2009)
	Taxaceae	Phomopsis sp. BKH 27	Leaves; Kangwondo forest, South Korea	PDA	418 µg/L	Dichloromethane	–	UV, IR, HPLC, LC-MS, NMR	Kumaran and Hur (2009)
		Fusarium arthrosporioides	Inner bark; Kang-Won province, Korea	PDA	131 mg/L	Chloroform/methanol (20:1, v/v)	–	TLC, RP-HPLC, LC-MS and NMR	Li et al. (2008)
Taxus mairei (Lemée & H. Lév.) S.Y. Hu	Taxaceae	Tubercularia sp. TF5	Inner bark; Fujian province, southeast China	PDA	–	Chloroform/methanol (10:1)	–	UV, TLC, HPLC, MS	Wang et al. (2000)
	Taxaceae	Fusarium maire K178	Bark; China	–	20 µg/L	Diethyl sulfate	–	HPLC, MS, CIEIA	Xu et al. (2006)
Taxus × media Rehder	Taxaceae	Cladosporium cladosporioides MD2	Inner bark	PDB	–	Chloroform, methanol (10:1)	DBAT	HPLC, NMR	Zhang et al. (2009)
	Taxaceae	Aspergillus terreus	Bark; University of Guelph Main Campus and Arboretum, Guelph, Canada	PDA	80 µg/L	Chloroform/methanol (9:1)	Alamar blue assay	LC-MS	Soliman et al. (2011)
	Taxaceae	Guignardia mangiferae HAA11	Bark, needles; Shanghai, China	PDB	720 ng/L	Ethyl acetate	ts, dbat, bapt	HPLC-MS	Xiong et al. (2013)
	Taxaceae	Fusarium proliferatum HBA29	Bark, needles; Shanghai, China	PDB	240 ng/L	Ethyl acetate	ts, dbat, bapt	HPLC-MS	Xiong et al. (2013)
	Taxaceae	Colletotrichum gloeosporioides TA67	Bark, needles; Shanghai, China	PDB	120 ng/L	Ethyl acetate	ts, dbat, bapt	HPLC-MS	Xiong et al. (2013)

Taxus wallichiana Zucc.	Taxaceae	*Sporormia minima*	Stem; Himalayan region of Nepal	PDA	15.7 ng/mL	Methylene chloride	–	TLC, LC-MS	Shrestha et al. (2001)
		Trichothecium sp.	Stem; Himalayan region of Nepal	PDA	165.7 ng/mL	Methylene chloride	–	TLC, LC-MS	Shrestha et al. (2001)
		Pestalotiopsis microspora	Stem; Himalayan region of Nepal	PDA	25.7 ng/mL	Methylene chloride	–	TLC, LC-MS	Shrestha et al. (2001)
	Taxaceae	*Pestalotiopsis microspora*	Inner bark; Himalaya, India	PDA or MID	60–70 µg/L	Methylene chloride	–	MS, NMR	Strobel et al. (1996)
Taxus wallichiana var. *mairei* (Lemée and H. Lév.) L.K. Fu and N. Li	Taxaceae	*Phoma medicaginis*	Bark; Taihang Mountain, Henan Province, China	PDB, spent culture, dry mycelium	1.215 mg/L (PDB), 0.936 mg/L (spent culture), 20 mg/kg (dry mycelium)	Chloroform and methanol (1:1)	–	HPLC, ESI-MS/MS	Zaiyou et al. (2017)
Aegle marmelos Correa ex Roxb	Rutaceae	*Bartalinia robillardoides* AMB-9	Leaves; Chennai, India	MID	187.6 µg/L	Methylene chloride	–	TLC, UV, HPLC	Gangadevi and Muthumary (2008b)
Calotropis gigantea (L.) R. Br.	Aapocynaceae	*Phoma* sp.	Leaves; Chennai, TN, India	MID	76.13 µg	Dichloromethane	–	HPLC, FTIR	Hemamalini et al. (2015)
Capsicum annuum L.	Solanaceae	*Colletotrichum capsici*	Fruit	MID	687 mg/L	Dichloromethane	ts	HPLC	Kumaran et al. (2011)
Cardiospermum halicacabum L.	Sapindaceae	*Pestalotiopsis pauciseta* CHP-11	Leaves	–	113.3 µg/L	Methylene chloride	–	HPLC	Gangadevi et al. (2008)
Citrus medica L.	Rutaceae	*Phyllosticta citricarpa*	Leaves; Chennai, India	MID	265 µg/L	Dichloromethane	–	HPLC, NMR	Kumaran et al. (2008a)
Corchorus olitorius L.		*Grammothele lineata*	Root, stem, leaf, flower, seed; University of Dhaka	PDB	382.2 µg/L	Ethyl acetate	–	TLC, HPLC, FTIR, LC-ESI-MS/MS	Das et al. (2017)
Cupressus sp.	Cupressaceae	*Phyllosticta spinarum*	Needle; Tamil nadu, South India	MID	235 µg/L	Dichloromethane	–	UV, IR,TLC, HPLC	Kumaran et al. (2008b)
Ginkgo biloba L.	Ginkgoaceae	*Phomopsis* sp. BKH 30	leaves; Kangwondo forest, South Korea	PDA	372 µg/L	dichloromethane	–	UV, IR, HPLC, LC-MS, NMR	Kumaran and Hur (2009)
	Ginkgoaceae	*Phoma betae*	Leaves; Konkuk University campus	MID	795 µg/L	15,900-fold	ts	TLC,UV,IR, HPLC, LC-MS, ^{1}H NMR	Kumaran et al. (2012)
Hibiscus rosa-sinensis L.	Malvaceae	*Phyllosticta dioscoreae*	Leaves; Tamil nadu, India	M1D	298 µg/L	Dichloromethane	–	HPLC	Kumaran et al. (2009a)
Justicia gendarussa Burm. f.	Acanthaceae	*Colletotrichum gloeosporioides* JGC-9	Leaves; Chennai, India	PDA	163.4 µg/L	Methylene chloride	–	HPLC	Gangadevi and Muthumary (2008a)

Continued

Table 8.1 List of endophytes producing paclitaxel from different host plant species—cont'd

Host	Family	Endophytic fungi	Isolated from	Paclitaxel produced by the test fungus in medium	Paclitaxel yield	Solvent used for Paclitaxel extraction	Reported genes	Detection methods	Reference/s
Larix leptolepis L.	Pinaceae	*Phomopsis* sp. BKH 35	Leaves; Kangwondo forest, South Korea	PDA	334 µg/L	Dichloromethane	–	UV, IR, HPLC, LC-MS, NMR	Kumaran and Hur (2009)
Michelia champaca L.	Magnoliaceae	*Chaetomium* sp. TBPJ-B	Needles; Chennai, India	PDA	77.23	Dichloromethane	–	HPLC, UV	Rebecca et al. (2012)
Moringa oleifera Lam.	Moringaceae	*Cladosporium oxysporum*	Leaves	M1D	550 µg/L	Dichloromethane	dbat	UV, HPLC, LC-MS, IR, NMR	Gokul Raj et al. (2015)
Morinda citrifolia L.	Rubiaceae	*Lasiodiplodia theobromae*	Leaves	MID	245 µg/L	Dichloromethane		UV, IR, HPLC, NMR, FAB-MS	Pandi et al. (2011)
Plantago major L.	Plantaginaceae	*Alternaria alternata, Colletotrichum gloesporioides, Glomerella cingulata, Nigrospora sphaerica*	–	PDA	5.24–4.4 µg/mL	Ethyl acetate extract	ts, dbat	UV, LC-MS	de Andrade et al. (2018)
Plectranthus amboinicus (Lour.) Spreng.	Lamiaceae	*Pestalotiopsis microspora* EF01	Leaves; Tamil Nadu, India	M1D	204.7 µg/L	Dichloromethane		TLC, UV	Rajendran et al. (2013)
Rhizophora annamalayana	Rhizophoraceae	*Fusarium oxysporum* RAEN-017	Leaves; Vellar estuary, Tamil Nadu, India	PDB	172.3 µg/L	Ethyl acetate	–	IR, TLC, HPLC	Elavarasi et al. (2012)
Salacia oblonga Wall. ex Wight & Arn.	Celastraceae	*Alternaria* spp., *Fusarium* sp., *Botryosphaeria rhodina, Trichoderma longibrachiatum, Lasiodiplodia theobromae, Aspergillus terreus, Armilaria* sp., *Phoma* sp., *Coriolopsis caperata, Phomopsis* sp.	Bark; Karnataka, India (Western ghat)	PDA	–		dbat, bapt	Genomic mining	Roopa et al. (2015)
Tarenna asiatica (L.) Kuntze ex K. Schum.	Rubiaceae	*Aspergillus oryzae*	Leaves; North Tamil nadu, India	M1D	95.04 µg/L	Dichloromethane	–	TLC, LC-MS, FTIR, UV-Vis	Suresh et al. (2020)
Taxodium distichum (L.) Rich.	Cupressaceae	*Pestalotiopsis microspora* CP-4	Outer bark; inland, the central coast, South Carolina	PDA	14–1487 ng/L	Methylene chloride	–	TLC, HPLC, MS	Li et al. (1996)
	Cupressaceae	*Aspergillus fumigatus* TXD105	bark	PDA, M1D, S7, FBM, REF, YPD	84.41 µg/L	Dichloromethane	–	*TLC, UV, HPLC*	Ismaiel et al. (2017)

Plant host	Family	Fungal species	Tissue/Location	Medium	Concentration	Solvent		Methods	References
Terminalia arjuna (Roxb. ex DC.) Wight & Arn.	Combretaceae	Alternaria tenuissima TER995	Bark	PDA, M1D, S7, FBM, REF, YPD	37.92 µg/L	Dichloromethane	–	TLC, UV, HPLC	Ismaiel et al. (2017)
	combretaceae	Alternaria brassicicola	Needles, Thapar Institute of Engineering and Technol-ogy campus, Patiala, India	PDA	140.8 µg/L	Ethyl acetate extract	–	UV, HPLC, FTIR, LC-ESI-MS	Gill and Vasundhara (2019)
	Combretaceae	Chaetomella raphigera TAC-15	Leaves; Chennai, India	MID liquid	79.6 µg/L	Methylene chloride	–	UV, IR, FAB-MS, NMR	Gangadevi and Muthumary (2009a, 2009b)
	Combretaceae	Pestalotiopsis terminaliae	Leaves; Chennai, India	MID	211.1 µg/L	Methylene chloride	–	UV, TLC, HPLC	Gangadevi and Muthumary (2009a, 2009b)
Torreya grandifolia Raf.	Taxaceae	Periconia sp.	Inner bark; Huangshan National, China	MID	30–821 ng/L	Methylene chloride	–	MS, NMR	Li et al. (1998)
Wollemia nobilis W. G. Jones, K. D. Hill and J. M. Allan	Araucariaceae	Pestalotiopsis guepinii W-1f-2	Stem; Wollemi National Park near, Sydney, Australia	MID	485 ng/L	Methylene chloride	–	UV, TLC, EIA	Strobel et al. (1997)
Wrightia tinctoria (Roxb.) R. Br.	Apocyanaceae	Phyllosticta tabernaemontanae	Leaves; Yercaud, Tamilnadu, India	M1D, PDB	M1D (461 µg/L), PDB (150 µg/L)	Dichloromethane	–	HPLC	Kumaran et al. (2009a)

DBAT, 10-deacetylbaccatin III-10-O-acetyl transferase; TS, taxadiene synthase; BAPT, C-13 phenylpropanoid side chain–CoA acyltransferase; ZUV, ultraviolet-visible spectroscopy; IR, infrared spectroscopy; MS, Mass spectrometry; TLC, thin layer chromatography; MPLC, medium-pressure liquid chromatography; HPLC, high-performance liquid chromatography; LC-MS, liquid chromatography; ESI-MS, electrospray ionization mass spectrometry; LC-MS/MS, liquid chromatography-tandem mass spectrometry; EIMS, electron impact mass spectrometry; FAB-MS, fast atom bombardment mass spectrometry; NMR, nuclear magnetic resonance spectroscopy; EIA, competitive inhibition enzyme immunoassay; PDA, potato dextrose agarose; PDB, potato-dextrose broth; YPD, yeast peptone dextrose; REF, reformativemedium; FBM, flask basal medium; MEA, malt extract agar; M1D, reformed liquid medium.

can also be produced from endophytes of Acanthaceae, Araucariaceae, Combretaceae, Cupressaceae, Ginkgoaceae, Lamiaceae, Magnoliaceae, Malvaceae, Moraceae, Pinaceae, Plantaginaceae, Podocarpaceae, Rhizophoraceae, Rubiaceae, Rutaceae, and Sapindaceae, Solanaceae. Most of the fungi from various families were isolated from leaves (28) and bark (24).

Taxonomically (Fig. 8.1), nearly all genera belong to Ascomycotina (90%), involving Sordariomycetes (44%), Dothideomycetes (33%), Eurotiomycetes (9%), and Leotiomycetes (1%) others belong to Basidiomycota (4%) involving Agaricomycetes, Zygomycota (1.5%) involving Mucoromycetes, and Deuteromycotina (3%). The genera contained two or more species of paclitaxel producing endophytic fungi, namely, *Alternaria, Aspergillus, Pestalotiopsis, Fusarium, Phomposis, Colletotrichum, Guignardia, Phyllosticta, Trichothecium.* These endophytic fungi could be used as a natural alternative source for the production of anticancer drug paclitaxel. Yields of paclitaxel by these fungi can be increased by genetics or biotechnological manipulation.

For the extraction of paclitaxel, various organic solvents were used, such as ethyl acetate, dichloromethane, and methyl alcohol. Screening of endophytic fungus derived paclitaxel determines by thin-layer chromatography (TLC), High performance liquid chromatography (HPLC), High Performance Liquid chromatography/Liquid chromatography-mass spectrometry (HPLC/LC-MS), ultraviolet (UV) immunity analysis (Kumar et al., 2019), Liquid chromatography electrospray ionization tandem mass spectrometric (LC-ESI-MS) (Caruso et al., 2000), Nuclear magnetic resonance (NMR), and so on. Among them, HPLC still the most popular method for determining the

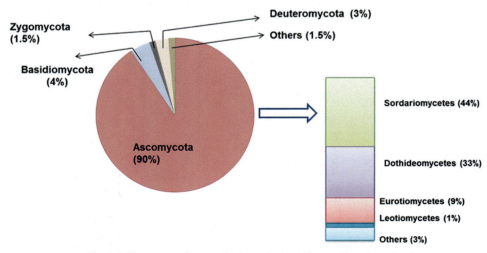

Fig. 8.1 Taxonomy of the endophytes isolated for paclitaxel production.

concentration and screening of paclitaxel (Deng et al., 2009; Gangadevi et al., 2008; Gangadevi and Muthumary, 2008a; Kumaran et al., 2009a; Mirjalili et al., 2012; Somjaipeng et al., 2015). *Aspergillus fumigatus, A. terreus, A. aculeatinus, A. niger, A. oryzae* were reported to be paclitaxel producers (Kumar et al., 2019). Kumar et al. (2019) found that *A. fumigatus* KU-837249 is a paclitaxel producer that hyper-produced 1.60 g/L of paclitaxel on S7 medium. Previous studies also supported paclitaxel production on the S7 medium (Garyali et al., 2013; Stierle et al., 1993). Other species *A. niger* (Zhao et al., 2009) and *A. aculeatinus* (Qiao et al., 2017) produced paclitaxel on PDA media and malt extract agar (MEA) media, respectively. *A. aculeatinus* producing relatively high original paclitaxel and also showed quick accumulation of paclitaxel that indicate the Tax-6 strain of *A. aculeatinus* could be a realistic choice of paclitaxel commercial production (Qiao et al., 2017). One recent study by Suresh et al. reported paclitaxel yield of 95.04 µg/L with *A. oryzae* from leaves of non-*Taxus* source *Tarenna asiatica* in synthetic reformed liquid medium (M1D) (Suresh et al., 2020). *Alternaria* sp., *Beauveria* sp., *Epicoccum* sp., *Fusarium* sp., *Mycelia sterilia, Stemphylium sedicola, Paraconiothyrium variabile, Cladosporium* sp. endophytes were isolated from *Taxus baccata*, which produced paclitaxel (Caruso et al., 2000; Mirjalili et al., 2012; Somjaipeng et al., 2015; Zaiyou et al., 2017). Among all the reports, *Paraconiothyrium variabile* fungus is the most promising one which yield the highest amount (1.75 µg/L) of paclitaxel from *T. baccata* on the M1D medium modified by adding glycerol, sorbitol, glucose, NaCl, and KCl. However, *Epicoccum nigrum* fungi isolated in the same study on the same media, but paclitaxel yield was low at about 1.32 µg/L (Somjaipeng et al., 2015). By the genomic DNA extraction from isolated *Stemphylium sedicola* fungi and PCR analysis, the presence of *Taxus* taxadiene synthase (ts) gene detected, that encodes the enzyme catalyzing the first committed step of taxol biosynthesis (Mirjalili et al., 2012). Zaiyou et al. (2017) observed paclitaxel content of *Phoma medicaginis* isolated from the bark of *T. wallichiana* var. *mairei* on whole PDB culture, spent culture, and dry mycelia culture, and resulted in paclitaxel content in the spent culture medium is the close to that in the whole culture, hence, paclitaxel synthesized in hypha is secreted into the culture medium. Li et al. (2008) reported paclitaxel yield of 131 mg/L with *Fusarium arthrosporioides* from the inner bark of *T. cuspida* in the PDA medium. The highest paclitaxel productivities were obtained in several previous studies using the MID medium composition, 461 µg/L paclitaxel from *Phyllosticta tabernaemontanae* (Kumaran et al., 2009b), 485 ng/L paclitaxel from *Pestalotiopsis guepinii* W-1f-2 (Strobel et al., 1997), 30–821 ng/L paclitaxel from *Periconia* sp. (Li et al., 1998), 79.6 µg/L paclitaxel from *Chaetomella raphigera* TAC-15 (Gangadevi and Muthumary, 2009a, b), and 211.1 µg/L paclitaxel from *Pestalotiopsis terminaliae*.

Phyllosticta tabernaemontanae is an produced 9.2×10^3 fold more paclitaxel than earlier reported fungus, *Taxomyces andreanae*, hence, *P. tabernaemontanae* is a promising candidate for paclitaxel production (Kumaran et al., 2009b). Ismaiel et al. Observed paclitaxel content in A. fumigatus *TXD105* and *Alternaria tenuissima TER995* on six types of media PDA, M1D, S7, FBM, REF, YPD, among the tried media M1D and FBM were found to be the most conductive fermentation medium for the production (Ismaiel et al., 2017).

8.5 Anticancer properties of endophytes-derived paclitaxel

Paclitaxel, as a microtubule-stabilizing agent induces cell cycle arrest and leads to apoptosis of the cell. The microtubule maintains the cytoskeleton structure and shape of the cell. β-tubulin is a subunit of the microtubule, binds to the specific site of paclitaxel and prevent depolymerization of paclitaxel (Schiff and Horwitz, 1980). Microtubule stabilization by paclitaxel induces G2/M-phase mitotic arrest (Jordan and Wilson, 2004; Sherline and Schiavone, 1977; Snyder et al., 2001), and finally undergoes towards apoptosis (Schiff et al., 1979).

Endophytic fungi-derived paclitaxel inhibit cell proliferation at low concentration by blocking mitosis. Kumaran et al. (2008a, b, 2009b, 2011) Gangadevi et al. (2008), and many other research findings support this (Table 8.2). They revealed application of low concentration 0.005 to 0.05 µmol/L of paclitaxel to cancer cells increases cell death by stabilizing the spindle fiber during mitosis, however increasing concentration of paclitaxel to cancer cell gradually decreased cell death. They showed apoptotic activity of fungal paclitaxel in the colon (H116), intestine (Int4070), lung (HL251), leukemia (HLK210), and breast (MCF-7, BT220) cancer cell lines (Gangadevi et al., 2008; Kumaran et al., 2009a, 2011). The cytotoxicity of paclitaxel was also assessed against BT 220, H116, Int 407, HL 251 and HLK 210 cells. Paclitaxel induced cell death with 0.005 to 0.05 µmol/L concentration. However, further increase in paclitaxel concentration from 0.5 µmol/L to 5 µmol/L cell death rate decreased significantly (Gangadevi et al., 2008). Pandi et al. (2011) observed fungal paclitaxel effects depend on incubation time and specificity of cancer cells by cell viability assay and revealed the increase in paclitaxel concentration from 300 to 600 µg significantly induced cell death in MCF-7 cell lines by apoptosis (Pandi et al., 2011). This study contrast with Ruckdeschel colleague's study in 1997, which stated paclitaxel-induced cell apoptosis with particular concentrations but higher incubation time (Ruckdeschel et al., 1997).

Table 8.2 List of in vitro study of endophytes derived paclitaxel anticancer properties.

Endophytic fungi	Host plant	Cell lines	Applied paclitaxel concentration	Incubation time, cytotoxicity assay	Outcome of in vitro study	Ic_{50}/ID_{50}	Reference/s
Tubercularia sp. TF5	Taxus mairei (Lemée & H. Lév.) S.Y.Hu	KB, P388	–	MTT assay	Induce tubulin polymerization	ID_{50} (infectious dose): 1/2400 (KB), 1/38400 (P388)	Wang et al. (2000)
Bartalinia robillardoides	Aegle marmelos (L.) Corrêa	BT220, H116, Int407, HLK 210, HL251	0.0005–5 μM	24, 48, 72 h; phosphatidylserine detection assay	Condensed, segmented nuclei degraded after 72 h, cells block in G2/M phase (0.005 μM)	–	Gangadevi and Muthumary (2007)
Bartalinia robil-lardoides	Aegle marmelos (L.) Corrêa	BT 220, H116, Int 407, HL 251, HLK 210	0.005–5 μM	24 h, 48 h, 72 h; apoptotic assay	Morphological changes of the cancer cells which	–	Gangadevi and Muthumary (2008b)
Pestalotiopsis pauciseta CHP-11	Cardiospermum halicacabum L.	BT 220, H116, Int 407, HL 251 and HLK 210	0.005–0.5 μmol/L	apoptotic assay	0.005–0.05 μmol/L: cell death increased 0.05–0.5 μmol/: cell death increased slightly 0.5–5 μmol/L: cell death decreased	–	Gangadevi et al. (2008)
Phyllosticta spinarum	Cupressus sp.	Int 407, HLK 210, BT220, HL 251	0.005 to 5 mM	48 h; apoptotic assay	0.005–0.05 μmol/L: cell death increased 0.05–0.5 μmol/: cell death increased slightly 0.5–5 μmol/L: cell death decreased		Senthil Kumaran et al. (2008)
Mucor rouxianus	Taxus chinensis (Pilg.) Rehder	liver carcinoma 7402	–	24 h	ED50 for the fungal taxol was $5.2 \pm 1.6 \times 10{-}3$ mg/mL		Miao et al. (2009)
Phyllosticta tabernaemontanae	Wrightia tinctoria (Roxb.) R. Br.	HLK210, H116, Int407, HL251, and BT220	0.005–0.5 μmol/L	48 h; apoptotic assay	0.005–0.05 μmol/L: cell death increased 0.05–0.5 μmol/: cell death increased slightly 0.5–5 μmol/L: cell death decreased	–	Kumaran et al. (2009b)
Lasiodiplodia theobromae	Morinda citrifolia L.	MCF-7	100 to 600 μg/ml	24 h, 48 h, 72 h; MTT assay	Cell viability decrease,	IC_{50}: 300 μg	Pandi et al. (2011)
Colletotrichum capsici	Capsicum annuum L.	MCF-7, HL 251 and HLK 210	0.005–0.5 μmol/L	48 h; apoptotic assay	0.005–0.05 μmol/L: cell death increased 0.05–0.5 μmol/L: cell death increased slightly 0.5–5 μmol/L: cell death decreased		Kumaran et al. (2011)

Continued

Endophytic fungi	Host plant	Cell lines	Applied paclitaxel concentration	Incubation time, cytotoxicity assay	Outcome of in vitro study	IC50/ID50	Reference/s
Didymostilbe sp. DF110	Taxus chinensis var. mairei	BEL7402	50 µg/mL	48 h	little cytotoxicity	–	Wang and Tang (2011)
Phoma betae	Ginkgo biloba L.	MCF-7, ATCC HTB-22, A549, ATCC CCL-185, T98G, ATTCC CRL-1690	0.005 to 0.05 µM	48 h, apoptotic assay	0.005–0.05 µmol/L: cell death increased 0.5–5 µmol/L: cell death decreased	–	Kumaran et al. (2012)
Pestalotiopsis pauciseta VM1	Tabebuia pentaphylla Hemsl.	MCF-7	100 µg to 700 µg	24 h, 48 h, 72 h; MTT assay	Shrink and spherical in shape	IC50: 350 µg/mL	Vennila et al. (2012)
Fusurium solani	Taxus celebica (Warb.) H.L. Li	HeLa, HepG2, Jurkat, Ovcar3, T47D	10 nM (T47D, JR4-Jurkat, HeLa), 100 nM (HepG2, Ovcar3)	24 h, 48 h; flow cytometry	DNA fragmentation (6 nM); Bcl2-overexpressing in J16 Jurkat cells, loss in mitochondrial membrane potential, activation of caspase-10 in JR4-Jurkat cells	IC50: 0.005 to 0.2 µM (JR4-Jurkat cells)	Chakravarthi et al. (2013)
Pestalotiopsis microspora EF01	Plectranthus amboinicus (Lour.) Spreng.	Hep G2	0.005 to 0.05 µM	MTT assay	–	–	Rajendran et al. (2013)
Cladosporium oxysporum	Moringa oleifera Lam.	HCT 15	1–7.5 µM	24 h; MTT assay	Cell shape become shrink and spherical	IC50: 3.5 µM	Gokul Raj et al. (2015)
Phoma sp.	Calotropis gigantea (L.) Dryand.	MCF 7	100 µg/mL	24 h; MTT assay	–	IC50: 50 µg/mL	Hemamalini et al. (2015)
Aspergillus fumigatus, Alternaria tenuissima	Taxodium distichum (L.) Rich., Terminalia arjuna (Roxb. ex DC.) Wight & Arn.	HepG-2, MCF-7 A-549 HEp-2, CHO-K1	0.39 µg/mL 0.78 mL^{-1} 1.56 µg/mL	24 h, MTT assay	Increasing paclitaxel concentration decrease cell proliferation	IC50: HEp-2(14.80 µg/mL); A-549 (3.04 µg/mL), MCF-7 (5.48 µg/mL), HepG-2 (5.78 µg/mL), CHO-K1 (11.80 µg/mL)	Ismaiel et al. (2017)
Grammothele lineata	Corchorus olitorius L.	HeLa	0.005 µM	24 h	35% cell death (0.005 µM)	–	Das et al. (2017)
Aspergillus oryzae	Tarenna asiatica (L.) Kuntze ex K.Schum.	NCI-H460	30 µg/mL	6, 12, 24, 48 h; MTT assay	Shrinkage of cancer cell, octagonal cells that transformed to sphere-shaped cells	IC50: 50 µg/mL	Suresh et al. (2020)

BT220, MCF-7, ATCC, HTB-22 (human breast cancer cell line), H116 (human colon cancer cell line), cell Int407 (human intestine cancer cell line), HL251, A549, ATCC CCL-185 (human lung cancer cell line), HLK 210 (human leukemia cell line), HeLa (human cervical carcinoma cell line), HepG2 (human liver carcinoma cell line): HepG2, Jurkat-JR4 (human leukemia T cell line), Ovcar-3 (human ovarian carcinoma cell line), T47D (human ductal breast epithelial tumor cell line), Jurkat-JR16 (Jurkat cells overexpressing Bcl2). glioblastoma cells (T98G, ATTCC CRL-1690).

Pestalotiopsis microspora EF01 endophytic fungi isolated from leaves of *Plectranthus amboinicus*, screened for production of paclitaxel, and resulted in 204.7 µg/L yield. This fungal derived paclitaxel showed cytotoxic activity in Hep G2 (human liver carcinoma) cell line (Rajendran et al., 2013). Vennila et al. (2012) analyzing result in their study that the effects of *Pestalotiopsis pauciseta* derived paclitaxel on MCF-7 breast cancer cell proliferation and showed fungal paclitaxel has very high in vitro cytotoxic effects on MCF-7 cell with IC_{50} 350 µg/mL in 48 h treatment. Fungal paclitaxel also changed cells shape polygonal to spherical and inhibited the development of consistent spindles in the metaphase stage ultimately lead to G2/M phase cell cycle arrest and apoptotic cell death (Vennila et al., 2012). Paclitaxel extracted from the *Phoma* sp. showed strong anti-proliferative activity with 100–1000 µg/mL concentration against human breast cancer (MCF 7) cell line detected by the MTT assay. 500 µg/mL concentration of fungal paclitaxel showed 50% cell viability with 48 h incubation (Hemamalini et al., 2015).

A. fumigatus, A. tenuissima, from *Taxodium distichum, Terminalia arjuna* producing paclitaxel, was identified based on the morphology of the fungal culture, mechanism of spores production, characteristics of the spores and molecular characterization. The fungal paclitaxel showed strong activity against human cancer cell lines (HepG-2, MCF-7, A-549, HEp-2, CHO-K1) by MTT assay (Ismaiel et al., 2017). The fungal paclitaxel inhibits HCT 15 cell growth with higher concentrations by 24 h treatment. They revealed that *Cladosporium oxysporum* fungus derived paclitaxel can be used as an alternative source of paclitaxel production (Gokul Raj et al., 2015). DNA fragmentation, mitochondrial membrane potential loss, activation of caspase-10 and apoptosis induced, by application of *Fusarium solani* derived paclitaxel from *T. celebica* (Chakravarthi et al., 2013). Fungal paclitaxel from *A. oryzae* treatment for 24 h and 48 h in NCI-H460 cells displayed alterations in the structure of cells and cells become sphere-shaped. This fungal derived paclitaxel significantly showed in vitro cytotoxic activity and induced apoptosis with 30 µg/mL concentration (Suresh et al., 2020).

Fungal paclitaxel treatment decreased the activity of matrix metalloproteinase (MMP9) which is known as tumor micro-environment regulators and initiates cancer progression (Abbas et al., 2017). Fungal derived paclitaxel treatment also increased caspase 3 activity by 2.5 folds that lead to initiate the cytotoxic activity in mice infected with EAC (Ehrlich ascites carcinoma) cells (Zein et al., 2019). Pandi et al. (2010), observed paclitaxel activity in Sprague dawley rats breast tissue. They treated (DMBA)-induced mammary glands of the rat with fungal paclitaxel (*Botryodiplodia theobromae*) isolated from Morinda citrifolia and showed increase in antioxidants (Superoxide dismutase:

SOD, catalase: CAT and Glutathione peroxidases: GPx) as compared to the control group. The result also revealed fungal paclitaxel increased levels of non-enzymic antioxidants (vitamin C, vitamin E and GSH) (Pandi et al., 2010). SOD is an intracellular enzyme and tumor suppressor protein that breaks down O_2 to H_2O_2 into cell and prevents tissue damage (Weydert et al., 2006). Catalase enzyme decomposes hydrogen peroxide (H_2O_2) to water and oxygen. GPx antioxidant reacts with H_2O_2 and prevents intracellular damage. Antioxidant (SOD and CAT) increases by paclitaxel in cancer cells blocks ROS. ROS have cell proliferative effects (Simon et al., 2000). COX-2 (cyclooxygenase-2) is an enzyme that promotes cancer cell growth (Pang et al., 2016). Fungal paclitaxel significantly decreases COX-2 in Sprague dawley rat and sensitize cancer cells to apoptosis (Pandi et al., 2010).

8.6 Conclusions

Plant endophytic fungi have been showing its importance for the isolation of novel natural bioactive products. This chapter mainly demonstrates the research progress on endophytes derived paclitaxel production and roles in cancer cure between 1993 and 2020. Paclitaxel has been already proven for its anticancer activity. All the endophytes species producing paclitaxel are taxonomically classified and the genera of *Aspergillus, Altenaria, Pestalotiopsis,* and *Paraconiothyrium* are the most promising candidate for the paclitaxel production. Most of the species of endophytic fungi producing paclitaxel belong to Ascomycota (90%), others are Basidiomycota (4%), Deuteromycota (3%), and Zygomycota (1.5%). The main source of paclitaxel is the *Taxus* species, apart from that *Podocarpus gracilior, Corylus mandshurica,* and *Corylus avellana* are natural sources of paclitaxel. Several biotechnological approaches are tried for paclitaxel production, however because of a large number of reaction steps, long incubation time, and low yields that limit its practicality. For the first time, the endophyte, *T. andeanae* was isolated from *T. brevifolia* for the production of paclitaxel in 1993. After that discovery, scientists were taking interest in endophytes derived paclitaxel production at lower cost, and yet a higher yield. Endophytic fungi derived paclitaxel proven its effectively in respective cancer cell lines (BT 220, H116, Int 407, MCF-7, HL 251 and HLK 210). Concentrations of paclitaxel, the incubation time of paclitaxel with cell, cell specificity are some factors that influence paclitaxel mediated apoptosis of cancer cells. Exceedingly, in vitro work reported on the anti-cancer activities of paclitaxel from endophytic fungi, only 2 in vivo animal studies have been completed so far (Pandi et al., 2010). Human cancer-related clinical trials need to be done in the future. However, new strategies of cultivation of endophytic fungi and isolation of paclitaxel will provide a promising chance to summons the supply of drugs for cancer that stops cancer cell growth.

References

Abbas, N.F., Shabana, M.E., Habib, F.M., Soliman, A.A., 2017. Histopathological and immunohistochemical study of matrix metalloproteinase-2 and matrix metalloproteinase-9 in breast carcinoma. J. Arab Soc. Med. Res. 12, 6–12.

Alurappa, R., Chowdappa, S., Narayanaswamy, R., Sinniah, U.R., Mohanty, S.K., Swamy, M.K., 2018. Endophytic fungi and bioactive metabolites production: An update. In: Patra, J.K., Das, G., Shin, H.-S. (Eds.), Microbial Biotechnology. Springer, Singapore, pp. 455–482.

de Andrade, H.F., de Araújo, L.C.A., dos Santos, B.S., Paiva, P.M.G., Napoleão, T.H., dos Santos Correia, M.T., de Oliveira, M.B.M., de Lima, G.M.S., Ximenes, R.M., da Silva, T.D., da Silva, G.R., da Silva, M.V., 2018. Screening of endophytic fungi stored in a culture collection for taxol production. Braz. J. Microbiol. 49, 59–63.

Bala, S., Uniyal, G.C., Chattopadhyay, S.K., Tripathi, V., Sashidhara, K.V., Kulshrestha, M., Sharma, R.P., Jain, S.P., Kukreja, A.K., Kumar, S., 1999. Analysis of taxol and major taxoids in Himalayan yew, *Taxus wallichiana*. J. Chromatogr. A 858, 239–244.

Bary, A., 1866. Morphologie und Physiologie der Pilze, Flechten und Myxomyceten. W. Engelmann.

Bestoso, F., Ottaggio, L., Armirotti, A., Balbi, A., Damonte, G., Degan, P., Mazzei, M., Cavalli, F., Ledda, B., Miele, M., 2006. In vitro cell cultures obtained from different explants of *Corylus avellana* produce Taxol and taxanes. BMC Biotechnol. 6, 1–11.

BoLin, L., QunHua, Z., Shan, L., RiQiang, C., 1995. Preliminary study of cell callus culture and taxol production of several *Taxus* speicies. J. Nanjing Univ. Nat. Sci. 31 (3), 424–429.

Caruso, M., Colombo, A.L., Fedeli, L., Pavesi, A., Quaroni, S., Saracchi, M., Ventrella, G., 2000. Isolation of endophytic fungi and actinomycetes taxane producers. Ann. Microbiol. 50, 3–14.

Chakravarthi, P.B.V.S.K., Das, P., Surendranath, K., Karande, A.A., Jayabaskaran, C., 2008. Production of paclitaxel by *Fusarium solani* isolated from *Taxus celebica*. J. Biosci. 33, 259–267.

Chakravarthi, B.V.S.K., Sujay, R., Kuriakose, G.C., Karande, A.A., Jayabaskaran, C., 2013. Inhibition of cancer cell proliferation and apoptosis-inducing activity of fungal taxol and its precursor baccatin III purified from endophytic *Fusarium solani*. Cancer Cell Int. 13, 1–11.

Choi, H.-K., Adams, T.L., Stahlhut, R.W., Kim, S.-I., Yun, J.-H., Song, B.-K., Kim, J.-H., Song, J.-S., Hong, S.-S., Lee, H.-S., 1999. Method for Mass Production of Taxol by Sem-Continuous Culture with *Taxus chinensis* Cell Culture. U.S. Pat. 5,871,979.

Croteau, R., Ketchum, R.E.B., Long, R.M., Kaspera, R., Wildung, M.R., 2006. Taxol biosynthesis and molecular genetics. Phytochem. Rev. 5, 75–97.

Danishefsky, S.J., Masters, J.J., Young, W.B., Link, J.T., Snyder, L.B., Magee, T.V., Jung, D.K., Isaacs, R.C.A., Bornmann, W.G., Alaimo, C.A., Coburn, C.A., Grandi, M.J.D., 1996. Total synthesis of baccatin III and taxol. J. Am. Chem. Soc. 118, 2843–2859. https://doi.org/10.1021/ja952692a.

Das, A., Rahman, M.I., Ferdous, A.S., Amin, A., Rahman, M.M., Nahar, N., Uddin, M.A., Islam, M.R., Khan, H., 2017. An endophytic basidiomycete, *Grammothele lineata*, isolated from *Corchorus olitorius*, produces paclitaxel that shows cytotoxicity. PLoS One 12, 1–17.

Debbab, A., Aly, A.H., Edrada-Ebel, R.A., Wray, V., Müller, W.E.G., Totzke, F., Zirrgiebel, U., Schächtele, C., Kubbutat, M.H.G., Wen, H.L., Mosaddak, M., Hakiki, A., Proksch, P., Ebel, R., 2009. Bioactive metabolites from the endophytic fungus *Stemphylium globuliferum* isolated from *Mentha pulegium*. J. Nat. Prod. 72, 626–631.

Demain, A.L., Vaishnav, P., 2011. 2011. Natural products for cancer chemotherapy. Microb. Biotechnol. 4 (6), 687–699.

Deng, B.W., Liu, K.H., Chen, W.Q., Ding, X.W., Xie, X.C., 2009. *Fusarium solani*, Tax-3, a new endophytic taxol-producing fungus from *Taxus chinensis*. World J. Microbiol. Biotechnol. 25, 139–143.

Denis, J.N., Greene, A.E., Guenard, D., Gueritte-Voegelein, F., Mangatal, L., Potier, P., 1988. Highly efficient, practical approach to natural taxol. J. Am. Chem. Soc. 110, 5917–5919. https://doi.org/10.1021/ja00225a063.

Denis, J.N., Greene, A.E., Serra, A.A., Luche, M.J., 1986. An efficient, enantioselective synthesis of the taxol side chain. J. Org. Chem. 51, 46–50. https://doi.org/10.1021/jo00351a008.

Doi, T., Fuse, S., Miyamoto, S., Nakai, K., Sasuga, D., Takahashi, T., 2006. A formal total synthesis of taxol aided by an automated synthesizer. Chem. Asian J. 1, 370–383. https://doi.org/10.1002/asia.200600156.

Elavarasi, A., Rathna, G.S., Kalaiselvam, M., 2012. Taxol producing mangrove endophytic fungi *Fusarium oxysporum* from *Rhizophora annamalayana*. Asian Pac. J. Trop. Biomed. 2, S1081–S1085.

ElSohly, H.N., Croom, E.M., Kopycki, W.J., Joshi, A.S., ElSohly, M.A., McChesney, J.D., 1995. Concentrations of taxol and related taxanes in the needles of different *Taxus* cultivars. Phytochem. Anal. 6, 149–156. https://doi.org/10.1002/pca.2800060307.

Fang, W., Wu, Y., Zhou, J., Chen, W., Fang, Q., 1993. Qualitative and quantitative determination of taxol and related compounds in *Taxus cuspidata* Sieb et Zucc. Phytochem. Anal. 4, 115–119.

Frense, D., 2007. Taxanes: perspectives for biotechnological production. Appl. Microbiol. Biotechnol. 73, 1233–1240. https://doi.org/10.1007/s00253-006-0711-0.

Gangadevi, V., Murugan, M., Muthumary, J., 2008. Taxol determination from *Pestalotiopsis pauciseta*, a fungal endophyte of a medicinal plant. Shengwu Gongcheng Xuebao 24, 1433–1438.

Gangadevi, V., Muthumary, J., 2007. Preliminary studies on cytotoxic effect of fungal taxol on cancer cell lines. J. Biotechnol. 6, 1382–1386.

Gangadevi, V., Muthumary, J., 2008a. Isolation of *Colletotrichum gloeosporioides*, a novel endophytic taxol-producing fungus from the leaves of a medicinal plant, *Justicia gendarussa*. Mycol. Balc. 5, 1–4.

Gangadevi, V., Muthumary, J., 2008b. Taxol, an anticancer drug produced by an endophytic fungus *Bartalinia robillardoides* Tassi, isolated from a medicinal plant, *Aegle marmelos* Correa ex Roxb. World J. Microbiol. Biotechnol. 24, 717–724.

Gangadevi, V., Muthumary, J., 2009a. A novel endophytic taxol-producing fungus *Chaetomella raphigera* isolated from a medicinal plant, *Terminalia arjuna*. Appl. Biochem. Biotechnol. 158, 675–684.

Gangadevi, V., Muthumary, J., 2009b. Taxol production by *Pestalotiopsis terminaliae*, an endophytic fungus of *Terminalia arjuna* (arjun tree). Biotechnol. Appl. Biochem. 52, 9–15.

Garyali, S., Kumar, A., Reddy, M.S., 2013. Taxol production by an endophytic fungus, *Fusarium redolens*, isolated from himalayan yew. J. Microbiol. Biotechnol. 23, 1372–1380.

Gill, H., Vasundhara, M., 2019. Isolation of taxol producing endophytic fungus *Alternaria brassicicola* from non-*Taxus* medicinal plant *Terminalia arjuna*. World J. Microbiol. Biotechnol. 35, 1–8.

Gokul Raj, K., Manikandan, R., Arulvasu, C., Pandi, M., 2015. Anti-proliferative effect of fungal taxol extracted from *Cladosporium oxysporum* against human pathogenic bacteria and human colon cancer cell line HCT 15. Spectrochim. Acta A Mol. Biomol. Spectrosc. 138, 667–674.

Guo, B.H., Wang, Y.C., Zhou, X.W., Hu, K., Tan, F., Miao, Z.Q., Tang, K.X., 2006. An endophytic Taxol-producing fungus BT2 isolated from *Taxus chinensis* var. *mairei*. Afr. J. Biotechnol. 5, 875–877.

Hao, X., Pan, J., Zhu, X., 2013. Taxol producing Fungi. In: Ramawat, K., Mérillon, J.M. (Eds.), Natural Products. Springer, Berlin, https://doi.org/10.1007/978-3-642-22144-6_124.

Hemamalini, V., Mukesh Kumar, D.J., Immaculate, A., Rebecca, N., Srimathi, S., Muthumary, J., Kalaichelvan, P.T., 2015. Isolation and characterization of taxol producing endophytic *Phoma* sp. from *Calotropis gigantea* and its anti-proliferative studies. J. Acad. Ind. Res. 3, 645.

Hirasuna, T.J., Pestchanker, L.J., Srinivasan, V., Shuler, M.L., 1996. Taxol production in suspension cultures of *Taxus baccata*. Plant Cell Tissue Organ Cult. 44, 95–102.

Hoffman, A., Shahidi, F., 2009. Paclitaxel and other taxanes in hazelnut. J. Funct. Foods 1, 33–37.

Holton, R.A., Biediger, R.J., Boatman, P.D., 1995. Semisynthesis of taxol and taxotere. In: Suffness, M. (Ed.), Taxol: Science and Applications. CRC Press, Boca Raton, pp. 3–25.

Holton, R.A., Kim, H.B., Somoza, C., Liang, F., Biediger, R.J., Boatman, P.D., Shindo, M., Smith, C.C., Kim, S., Nadizadeh, H., Suzuki, Y., Tao, C., Vu, P., Tang, S., Zhang, P., Murthi, K.K., Gentile, L.N., Liu, J.H., 1994b. First total synthesis of taxol. 2. Completion of the C and D rings. J. Am. Chem. Soc. 116, 1599–1600.

Holton, R.A., Somoza, C., Kim, H.B., Liang, F., Biediger, R.J., Boatman, P.D., Shindo, M., Smith, C.C., Kim, S., Nadizadeh, H., Suzuki, Y., Tao, C., Vu, P., Tang, S., Zhang, P., Murthi, K.K., Gentile, L.N., Liu, J.H., 1994a. First total synthesis of taxol. 1. Functionalization of the B ring. J. Am. Chem. Soc. 116, 1597–1598. https://doi.org/10.1021/ja00083a066.

Ibrahim, M., Sieber, T.N., Schlegel, M., 2017. Communities of fungal endophytes in leaves of *Fraxinus ornus* are highly diverse. Fungal Ecol. 29, 10–19.

Ismaiel, A.A., Ahmed, A.S., Hassan, I.A., 2017. Production of paclitaxel with anticancer activity by two local fungal endophytes, *Aspergillus fumigatus* and *Alternaria tenuissima*. Appl. Microbiol. Biotechnol. 101 (14), 5831–5846.

Jalgaonwala, R.E., Mohite, B.V., Mahajan, R.T., 2011. A review: natural products from plant associated endophytic fungi. J. Microbiol. Biotechnol. Res. 1, 21–32.

Jia, M., Chen, L., Xin, H.L., Zheng, C.J., Rahman, K., Han, T., Qin, L.P., 2016. A friendly relationship between endophytic fungi and medicinal plants: a systematic review. Front. Microbiol. 7, 906.

Jordan, M.A., Wilson, L., 2004. Microtubules as a target for. Nat. Rev. Cancer 4, 253–265.

Kasaei, A., Mobini-Dehkordi, M., Mahjoubi, F., Saffar, B., 2017. Isolation of taxol-producing endophytic fungi from Iranian yew through novel molecular approach and their effects on human breast cancer cell line. Curr. Microbiol. 74, 702–709.

Ketchum, R.E.B., Gibson, D.M., 1996. Paclitaxel production in suspension cell cultures of *Taxus*. Plant Cell Tissue Org. Cult. 46, 9–16. https://doi.org/10.1007/BF00039691.

Khosroushahi, A.Y., Valizadeh, M., Ghasempour, A., Khosrowshahli, M., Naghdibadi, H., Dadpour, M.R., Omidi, Y., 2006. Improved taxol production by combination of inducing factors in suspension cell culture of *Taxus baccata*. Cell Biol. Int. 30, 262–269. https://doi.org/10.1016/j.cellbi.2005.11.004.

Kumar, P., Singh, B., Thakur, V., Thakur, A., Thakur, N., Pandey, D., Chand, D., 2019. Hyper-production of taxol from *Aspergillus fumigatus*, an endophytic fungus isolated from *Taxus* sp. of the Northern Himalayan region. Biotechnol. Rep. 24, e00395.

Kumaran, R.S., Choi, Y.K., Lee, S., Jeon, H.J., Jung, H., Kim, H.J., 2012. Isolation of taxol, an anticancer drug produced by the endophytic fungus, *Phoma betae*. Afr. J. Biotechnol. 11, 950–960. https://doi.org/10.5897/AJB11.1937.

Kumaran, R.S., Hur, B., 2009. Screening of species of the endophytic fungus *Phomopsis* for the production of the anticancer drug taxol. Biotechnol. Appl. Biochem. 54, 21–30.

Kumaran, R.S., Jung, H., Kim, H.J., 2011. In vitro screening of taxol, an anticancer drug produced by the fungus, *Colletotrichum capsici*. Eng. Life Sci. 11, 264–271.

Kumaran, R.S., Muthumary, J., Hur, B.K., 2008a. Production of taxol from *Phyllosticta spinarum*, an endophytic fungus of *Cupressus* sp. Eng. Life Sci. 8, 438–446.

Kumaran, R.S., Muthumary, J., Hur, B.K., 2008b. Taxol from *Phyllosticta citricarpa*, a leaf spot fungus of the angiosperm *Citrus medica*. J. Biosci. Bioeng. 106, 103–106.

Kumaran, R.S., Muthumary, J., Hur, B.K., 2009b. Isolation and identification of an anticancer drug, taxol from *Phyllosticta tabernaemontanae*, a leaf spot fungus of an angiosperm, *Wrightia tinctoria*. J. Microbiol. 47, 40–49.

Kumaran, R.S., Muthumary, J., Kim, E.K., Hur, B.K., 2009a. Production of taxol from *Phyllosticta dioscoreae*, a leaf spot fungus isolated from *Hibiscus rosa-sinensis*. Biotechnol. Bioprocess Eng. 14, 76–83.

Kusari, S., Spiteller, M., 2011. Are we ready for industrial production of bioactive plant secondary metabolites utilizing endophytes? Nat. Prod. Rep. 28, 1203–1207.

Li, C.T., Li, Y., Wang, Q.J., Sung, C.K., 2008. Taxol production by *Fusarium arthrosporioides* isolated from yew, *Taxus cuspidata*. J. Med. Biochem. 27, 454–458.

Li, J.Y., Sidhu, R.S., Ford, E.J., Long, D.M., Hess, W.M., Strobel, G.A., 1998. The induction of taxol production in the endophytic fungus—*Periconia* sp. from *Torreya grandifolia*. J. Ind. Microbiol. Biotechnol. 20, 259–264.

Li, J.Y., Strobel, G., Sidhu, R., Hess, W.M., Ford, E.J., 1996. Endophytic taxol-producing fungi from bald cypress, *Taxodium distichum*. Microbiology 142, 2223–2226.

Liu, K., Ding, X., Deng, B., Chen, W., 2009. Isolation and characterization of endophytic taxol-producing fungi from *Taxus chinensis*. J. Ind. Microbiol. Biotechnol. 36, 1171–1177.

Luo, F., Fei, X., Tang, F., Li, X., 2011. Simultaneous determination of paclitaxel in hazelnut by HPLC-MS/MS. For. Res. 24 (6), 779–783.

Ma, H., Lu, Z., Liu, B., Qiu, Q., Liu, J., 2013. Transcriptome analyses of a Chinese hazelnut species *Corylus mandshurica*. BMC Plant Biol. 13, 152. https://doi.org/10.1186/1471-2229-13-152.

Majumder, A., Jha, S., 2009. Biotechnological approaches for the production of potential anticancer leads podophyllotoxin and paclitaxel: an overview. J. Biol. Sci. 1 (1), 46–69.

Mamadalieva, N.Z., Mamedov, N.A., 2020. *Taxus brevifolia* a high-value medicinal plant, as a source of taxol. In: Máthé, Á. (Ed.), Medicinal and Aromatic Plants of North America. Medicinal and Aromatic Plants of the World. vol 6. Springer, Cham, https://doi.org/10.1007/978-3-030-44930-8_9.

Miao, Z., Wang, Y., Yu, X., Guo, B., Tang, K., 2009. A new endophytic taxane production fungus from *Taxus chinensis*. Appl. Biochem. Microbiol. 45, 81–86.

Mirjalili, M.H., Farzaneh, M., Bonfill, M., Rezadoost, H., Ghassempour, A., 2012. Isolation and characterization of *Stemphylium sedicola* SBU-16 as a new endophytic taxol-producing fungus from *Taxus baccata* grown in Iran. FEMS Microbiol. Lett. 328, 122–129.

Mukaiyama, T., Shiina, I., Iwadare, H., Saitoh, M., Nishimura, T., Ohkawa, N., Sakoh, H., Nishimura, K., Tani, Y., Hasegawa, M., Yamada, K., Saitoh, K., 1999. Asymmetric total synthesis of taxol. Chem. Eur. J. 5, 121–161. https://doi.org/10.1002/(SICI)1521-3765(19990104).

Mukherjee, S., Ghosh, B., Jha, T.B., Jha, S., 2002. Variation in content of taxol and related taxanes in Eastern Himalayan populations of *Taxus wallichiana*. Planta Med. 68, 757–759. https://doi.org/10.1055/s-2002-33808.

Nadeem, M., Rikhari, H.C., Kumar, A., Palni, L.M.S., Nandi, S.K., 2002. Taxol content in the bark of Himalayan yew in relation to tree age and sex. Phytochemistry 60, 627–631. https://doi.org/10.1016/S0031-9422(02)00115-2.

Ne'meth-Kiss, V., Forga'cs, E., Cserha'ti, T., Schmidt, G., 1996. Taxol content of various *Taxus* species in Hungary. J. Pharm. Biomed. Anal. 14, 997–1001. https://doi.org/10.1016/0731-7085(95)01682-1.

Nicolaou, K.C., Yang, Z., Liu, J.J., Ueno, H., Nantermet, P.G., Guy, R.K., Claiborne, C.F., Renaud, J., Couladouros, E.A., Paulvannan, K., Sorensen, E.J., 1994. Total synthesis of taxol. Nature 367, 630–634. https://doi.org/10.1038/367630a0.

Nisa, H., Kamili, A.N., Nawchoo, I.A., Shafi, S., Shameem, N., Bandh, S.A., 2015. Fungal endophytes as prolific source of phytochemicals and other bioactive natural products: a review. Microb. Pathog. 82, 50–59.

Ojima, I., Habus, I., Zhao, M., Georg, G.I., Jayasinghe, L.R., 1991. Efficient and practical asymmetric synthesis of the taxol C-13 side chain, N-benzoyl-(2R, 3S)-3-phenylisoserine, and its analogs via chiral 3-hydroxy-4-aryl-. beta.-lactams through chiral ester enolate-imine cyclocondensation. J. Org. Chem. 56 (5), 1681–1683.

Ojima, I., Habus, I., Zhao, M., Zucco, M., Park, Y.H., Sun, C.M., Brigaud, T., 1992. New and efficient approaches to the semisynthesis of taxol and its C-13 side chain analogs by means of b-lactam synthon method. Tetrahedron 48, 6985–7012. https://doi.org/10.1016/S0040-4020(01)91210-4.

Pandi, M., Kumaran, R.S., Choi, Y.K., Kim, H.J., Muthumary, J., 2011. Isolation and detection of taxol, an anticancer drug produced from *Lasiodiplodia theobromae*, an endophytic fungus of the medicinal plant Morinda citrifolia. Afr. J. Biotechnol. 10, 1428–1435.

Pandi, M., Manikandan, R., Muthumary, J., 2010. Anticancer activity of fungal taxol derived from *Botryodiplodia theobromae* pat., an endophytic fungus, against 7, 12 dimethyl benz (a) anthracene (DMBA)-induced mammary gland carcinogenesis in Sprague dawley rats. Biomed. Pharmacother. 64 (1), 48–53.

Pang, L.Y., Hurst, E.A., Argyle, D.J., 2016. Cyclooxygenase-2: a role in cancer stem cell survival and repopulation of cancer cells during therapy. Stem Cells Int. 2016, 2048731. https://doi.org/10.1155/2016/2048731.

Parc, G., Canaguier, A., Landre´, P., Hocquemiller, R., Chriqui, D., Meyer, M., 2002. Production of taxoids with biological activity by plants and callus culture from selected *Taxus* genotypes. Phytochemistry 59, 725–730. https://doi.org/10.1016/S0031-9422(02)00043-2.

Patel, R.N., 1998. Tour de paclitaxel: biocatalysis for semisynthesis. Annu. Rev. Microbiol. 52, 361–395.

Patel, R.N., Banerjee, A., Nanduri, V.V., 2000. Enzymatic acetylation of 10-deacetylbaccatin III to baccatin III by C-10 deacetylase from *Nocardioides luteus* SC 13913. Enzym. Microb. Technol. 27, 371–375.

Phisalaphong, M., Linden, J.C., 1999. Kinetic studies of paclitaxel production by *Taxus canadensis* cultures in batch and semicontinuous with total cell recycle. Biotechnol. Prog. 15, 1072–1077.

Poupat, C., Hook, I., Gue'ritte, F., Ahond, A., Gue'nard, D., Adeline, M.T., Wang, X.P., Dempsey, D., Breuillet, S., Potier, P., 2000. Neutral and basic taxoid contents in the needles of *Taxus* species. Planta Med. 66, 580–584. https://doi.org/10.1055/s-2000-8651.

Prasad, V., De Jesús, K., Mailankody, S., 2017. The high price of anticancer drugs: origins, implications, barriers, solutions. Nat. Rev. Clin. Oncol. 14 (6), 381–390.

Qiao, W., Ling, F., Yu, L., Huang, Y., Wang, T., 2017. Enhancing taxol production in a novel endophytic fungus, *Aspergillus aculeatinus* Tax-6, isolated from *Taxus chinensis* var. *mairei*. Fungal Biol. 121, 1037–1044.

Rai, M., Rathod, D., Agarkar, G., Dar, M., Brestic, M., Pastore, G.M., Junior, M.R., 2014. Fungal growth promotor endophytes: a pragmatic approach towards sustainable food and agriculture. Symbiosis 62 (2), 63–79.

Rajendran, L., Rajagopal, K., Subbarayan, K., 2013. Efficiency of fungal taxol on human liver carcinoma cell lines. Am. J. Res. Commun. 1, 112–121.

Rao, K.V., Bnakuni, R.S., Juchum, J., Davies, R.M., 1996. A large scale process for paclitaxel and other taxanes from the needles of *Taxus x media Hicksii* and *Taxus floridana* using reverse phase column chromatography. J. Liq. Chromatogr. Relat. Technol. 19, 427–447.

Rebecca, A., Hemamalini, V., Kumar, D., 2012. Endophytic *Chaetomium* sp. from *Michelia champaca* L. and its taxol production. Jairjp. Com, J. Acad. Ind. Res. 1, 68–72.

Remesh, A., 2017. Toxicities of anticancer drugs and its management. Int. J. Basic Clin. Pharmacol. 1, 2–12. https://doi.org/10.5455/2319-2003.ijbcp000812.

Rency, A.S., Pandian, S., Kasinathan, R., Satish, L., Swamy, M.K., Ramesh, M., 2019. Hairy root cultures as an alternative source for the production of high-value secondary metabolites. In: Swamy, M.K., Akhter, M.S. (Eds.), Natural Bio-Active Compounds. Springer, Singapore, pp. 237–264.

Roberts, S.G., Shuler, M.L., 1997. Large-scale plant cell culture. Curr. Opin. Biotechnol. 8, 154–159. https://doi.org/10.1016/S0958-1669(97)80094-8.

Rodriguez, R.J., White, J.F., Arnold, A.E., Redman, R.S., 2009. Fungal endophytes: diversity and functional roles: Tansley review. New Phytol. 182 (2), 314–330.

Roopa, G., Madhusudhan, M.C., Sunil, K.C.R., Lisa, N., Calvin, R., Poornima, R., Zeinab, N., Kini, K.R., Prakash, H.S., Geetha, N., 2015. Identification of taxol-producing endophytic fungi isolated from *Salacia oblonga* through genomic mining approach. J. Genet. Eng. Biotechnol. 13, 119–127.

Ruckdeschel, K., Roggenkamp, A., Lafont, V., Mangeat, P., Heesemann, J., Rouot, B., 1997. Interaction of *Yersinia enterocolitica* with macrophages leads to macrophage cell death through apoptosis. Infect. Immun. 65, 4813–4821.

Ruiz-Sanchez, J., Flores-Bustamante, Z.R., Dendooven, L., Favela-Torres, E., Soca-Chafre, G., Galindez-Mayer, J., Flores-Cotera, L.B., 2010. A comparative study of Taxol production in liquid and solid-state fermentation with *Nigrospora* sp. a fungus isolated from *Taxus globosa*. J. Appl. Microbiol. 109, 2144–2150.

Satish, L., Shamili, S., Yolcu, S., Lavanya, G., Alavilli, H., Swamy, M.K., 2020. Biosynthesis of secondary metabolites in plants as influenced by different factors. In: Swamy, M.K. (Ed.), Plant-Derived Bioactives. Springer, Singapore, pp. 61–100.

Schiff, P.B., Fant, J., Horwitz, S.B., 1979. Promotion of microtubule assembly in vitro by taxol. Nature 277, 665–667.

Schiff, P.B., Horwitz, S.B., 1980. Taxol stabilizes microtubules in mouse fibroblast cells. Proc. Natl. Acad. Sci. U. S. A. 77, 1561–1565.

Senthil Kumaran, R., Muthumary, J., Hur, B.K., 2008. Production of taxol from Phyllosticta spinarum, an endophytic fungus of Cupressus sp. Eng. Life Sci. 8 (4), 438–446.

Sherline, P., Schiavone, K., 1977. Immunofluorescence localization of proteins of high molecular weight along intracellular microtubules. Science 198, 1038–1040.

Shrestha, K., Strobel, G.A., Shrivastava, S.P., Gewali, M.B., 2001. Evidence for paclitaxel from three new endophytic fungi of Himalayan yew of Nepal. Planta Med. 67, 374–376.

Simon, H.U., Haj-Yehia, A., Levi-Schaffer, F., 2000. Role of reactive oxygen species (ROS) in apoptosis induction. Apoptosis 5, 415–418.

Snyder, J.P., Nettles, J.H., Cornett, B., Downing, K.H., Nogales, E., 2001. The binding conformation of Taxol in β-tubulin: a model based on electron crystallographic density. Proc. Natl. Acad. Sci. U. S. A. 98, 5312–5316.

Soliman, S.S.M., Tsao, R., Raizada, M.N., 2011. Chemical inhibitors suggest endophytic fungal paclitaxel is derived from both mevalonate and non-mevalonate-like pathways. J. Nat. Prod. 74, 2497–2504.

Somjaipeng, S., Medina, A., Kwaśna, H., Ordaz Ortiz, J., Magan, N., 2015. Isolation, identification, and ecology of growth and taxol production by an endophytic strain of *Paraconiothyrium variabile* from English yew trees (*Taxus baccata)*. Fungal Biol. 119, 1022–1031.

Stahlhut, R., Park, G., Petersen, R., Ma, W., Hylands, P., 1999. The occurrence of the anti-cancer diterpene taxol in *Podocarpus gracilior* Pilger (Podocarpaceae). Biochem. Syst. Ecol. 27, 613–622.

Stierle, A., Strobel, G., Stierle, D., 1993. Taxol and taxane production by *Taxomyces andreanae*, an endophytic fungus of Pacific yew. Science 260 (5105), 214–216.

Strobel, G.A., 2014. Methods of discovery and techniques to study endophytic fungi producing fuel-related hydrocarbons. Nat. Prod. Rep. 31 (2), 259–272.

Strobel, G., 2018. The emergence of endophytic microbes and their biological promise. J. Fungi 4 (2), 57. https://doi.org/10.3390/jof4020057.

Strobel, G.A., Hess, W.M., Li, J.Y., Ford, E., Sears, J., Sidhu, R.S., Summerell, B., 1997. *Pestalotiopsis guepinii*, a taxol-producing endophyte of the wollemi pine, *Wollemia nobilis*. Aust. J. Bot. 45, 1073–1082.

Strobel, G., Yang, X., Sears, J., Kramer, R., Sidhu, R.S., Hess, W.M., Young, B., 1996. Taxol from *Pestalotiopsis microspora*, an endophytic fungus of *Taxus wallachiana*. Microbiology 142, 435–440.

Sturz, A.V., Nowak, J., 2000. Endophytic communities of rhizobacteria and the strategies required to create yield enhancing associations with crops. Appl. Soil Ecol. 15, 183–190.

Suresh, G., Kokila, D., Suresh, T.C., Kumaran, S., Velmurugan, P., Vedhanayakisri, K.A., Sivakumar, S., Ravi, A.V., 2020. Mycosynthesis of anticancer drug taxol by *Aspergillus oryzae*, an endophyte of *Tarenna asiatica*, characterization, and its activity against a human lung cancer cell line. Biocatal. Agric. Biotechnol. 24, 101525.

Swamy, M.K., Akhtar, M.S., Sinniah, U.R., 2016. Response of PGPR and AM fungi toward growth and secondary metabolite production in medicinal and aromatic plants. In: Akhtar, M.S. (Ed.), Plant, Soil and Microbes. Springer, Cham, pp. 145–168.

Tabata, H., 2004. Paclitaxel production by plant-cell-culture technology. Adv. Biochem. Eng. Biotechnol. 87, 1–23. https://doi.org/10.1007/b13538.

Uzma, F., Mohan, C.D., Hashem, A., Konappa, N.M., Rangappa, S., Kamath, P.V., Singh, B.P., Mudili, V., Gupta, V.K., Siddaiah, C.N., Chowdappa, S., 2018. Endophytic fungi—alternative sources of cytotoxic compounds: a review. Front. Pharmacol. 9, 309. https://doi.org/10.3389/fphar.2018.00309.

Van Rozendaal, E.L.M., Lelyveld, G.P., Van Beek, T.A., 2000. Screening of the needles of different yew species and cultivars for paclitaxel and related taxoids. Phytochemistry 53, 383–389. https://doi.org/10.1016/S0031-9422 (99)00094-1.

Vennila, R., Kamalraj, S., Muthumary, J., 2012. In vitro studies on anticancer activity of fungal taxol against human breast cancer cell line MCF-7 cells. Biomed. Aging Pathol. 2, 16–18.

Wang, J., Li, G., Lu, H., Zheng, Z., Huang, Y., Su, W., 2000. Taxol from *Tubercularia* sp. strain TF5, an endophytic fungus of *Taxus mairei*. FEMS Microbiol. Lett. 193, 249–253.

Wang, Y., Tang, K., 2011. A new endophytic taxol- and baccatin III-producing fungus isolated from *Taxus chinensis* var. *mairei*. Afr. J. Biotechnol. 10, 16379–16386.

Wastewater, X., Rodr, S., 2016. Unraveling the chemical interactions of fungal endophytes for exploitation as microbial factories. In: Purchase, D. (Ed.), Fungal Applications in Sustainable Environmental Biotechnology. Springer, pp. 3–28.

Weydert, C.J., Waugh, T.A., Ritchie, J.M., Iyer, K.S., Smith, J.L., Li, L., Spitz, D.R., Oberley, L.W., 2006. Overexpression of manganese or copper-zinc superoxide dismutase inhibits breast cancer growth. Free Radic. Biol. Med. 41, 226–237.

Wheeler, N.C., Jech, K., Masters, S., Brobst, S.W., Alvarado, A.B., Hoover, A.J., Snader, K.M., 1992. Effects of genetic, epigenetic, and environmental factors on taxol content in *Taxus brevifolia* and related species. J. Nat. Prod. 55, 432–440. https://doi.org/10.1021/np50082a005.

Wilson, D., 1995. Endophyte: the evolution of a term, and clarification of its use and definition. Oikos 73, 274–276.

Witherup, K.M., Look, S.A., Stasko, M.W., Ghiorzi, T.J., Muschik, G.M., Cragg, G.M., 1990. *Taxus* spp. Needles contain amounts of taxol comparable to the bark of *Taxus brevifolia*: analysis and isolation. J. Nat. Prod. 53, 1249–1255.

Wu, J., Lin, L., 2003. Enhancement of taxol production and release in *Taxus chinensis* cell cultures by ultrasound, methyl jasmonate and in situ solvent extraction. Appl. Microbiol. Biotechnol. 62, 151–155. https://doi.org/10.1007/s00253-003-1275-x.

Xiong, Z.Q., Yang, Y.Y., Zhao, N., Wang, Y., 2013. Diversity of endophytic fungi and screening of fungal paclitaxel producer from Anglojap yew, *Taxus x media*. BMC Microbiol. 13, 71.

Xu, F., Tao, W., Cheng, L., Guo, L., 2006. Strain improvement and optimization of the media of taxol-producing fungus *Fusarium maire*. Biochem. Eng. J. 31, 67–73.

Yukimune, Y., Tabata, H., Higashi, Y., Hara, Y., 1996. Methyl jasmonate-induced over-production of paclitaxel and Baccatin III in *Taxus* cell suspension cultures. Nat. Biotechnol. 14, 1129–1132.

Zaiyou, J., Li, M., Xiqiao, H., 2017. An endophytic fungus efficiently producing paclitaxel isolated from *Taxus wallichiana* var. *mairei*. Medicine 96, 8–11.

Zein, N., Aziz, S.W., El-Sayed, A.S., Sitohy, B., 2019. Comparative cytotoxic and anticancer effect of Taxol derived from *Aspergillus terreus* and *Taxus brevifolia*. Biosci. Res. 16, 367–374.

Zhang, C., Fevereiro, P.S., 2007. The effect of heat shock on paclitaxel production in *Taxus yunnanensis* cell suspension cultures: role of abscisic acid-pretreatment. Biotechnol. Bioeng. 96, 506–514. https://doi.org/10.1002/bit.21122.

Zhang, P., Zhou, P.P., Yu, L.J., 2009. An endophytic taxol-producing fungus from *Taxus media*, *Cladosporium cladosporioides* MD2. Curr. Microbiol. 59, 227–232.

Zhao, K., Ping, W., Li, Q., Hao, S., Zhao, L., Gao, T., Zhou, D., 2009. *Aspergillus niger* var. *taxi*, a new species variant of taxol-producing fungus isolated from *Taxus cuspidata* in China. J. Appl. Microbiol. 107, 1202–1207.

Zhou, X., Zhu, H., Liu, L., Lin, J., Tang, K., 2010. A review: recent advances and future prospects of taxol-producing endophytic fungi. Appl. Microbiol. Biotechnol. 86, 1707–1717.

9

Metabolic engineering strategies to enhance the production of anticancer drug, paclitaxel

Lakkakula Satish[a,b], Yolcu Seher[i], Kasinathan Rakkammal[d], Pandiyan Muthuramalingam[d,h], Chavakula Rajya Lakshmi[e], Alavilli Hemasundar[c], Kakarla Prasanth[b], Sasanala Shamili[b], Mallappa Kumara Swamy[f], Malli Subramanian Dhanarajan[g], and Manikandan Ramesh[d]

[a]Department of Biotechnology Engineering, Ben-Gurion University of Negev, Beer Sheva, Israel, [b]French Associates Institute for Agriculture and Biotechnology of Drylands, The Jacob Blaustein Institutes of Desert Research, Ben-Gurion University of the Negev, Sede-Boqer Campus, Beer Sheva, Israel, [c]Department of Life Science, Sogang University, Seoul, South Korea, [d]Department of Biotechnology, Alagappa University, Science Campus, Karaikudi, TN, India, [e]Department of Chemistry, Vishnu Institute of Technology, Bhimavaram, AP, India, [f]Department of Biotechnology, East West First Grade College, Bengaluru, Karnataka, India, [g]Department of Biochemistry and Biotechnology, Jeppiaar College of Arts and Science, Padur, TN, India, [h]Department of Biotechnology, Sri Shakthi Institute of Engineering and Technology, Coimbatore, TN, India, [i]Faculty of Engineering and Natural Sciences, Sabanci University, Istanbul, Turkey

9.1 Introduction

Paclitaxel is an important diterpenoid with a taxane skeleton that is isolated from *Taxus* species (yew tree) and used as a component of anticancer drugs (Croteau et al., 2006). The U.S. Food and Drug Administration (FDA) has approved paclitaxel for treating many cancer types, such as breast, lung and ovarian cancers and Kaposi's sarcoma. Furthermore, it is also known as a crucial agent for arterial stents to prevent the formation of scar tissues. As yew trees grow slowly and have low levels of paclitaxel (0.01%–0.03% of the dry weight), it is

Paclitaxel. https://doi.org/10.1016/B978-0-323-90951-8.00003-5
Copyright © 2022 Elsevier Inc. All rights reserved.

a very slow and difficult process to extract paclitaxel from the trees on a commercial scale (Waugh et al., 1991).

Currently, paclitaxel is also being evaluated for treating diseases that are not related with cancers requiring stabilization of microtubules and the evasion of cell proliferation and angiogenesis, for instance, Alzheimer's and psoriasis (Malik et al., 2011; Vidal-Limon et al., 2018). However, paclitaxel exhibits drawbacks, for example, lower solubility, adverse side effects. To overcome these problems, several semisynthetic derivatives have been established and are made available in the market. As the demand for paclitaxel is continuously growing, identification and utilization of novel sources or the improvement of existing systems of its production is required (Onrubia et al., 2013a; Vidal-Limon et al., 2018). Though many systems have been explained for the total synthesis of paclitaxel, truncated yields hamper their use at a commercial level (Guo et al., 2006). Nevertheless, the paclitaxel derivatives produced through a semisynthetic approach using baccatin III or other precursors obtained from *Taxus* species has been found to be feasible, commercially with a little success rate. Recently, a great number of yew trees have been destroyed by humans for the extraction of this important drug. Therefore, biotechnological approaches of producing paclitaxel or its precursors from different *Taxus* species remains as the major and sustainable source at present, and majority of its global supply is fulfilled by the leading commercial biotech company, Phyton Biotech, which employs its proprietary "green" Plant Cell Fermentation Technology (PCF®) to produce API-grade paclitaxel (https://phytonbiotech.com/). Alternatively, to improve the efficiency of the paclitaxel treatment, Abraxis BioScience (a biotechnology company) has developed albumin-bound paclitaxel, which is available with the trade name of abraxane (nab-paclitaxel) as an alternative formulation. Abraxane is basically a paclitaxel molecule attached to albumin nanoparticles. In addition, the elicitation and metabolic engineering strategies is useful for improved production of many biopharmaceutical products, and hence can be employed for the increasing paclitaxel production (Ye and Bhatia, 2012). For instance, abiotic and biotic elicitors are generally utilized to enhance the production of pharmaceutical terpenoids. Several research groups have identified few effective elicitors, including coronatine, MJ, SA and abscisic acid to induce paclitaxel accumulation in suspension cell lines (Lu et al., 2016). The biotic elicitor MJ, which regulates plant defense was found as a very effective molecule to induce overproduction of paclitaxel in *Taxus* cell suspension cultures. In addition to elicitors, higher paclitaxel production was observed in bioreactors through co-culturing of *Taxus* cells with its endophytic fungi (Ramirez-Estrada et al., 2016). In previous studies, many genes and enzymes involved in the biosynthetic pathways of taxanes have been partially identified in *Taxus* species tissues and cells. Overexpression of genes involved in paclitaxel biosynthesis is another remarkable strategy to enhance the production of this target drug (Kundu et al., 2017; Li et al., 2019). The production of plant cell lines can be established from either the transgenic whole-plant or by direct transformation of cultured plant cells with the target genes. Hairy root culture, adventitious root culture, and callus culture approaches can be effective in producing specific biopharmaceuticals. Scale-up production and downstream purification methodologies adapted from other commercialized cell-based bioproduction systems can be used in delivering plant-made pharmaceuticals and proteins to the marketplace (Fig. 9.1). In this chapter, molecular engineering strategies, such as elicitors, overexpression of genes involved in paclitaxel biosynthesis, and ectopic expression to produce higher amounts of paclitaxel are discussed.

9.2 Historical perspective of paclitaxel

Taxanes are a significant new class of anticancer agents that apply their cytotoxic impacts through a unique mechanism. Paclitaxel is a gathering of the taxane class of drugs, and it was the first taxane to enter clinical preliminaries and to get the FDA endorsement (Ojima et al., 2016). It is a natural compound acquired from *T. brevifolia*, also called the yew tree, which is a slow-developing and uncommon evergreen found in the old-growth forests region of the Pacific North-west. Paclitaxel, the brand name called taxol is a strong cytotoxic diterpene obtained from yew trees, and is a hydrophobic mitotic inhibitor with a powerful anticancer effect (Zhu and Chen, 2019). Today yew is one of the most generally utilized anticancer agents on the planet. Furthermore, it is utilized in various cancer therapies, including lobular carcinoma, lung and ovarian cancers, and Kaposi's sarcoma (Appendino, 2002).

Samples of the Pacific yew's bark were first collected in 1962 by Arthur Barclay (an expert in botany) as a part of a cooperative plant-screening package tracked by the National Cancer Institute and the U.S. Department of Agriculture to discover natural products that may treat cancers. They were then sent to Wisconsin Alumni Research Foundation so as to plan crude extracts to be tried on oral epidermoid carcinoma cell culture – a cell line developed from a human disease. After 2 years, Drs. Monroe Wall and Dr. Mansukh Wani, along with their associates at the Research Triangle Institute's Natural Product Laboratory in Research Triangle Park of North Carolina found that extracts from this bark indicated significant cytotoxic action (Perdue and Hartwell, 1969; Wall and Wani, 1995; Weaver, 2014). Monroe Wall had chosen to name the molecule as "taxol" in light of the fact that it was certainly believed to be an alcohol, and furthermore on the grounds that it was a habitual way to name an established molecule after the genus of the originating plant (https://www.news-medical.net/health/Paclitaxel-History.aspx).

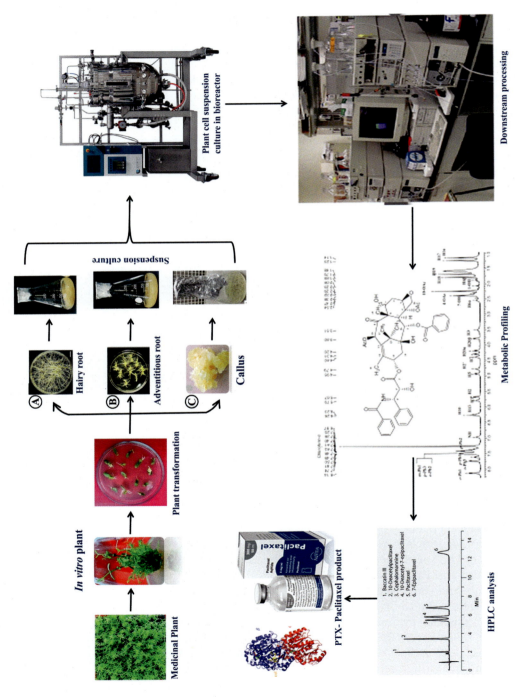

Fig. 9.1 Schematic representing the plant cell suspension culture process for the production of paclitaxel by plant metabolic engineering.

Additional examples of bark were gathered and sent to Wall's gathering for identification and subsequent purification of the active component in 1965. Despite the fact that the confinement of paclitaxel in unadulterated structure took quite a while, the chemical structure was at last published in 1971 (Wani et al., 1971). The antitumor action of paclitaxel was affirmed in 1977 in the mouse melanoma B16 model, after which it was chosen as a candidate drug for clinical turn of events. Likewise, significant antitumor activity was observed in animal models against MX-1 mammary, LX-1 lung and CX-1 colon cancers. In the exact year a request for 7000 pounds of bark was made, which implied relinquishing 1500 trees scattered more than a great many sections of land of old-growth forest in the Pacific Northwest (https://www.news-medical.net/health/Paclitaxel-History.aspx).

In 1978, an expert from the Albert Einstein College of Medicine of Yeshiva University reached Monroe Wall and requested certain radio- labeled paclitaxel so as to carry out investigations to understand its mechanisms of action. It was observed that paclitaxel prevented cell division through a fascinating mechanism, i.e., encouraging the advancement of microtubules (cell's ultrafine fibers). While, the past molecules destroyed cancerous cells by forestalling the establishment of microtubules, and in this manner hindering the division, the increased creation of microtubules disrupts appropriate coordination of cell division (https://www.news-medical.net/health/Paclitaxel-History.aspx). Fig. 9.2 illustrates the general timeline of the historical evolution of taxol as a drug from the identification to the FDA approval.

National Cancer Institute started stage I clinical preliminaries of Taxol against various cancer types in 1984. Interest for Taxol spiked in 1989 after agents at Johns Hopkins announced that the drug produced incomplete or complete responses in 30% of patients with cutting edge ovarian cancer (Goodman and Walsh, 2001; Walsh and Goodman, 2002). The FDA affirmed Taxol for the therapy for ovarian cancer in December 1992. Experts likewise tested the effectiveness of paclitaxel as a therapy for advanced breast cancer. Ensuing clinical preliminaries found that the drug was successful against this sickness, and in 1994 the FDA endorsed paclitaxel for use against breast cancer. The principal organization to accomplish the huge scope creation of paclitaxel was Polysciences, Inc. Clinical initiations became possible when a technique was inferred to extract a precursor of paclitaxel, i.e., 10-deacetyl-baccatin III from the normal yew *T. baccata*. The precursor was then changed over by the compound union to taxol. As of now, a cell culture strategy created by phyton catalytic is utilized by Bristol-Myers Squibb (BMS) to produce the drug (Weaver, 2014).

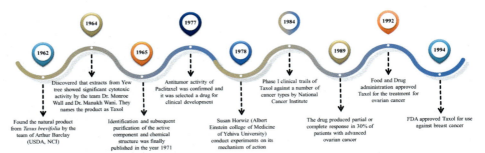

Fig. 9.2 Timeline of the historical evolution of Taxol drug (Identification to FDA approval).

9.3 Metabolic engineering strategies for paclitaxel production

Plants constitute numerous metabolic pathways that are essential for the synthesis of bio-active compounds. As an alternative approach, metabolic engineering is useful to unravel and improve the biosynthesis of important metabolites in plants. To prompt increased synthesis of metabolites, various biotechnological practices have been employed in plants. Some examples include selection of high-yielding cell lines, optimization of culture media, elicitation of cell cultures, application of bioreactors to produce metabolites in large scale, immobilization of cells/enzymes, addition of precursors to culture media, and biotransformation (Parsaeimehr et al., 2011; Sharma et al., 2019) (Fig. 9.2). In recent times, developments in gene editing approaches have significantly helped to exploit the abilities of plant cells to secrete higher levels of bioactive metabolites (Woo et al., 2015; Sharma et al., 2019).

Paclitaxel is a strong plant derived chemotherapeutic agent which can effectively promote the microtubule stabilization and promote mitosis inhibition by blocking the microtubule depolymerization (Zhu and Chen, 2019; Coelho et al., 2020). Hence, it can be served as a potential therapeutic agent for the treatment against wide array of cancers (Lenka et al., 2012). Being an important anticancer drug, paclitaxel utilization in cancer therapy is raising day by day. On the other hand, the paclitaxel scarcity is alarmingly raising due to continuous deforestation of Yew trees. Additionally, chemical synthesis of taxol entails ample amounts of financial input and labor intensive which adds an additional layer of complexity for its manufacturing and scaling-up process (Li et al., 2012). Hence, researchers across the globe continuously questing for the innovative ways to balance the demand and supply (Coelho et al., 2020). Starting from the Geranylgeranyl pyrophosphate (GGPP), which is the common diterpenoid precursor until to the paclitaxel biosynthesis it requires 19 biochemical steps (Walker and Croteau, 2001; Jennewein et al., 2004). Hence, overexpressing terpenoid biosynthetic pathway genes in native or ectopic system is one of the attractive approaches for synthesizing and scaling-up the scarcely available terpenoid components or intermediates (Lu et al., 2016). Furthermore, in a recent dated review Sabzehzari et al. hypothesized to increase the paclitaxel contents where two modes can be adopted (Sabzehzari et al., 2020). The first mode would be, by producing the paclitaxel through chemical, microbial or the plant sources and the another would be producing it through altering the gene expression of paclitaxel metabolic pathway genes (Sabzehzari et al., 2020).

9.3.1 Modulating the metabolic engineering strategy

9.3.1.1 Increasing the precursor access

The Isopentenyl pyrophosphate (IPP) and Geranylgeranyl pyrophosphate (GGPP) are the general precursors for the taxol biosynthetic pathway and were considered as a promising targets to enhance the availability of precursors (Chatzivasileiou et al., 2019). In a recent report, researchers have introduced the isopentenol utilization pathway (IUP) in *E. coli* (Chatzivasileiou et al., 2019). Compared to other intricate metabolic pathways, such as MVA (Mevalonic acid) and MEP (2-*C*-methyl-D-erythritol-4-phosphate), IUP pathway enhancement must be much faster and easier (Sabzehzari et al., 2020). Engineering the IUP approach found to ease the IPP production continuously by phosphorylating the prenol or isoprenol (Chatzivasileiou et al., 2019) and this method could be useful for generating high value secondary metabolites for therapeutic applications.

9.3.1.2 Modulating (enhancing or decreasing) the effectors

Enhancing the expression of positive regulators and knocking down the negative regulator expression of taxol biosynthetic genes is an alluring approach to scale-up the taxol contents (Sabzehzari et al., 2020). Specifically, taxol accumulation can be enhanced if we overexpress genes which catalyze hydroxylation and acetylation steps (Wang et al., 2019). Wang et al. empirically proved threefold higher presence of taxol contents in "Jinxishan" (JXS) cultivar which belongs to *Taxus mairei* when compared to its WT control due to higher hydroxylation and acetylation of taxol biosynthetic intermediates including 2-debenzoyl-7, 13-diacetylbaccatin III-2-O-benzoyl-transferase (DBBT), taxadiene 5-alpha hydroxylase (T5H), 5-alpha-taxadienol- 10-beta-hydroxylase

(T10OH), and taxadienol acetyltransferase (TAT) (Wang et al., 2019). In addition to this, while comparing the high and low taxol accumulated plant RNA-seq profile suggests that, the transcription factors including WRKY, MYB and ERF overexpression can also make a mark in taxol accumulation due to their differential regulation of genes (Wang et al., 2019). Taxadiene synthase (TS) is one of the most vital and favorite biosynthetic gene for the taxol biosynthesis. It was first engineered in *E. coli* by Huang et al. (2001) by fusing it with thioredoxin coupled system which further enhanced the protein solubility. The final recombinant TS protein resembled the native TS enzyme in terms of its mobility in SDS-PAGE (Huang et al., 2001). Together, using this thioredoxin fusion approach the overexpression of positive factors such as TS can be achieved without losing its native physical and enzymatic properties.

9.3.1.3 Regulation of transcriptional factors

Being their vital roles in various biological metabolic paths, the transcriptional factors considered as a vital instruments to engineer the metabolic flux in plants (Broun, 2004). Nevertheless, up- or downregulation of transcriptional factors could finally impact the metabolite accumulation. For instance, Lenka et al. (2015) established the negative regulatory roles of *TcJAMYC* transcription factors on paclitaxel biosynthetic gene expression in the *Taxus cuspidata* cultured cells. Similarly, Ethylene Responsive Factor (ERF) transcription factors including ERF114, ERF018, and ERF016 can modulate as both positive or negative regulators in the paclitaxel biosynthesis pathway (Sabzehzari et al., 2020). Zhang et al. (2015) also reported the similar activity in tfs including the TcERF12 and TcERF15, dual role as positive as well as negative regulators over the *TASY* gene of paclitaxel biosynthesis in *T. chinensis*. Together, modulating the transcriptional factors can concede an interesting output for taxol metabolism in plants.

9.3.1.4 Dual production of Phyto-metabolites

Another interesting approach is the deriving of two important phytochemical intermediates from a single plant reservoir. Using *Agrobacterium*-mediated gene transfer, a research group has transformed taxadiene synthase (TXS) gene to *Artemisia annua* (Li et al., 2015). Overproduction of TXS does not cause any yield penalty and growth reduction in the transgenic lines. Moreover, at the same time another important secondary metabolite, artemisinin was stored in the leaves and TXS was accumulated in the stem of the TXS overexpressing lines (Li et al., 2015). In summary, using this strategy we can produce multiple secondary metabolites in a single plant without any trade-off on the plant normal growth and development (Li et al., 2015).

9.3.2 Metabolic engineering in heterologous systems

Production of paclitaxel or its precursors through the heterologous hosts is considered as a fascinating approach compared to the chemical synthesis or microbial based production (Li et al., 2019). To date, few studies have been carried out to produce paclitaxel heterologously, i.e., using model organisms, such as *Escherichia coli*, *Saccharomyces cerevisiae*, and *Arabidopsis thaliana* (Besumbes et al., 2004; DeJong et al., 2006; Ajikumar et al., 2010). Taxadiene (a basic intermediate in paclitaxel biosynthesis) was synthesized in *E. coli* by overexpressing of taxadiene synthase (TS), geranylgeranyl diphosphate synthase (GGPPS), and Isopentenyl diphosphate (IPP) isomerase genes (Huang et al., 2001). Likewise, researchers produced taxadiene in *E. coli* T2 and T4 mutant strains by the engineering of mevalonate pathway (Bian et al., 2017). Also, they developed an *Agrobacterium*-mediated transformation approach to verify the strength of heterologous promoters in *Alternaria alternata* TPF6. Later, the taxadiene-expressing platform was transformed into *A. alternata* TPF6, and engineered the mevalonate pathway by introducing the plant taxadiene-yielding gene. Markedly, by co-overexpressing of isopentenyl diphosphate isomerase, taxadiene synthase (TS), and 3-hydroxy-3-methylglutaryl-CoA reductase (tHMG1), the researchers detected about $61.9 \mu g L^{-1}$ taxadiene in the engineered fungal strain, GB127 (Bian et al., 2017). In *E. coli*, heterologous protein interdependency was optimized to produce paclitaxel via P450-mediated taxol precursor biosynthesis. About $570 mg L^{-1}$ of paclitaxel was produced by this optimized approach (Biggs et al., 2016). As a suitable microbial host, *Bacillus subtilis* 168 was used as a cell factory to produce paclitaxel by expressing the plant-derived TS enzyme (Abdallah et al., 2020). In addition, eight enzymes involved in the biosynthesis pathway were overexpressed to upsurge the flux of the precursor, geranylgeranyl pyrophosphate. It was accomplished via constructing a biosynthetic operon containing the *B. subtilis* genes coding for MEP pathway (dxs, ispD, ispF, ispH, ispC, ispE, ispG) along with farnesyl diphosphate synthase (ispA). The total amount of taxadiene obtained by this strain was more than $17 mg L^{-1}$, which is 82-fold higher than that found in the *B. subtilis* wild strains. The study also revealed that *crtE* gene expression is vital for generating adequate GGPP. Ajikumar et al. (2010) reported a metabolic pathway engineering tactic to succeed in enhancing taxadiene titres in an engineered *E. coli* strain. The authors reported about $1 g L^{-1}$ (~ 15,000-fold) of taxadiene production in the host system.

For the first time, paclitaxel was produced in the yeast cells by introducing heterologous genes that encode enzymes involving in the initial part of the taxoid biosynthetic pathway and isoprenoid pathway (Engels et al., 2008). According to them, the expression of *T. chinensis* TS gene alone failed to enhance taxadiene quantity due to inadequate levels of GGPPS. Likewise, co-expression of TS and GGPPS synthase was also unsuccessful in increasing the levels, which may be attributed to steroid-linked negative feedback. Interestingly, expression of a truncated form of 3-hydroxyl-3-methylglutaryl-CoA (HMG-CoA) reductase isoenzyme 1 and a mutant regulatory protein (UPC2-1) allowed the uptake of steroid under aerobic conditions, and resulted increased taxadiene yield, i.e., up to 50%. Besumbes et al. (2004) reported the production of nearly 600 ng of taxadiene per g of dry weight (DW) in *Arabidopsis thaliana* by the overexpression of TASY and GGPP synthase genes, but affected the growth of plant, due to taxadiene amassing.

Few reports have previously demonstrated the production of paclitaxel or its derivatives using *Nicotiana* spp. (Li et al., 2012; Rontein et al., 2008). However, the accumulation of taxadiene amounts were detected to be lower in those approaches. In order to attain higher taxadiene amounts, recently Li et al. (2019) devised an experiment by coupling chloroplast metabolism with Isoprenoid pool enhancement, which successfully accumulated higher amount of 2 taxol derivatives, including taxadiene and taxadiene-5α-ol in *N. benthamiana* leaves. The taxadiene and taxadiene-5α-ol and cytochrome P450 reductase (CPR) in the chloroplast linked with elevated isoprenoid precursors were reportedly enhanced the taxol metabolite accumulation (Li et al., 2019).

Generally, these scientific efforts of biosynthesizing the paclitaxel in heterologous host organisms signifies that paclitaxel biosynthesis is an extremely complex pathway, and requires a lot more studies in this regard to completely understand about its mechanisms.

9.3.3 Plant cell cultures and elicitation approaches to enhance paclitaxel production

Since the paclitaxel amount is very low in *Taxus* spp. (0.01%–0.03% dry weight), alternative ways including tissue cultures, chemical synthesis and microbial fermentation by endophytic fungi have been used to produce a higher amount of this valuable anticancer drug (Somjaipeng et al., 2016). Plant cell cultures using different *Taxus* species (*T. cuspidata, T. yunnanensis, T. canadensis, T. globosa, T. baccata,* and *T. chinensis*) have been found successful in obtaining relatively a higher level of paclitaxel (reviewed in Liu et al., 2016). Further, the addition of different elicitors to the *Taxus* cell cultures has been found very advantageous. The molecules that activate secondary metabolism are called elicitors (Dörnenburg and Knorr, 1995). A great number of elicitors and signal molecules, such as fungal elicitors, silver ion, chitosan, benzoic acid, methyl jasmonate (MJ), and arachidonic acid were used to enhance the production of paclitaxel (Zhang et al., 2000; Yukimune et al., 1996). Elicitation, which is used to activate the secondary metabolism pathway (Lu et al., 2016), remarkably enhanced the paclitaxel concentrations. Besides, restructuring metabolic networks with recombinant DNA methods increases the production of metabolites and protein products (Bailey, 1991). We need to use elicitors for a large-scale biosynthesis of secondary metabolites even in transgenic plants-overexpressing key genes. Therefore, searching more efficient elicitors is required for inducing the biosynthesis of secondary metabolites in plant cell cultures (Naik and Al-Khayri, 2016). Different elicitors either alone or together with other compounds used for enhancement of paclitaxel production are presented in Table 9.1. In this section, we will summarize and discuss the studies that are related to the elicitor applications and genetic manipulations of paclitaxel biosynthesis along with elicitation for enhancement of paclitaxel production.

In plant cells, MJ treatment induces transcriptional reprogramming to regulate the biosynthesis of metabolites (Fonseca et al., 2009). MJ is the most effective elicitor (Zhang et al., 2000), which has been largely used to stimulate the biosynthesis of paclitaxel (Ketchum et al., 1999). It has been reported that the elicitation with $200\,\mu M$ of MJ for 7 days quickly triggered to produce paclitaxel and other compounds in *T. cuspidata* cell suspension cultures. The researchers found that the maximal amount of paclitaxel was $23.4\,mg\,L^{-1}$ after 5 days of elicitation in *T. cuspidata* cell line C93AD (Ketchum et al., 1999). In laboratory conditions, Ketchum et al. (1999) managed to maintain the production of moderate levels of paclitaxel for 2 years in *Taxus* cultures. In another study, Yukimune et al. (1996) tested three *Taxus* species (*T. media, T. baccata* and *T. brevifolia*), and found that *T. media* displayed the highest paclitaxel levels in a culture medium containing $100\,\mu M$ of MJ (Yukimune et al., 1996).

Similarly, addition of MJ in the suspension cultures of *T. media* caused to produce the highest paclitaxel yield $(2.09\,mg\,L^{1})$ (Cusidó et al., 2002). Immobilized *T. baccata* cells in Ca^{2+}-alginate beads elicited with MJ were shown to produce $2.71\,mg\,L^{-1}$ of paclitaxel per day in a stirred bioreactor (Bentebibel et al., 2005). Although most studies focused on a single elicitor, in a previous research, the effects of the combination of elicitors on the production of paclitaxel were investigated in *T. chinensis* cells. When compared with individual elicitors, they found a significant increase in the production of paclitaxel $(25.4\,mg\,L^{1})$ in *T. chinensis* suspension cells cultured in a combined medium containing $50\,mg\,L^{-1}$chitosan, $60\,\mu M$ MJ and $60\,\mu M$ Ag^{+}. The highest paclitaxel concentration was recorded on day 10 after elicitation. Among single

Table 9.1 Different elicitors either alone or together with other compounds used for enhancement of paclitaxel production.

Type of elicitors	Plant species	Cell culture approach	Paclitaxel yield	Reference(s)
Benzoic acid + Phenylalanine	*Taxus cuspidata*	Cell suspension culture	10 μg g^1 DW	Fett-neto et al. (1994)
Vanadyl sulfate (VSO$_4$)	*Taxus media*	Cell suspension culture	110.3 mg L^1	Yukimune et al. (1996)
		Two stage culture	21.5 mg L^1	Onrubia et al. (2013b)
Methyl jasmonate (MJ)	*Taxus baccata*	Cell suspension culture	13.1 μg g^1 DW	Cusidó et al. (1999)
	Taxus canadensis	Cell suspension culture	23.4 mg L^1	Ketchum et al. (1999)
	Taxus baccata	Cell suspension culture	2.09 mg L^{-1}	Cusidó et al. (2002)
MJ + Chitosan + Ag$^+$	*Taxus chinensis*	Cell suspension culture	25.4 mg L^{-1}	Zhang et al. (2000)
MJ + Ca^{2+}-algi-nate beads	*Taxus baccata*	Cell suspension culture	2.71 mg L^{-1}	Bentebibel et al. (2005)
SA + Mevastatin (MVS)	*Taxus chinensis* var. *mairei*	Cell suspension culture	1.626 mg g^{-1} DW	Wang et al. (2007)
Salicylic acid (SA)	*Taxus chinensis* var. *mairei*	Cell suspension culture	1.581 mg g^{-1} DW	Wang et al. (2007)
	Taxus baccata	–	14.7 μg L^{-1}	Somjaipeng et al. (2016)
	Taxodium mucronatum	Fungal culture	625.47 μg L^{-1}	Subban et al. (2019)
Fungal elicitors	*Taxus chinensis* var. *mairei*	Cell suspension culture	25.63 mg L^{-1}	Li et al. (2009)
	Taxus baccata	Cell suspension culture	291.5 μg L^{-1}	Salehi et al. (2018)
	Corylus avellana	Cell suspension culture	404.5 μg L^{-1}	Salehi et al. (2019)
	Taxus baccata	Fungal culture	116.19 μg L^{-1}	El-Bialy and El-Bastawisy (2020)
Benzoic acid	*Corylus avellana*	Cell suspension culture	15.69 μg g^{-1} DW	Bemani et al. (2012)
Coronatine (Cor)	*Taxus media*	Two stage culture	77.46 mg L^{-1}	Onrubia et al. (2013b)
Squalestatin	*Taxus baccata*	Cell suspension culture	2.47 mg L^{-1}	Amini et al. (2014)
VSO$_4$ + Phe + fructose	*Corylus avellana*	Cell suspension culture	4.2 μg g^{-1} DW	Rahpeyma et al. (2015)
SA + NaOAc + Cu^{2+}	*Taxus chinensis* var. *mairei*	Fungal culture	1337.56 μg L^{-1}	Qiao et al. (2017)
MBCD + fungal elicitors	*Corylus avellana*	Cell suspension culture	402.4 μg L^{-1}	Farhadi et al. (2020)

eliciters, MJ was shown to result in the highest amount of paclitaxel ($13.4\,mg\,L^1$) (Zhang et al., 2000). The combination of chitosan and MJ were speculated to have synergistic accumulative effects on the biosynthesis of paclitaxel. In conclusion, combined eliciters could be used as a good strategy to induce a large-scale paclitaxel production in *Taxus* cell cultures (Zhang et al., 2000). When the researchers added MJ together with chitin and chitosan into the suspension cultures of *T. cuspidata*, paclitaxel biosynthesis was increased as compared to single eliciter treatments (Linden and Phisalaphong, 2000).

Using fungal eliciters is the most efficient strategy for secondary metabolite biosynthesis in plant in vitro cultures (Yuan et al., 2002) and fungal eliciters were used to enhance paclitaxel production in *Taxus* for the first time in the early 1990s (Strobel et al., 1992). Endophytic fungi which have key functions in the production of paclitaxel in plants, form the conserved molecules called microbe-associated molecular patterns (MAMPs), and the receptors on plant cell surface recognize MAMPs, inducing plant defense (Newman et al., 2013). When endophytic fungi were added to plant cell culture system, plant defense- related secondary metabolites, including terpenoids were reported to be accumulated in plant cells (Li et al., 2009). A previous study indicated that the cells of *T. chinensis* var. *mairei* responded to the eliciters from *Fusarium oxysprum* through changing the general phenylpropanoid pathway and affecting the taxol biosynthesis (Yuan et al., 2002). The eliciters derived from *F. oxysprum* brought about cell apoptosis in suspension cultures of *T. chinensis* var. *mairei* in the late exponential stage, resulting in alterations in cell structure, plasmalemma permeability, general phenylpropanoid pathway and a dramatic increase in taxol synthesis. In conclusion, the researchers revealed a close relationship between cell apoptosis and taxol production in *T. chinensis* var. *mairei* culture exposed to fungal eliciters (Yuan et al., 2002). Eight fungal endophytes that belong to six genera, *Acremonium, Colletotrichum, Fusarium, Nodulisporium, Paecilomyces,* and *Periconia* were isolated from ecologically altered *Taxus baccata* grown in Egypt, and among them, only three fungal isolates were recorded to have a moderate capability for paclitaxel production (El-Bialy and El-Bastawisy, 2020). Among three endophytes, *Acremonium* species produced the highest level of paclitaxel ($116.19\,\mu g\,L^{-1}$) in M1D medium. Other endophytic fungi, *Colletotrichum,* and *Fusarium* species produced 109.5 and $41.53\,\mu g\,L^{-1}$ paclitaxel, respectively (El-Bialy and El-Bastawisy, 2020). To ensure the production stability of paclitaxel, the authors tested whether paclitaxel production changed with time (45, 90, 135 and 180 d) by different fungal endophytes. They found that *Fusarium* species could not synthesize paclitaxel, whereas *Acremonium* species and *Colletotrichum* species decreased the productivity by half and two thirds, respectively, after 6 months (El-Bialy and El-Bastawisy, 2020).

Li et al. (2009) reported that the co-culture of *Taxus chinensis* var. *mairei* suspension cells and its endophytic fungus, *Fusarium mairei* was successfully done to produce paclitaxel in a 20 L co-bioreactor. Co-culture system which was shown as an influential strategy for enhancement of paclitaxel production, produced 25.63 mg L^{-1} of paclitaxel within 15 days, and this amount was 38-fold higher than the uncoupled culture system in a bioreactor. They found that gibberellic acid from *F. mairei* induced the growth of *Taxus* cells (Li et al., 2009). Importantly, co-culture systems under optimum conditions not only significantly increased paclitaxel production but also reduced the culture time.

In vitro *Corylus avellana* (common hazel) culture is known as a promising and low-cost source for production of paclitaxel. There are some advantages of paclitaxel production with in vitro hazel culture. For instance, in vitro *Taxus* cultivation is much harder than hazel culture and because hazel is a member of Dicotyledonus plants, *Agrobacterium* transformation can be easily used for genetic manipulation as compared to Gymnosperms, including *Taxus* spp. (Miele et al., 2012; Salehi et al., 2019). In a recent work, elicitors derived from two endophytic fungal strains, such as *Chaetomium globosum* and *Paraconiothyrium brasiliense* were isolated from *T. baccata* and *C. avellana*, respectively. This was the first report shown the positive effect of fungal elicitors on the production of paclitaxel in *C. avellana* cell suspension culture (Salehi et al., 2018). The maximal paclitaxel concentration was obtained on day 17 in cell culture cycle by adding 10% (*v/v*) cell extract of *C. globosum*, whereas *P. brasiliense* elicitors from *C. avellana* caused lower paclitaxel production (Salehi et al., 2018). In another research, Salehi et al. (2019) studied whether paclitaxel production was affected by co-culturing of *C. avellana* cells with its endophytic fungus, *Epicoccum nigrum* strain YEF2. They found that co-culture of *C. avellana* cells and *E. nigrum* led to an improvement of paclitaxel production in comparison to monocultures of *C. avellana* cells and *E. nigrum* (Salehi et al., 2019). In the study, three important factors, such as fungal inoculum amount, co-culture establishment time and co-culture period were tested to achieve the highest amount of paclitaxel in the co-culture system. On day 13, the maximum paclitaxel concentration was 404.5 µg L^{-1} using by 3.2% (*v/v*) of *E. nigrum* mycelium suspension. As a result, the interaction between hazel cells and *E. nigrum* allows higher paclitaxel production (Salehi et al., 2019). Recently, it has been indicated that the combined treatment of whole fungal elicitors, cell extract and culture filtrate from endophytic fungus *Camarosporomyces flavigenus* resulted in higher paclitaxel biosynthesis than the individual elicitor treatment. The highest paclitaxel level was observed on day 21 of cell culture cycle (Salehi et al., 2020).

Vanadyl sulfate (VSO$_4$) has been shown to be an effective abiotic elicitor in culture media of different plant species (Cusidó et al., 1999).

In some reports, phenylalanine (Phe) which is one of the precursors in the taxol biosynthesis, was used together with VSO_4 in culture media to enhance the production of taxol and its natural precursor baccatin III. For example, callus growth enhanced in *T. baccata* medium supplemented with 1 mM of phenylalanine (Phe) or 0.05 and 0.1 mM of VSO_4. Under same conditions without Phe in the *T. baccata* cell suspension culture medium, the vanadyl sulfate considerably increased the levels of paclitaxel from $5.2 \mu g g^{1}$ DW to $13.1 \mu g g^{-1}$ DW. Also, the presence of VSO_4 increased the release of paclitaxel and baccatin III (120% and 97%, respectively) from the cells into the culture medium (Cusidó et al., 1999). In a research published by Rahpeyma et al. (2015), *C. avellana* cells were treated with $3 \mu M$ Phe, 0.05 and 0.1 mM VSO_4 and 3% fructose, and it has been revealed that the taxol content increased during culture period and the highest levels of taxol ($4.2 \mu g g^{-1}$ DW) was recorded on day 10. The combination of sugar, VSO_4 and Phe affected the paclitaxel production and biomass accumulation (Rahpeyma et al., 2015).

Cyclodextrin is known as an important agent/elicitor inducing secondary metabolite biosynthesis in plant cell cultures, and it makes the secretion of secondary metabolites easier from cells to culture media (Kashani et al., 2018; Farhadi et al., 2020). The synergistic effect of cyclodextrin and MJ on taxane production was reported in *Taxus x media* cell cultures. Because the low amount of paclitaxel was detected in control conditions of *T. x media* cell culture, the application of elicitors in culture medium is an effective biotechnological strategy for activation of genes involved in taxane metabolic pathway and an increase in the production of paclitaxel and related taxanes (Sabater-Jara et al., 2014). The key genes that encode rate-limiting enzymes in the paclitaxel synthesis pathway were identified in *T. baccata* cell cultures. Additionally, gene transcript levels were found to be enhanced into the culture medium containing methyl-β-cyclodextrin (MBCD) (Kashani et al., 2018). When two elicitors, MJ and cyclodextrin were both added to the culture medium, taxol levels were recorded 55-fold higher than the nonelicited cultures. Moreover, with using two elicitors, retroinhibition processes and taxane toxicity were limited for the producer cells (Sabater-Jara et al., 2014). In a very recent study, the elicitors from endophytic fungus *Coniothyrium palmarum* isolated from *T. baccata*, were tested alone or in combination with MBCD in terms of paclitaxel production in the cell suspension cultures of *C. avellana* (Farhadi et al., 2020). The results showed that the cell wall of *C. palmarum* was the most effective fungal elicitor for paclitaxel biosynthesis. Furthermore, MBCD alone and MBCD + fungal elicitors both stimulated a higher amount of paclitaxel secretion. 2.5% (v/v) of *C. palmarum* on day 17 and 50 mM of MBCD helped synergistically improve paclitaxel level to $402.4 \mu g L^{-1}$, and 78.6% of total concentration was secreted into the culture medium (Farhadi et al., 2020).

Coronatine (Cor) which is a bacterial phytotoxin produced by pathogen *Pseudomonas syringae* (Bender et al., 1999), acts as an elicitor of secondary metabolism in plants through inducing the jasmonate signaling pathway (Zhao et al., 2003). The effects of Cor and MJ on the production of paclitaxel and related taxanes were examined in *Taxus media* cell cultures. For that purpose, 1 µM of Cor and 100 µM of MJ were supplemented into the culture medium and a significant increase in total taxane production was observed in the cell suspension (Onrubia et al., 2013b). Total taxane content was $21.48 \, mg \, L^{-1}$ in MJ-supplemented medium on day 12 and $77.46 \, mg \, L^{-1}$ in Cor-supplemented medium on day 16. Besides, transcript levels of *taxane 13α-hydroxylase* (*T13αH*) and *taxane-2α-O-benzoyl transferase* (*TBT*) genes involved in paclitaxel biosynthetic pathway remarkably increased after elicitation (Onrubia et al., 2013b).

SA is an effective elicitor that has been used to increase taxol biosynthesis in *Pestalotiopsis microspora* isolated from the bark of *Taxodium mucronatum*. It was shown that SA elevated phenylalanine ammonia-lyase (PAL) activity and the levels of taxol and caused cell death in *Taxus* suspension cultures (Wang et al., 2007). The presence of 300 µM SA caused the production of paclitaxel ($625.47 \, µg \, L^1$) which is 45-times higher than control conditions (Subban et al., 2019). Similarly, the endophytic fungus *Paraconiothyrium variabile* produced a higher amount of paclitaxel ($14.7 \, µg \, L^{-1}$) when SA was added to its growth medium (Somjaipeng et al., 2016). Wang et al. (2007) reported that adding $20 \, mg \, L^1$ SA and $100 \, nmol \, L^{-1}$ mevastatin (MVS) into suspension cultures of *Taxus chinensis* var. *mairei* induced the increments of taxol contents ($1.626 \, mg \, g^{-1}$ DW). When the culture medium was supplemented only with SA, the taxol content was $1.581 \, mg \, g^{-1}$ DW, which is almost equal to the combined treatments of SA and MVS (Wang et al., 2007). In another work, the endophytic fungus *Aspergillus aculeatinus* Tax-6 isolated from *T. chinensis* var. *mairei* could produce taxol in potato dextrose agar liquid medium containing $10 \, mg \, L^{-1}$ SA, $8 \, g \, L^{-1}$ NaOAc (sodium acetate) and $0.1 \, mg \, L^{-1}$ Cu^{2+}. Sodium acetate is a precursor of taxol biosynthesis (Zamir et al., 1992), Cu^{2+} could increase oxidase activity and catalyze the production of taxol. In this medium, taxol content improved from 334.92 to $1337.56 \, µg \, L^{-1}$ (Qiao et al., 2017).

Benzoic acid is a precursor of the plant taxol (Fett-Neto et al., 1994), and it was also used to obtain elevated levels of paclitaxel. Fett-Neto et al. (1994) examined whether the taxol accumulation is affected by different amounts of benzoic acid, phenylalanine, *N*-benzoylglycine, serine, glycine, alanine, and 3-amino-3-pheny-propionic acid in *T. cuspidata* cell suspensions, and in developing callus cultures of *T. cuspidata*. Among different compounds, the maximum taxol levels ($10 \, µg \, g^1$ DW) were detected in the presence of benzoic acid and

Phe (Fett-Neto et al., 1994). Twofold elevation was recorded in fungal taxol content when some plant defense compounds such as SA, and benzoic acid were added into the culture medium (Soliman and Raizada, 2013). Bemani et al. (2012) found that the paclitaxel content in *Corylus avellana* cells supplemented with benzoic acid was fourfold higher than the control cells. Total paclitaxel content in *C. avellana* cells treated with 1 mM benzoic acid was recorded 15.69 μg g^{-1} DW (Bemani et al., 2012).

Another fungal elicitor, squalestatin was found to enhance the production of paclitaxel and baccatin III, significantly. A remarkable increase in paclitaxel levels (2.47 mg L^{-1}) was detected under 0.1 and 1 μM squalestatin treatments as compared to the control (Amini et al., 2014).

In addition to elicitor applications, metabolic engineering techniques together with elicitation are also used for enhancement of paclitaxel production. Although it is known that the MJ treatments induce the production of paclitaxel in *Taxus* cells, little is known about how MJ mediates the paclitaxel biosynthesis pathway. In a research work, promoters of *TASY, T5αH, DBAT, DBBT, PAM, BAPT,* and *DBTNBT* genes, encoding enzymes involved in paclitaxel biosynthesis were isolated and the roles of three MJ-regulated bHLH transcription factors (*TcJAMYC1, TcJAMYC2,* and *TcJAMYC4*) were investigated in *T. cuspidata* (Lenka et al., 2015). The promoters of genes were reported to be induced by MJ, and the expression of genes enhanced within 6 h and reached to the maximum levels within 12–18 h after elicitation (Lenka et al., 2015). In a previous report, *Taxus x media* var. *hicksii* transgenic root lines harboring the *taxadiene synthase* gene of *T. baccata* were elicited with diverse elicitors, and the combination of MJ, and Phe showed the highest level of total paclitaxel (10.78 mg L^{-1}) at the first week of treatments (Sykłowska-Baranek et al., 2015). Hoa et al. (2005) reported that transgenic cells overexpressing *10-deacetyl baccatin III-10-O-acetyl transferase (DBAT)* gene involved in paclitaxel biosynthesis pathway, increased paclitaxel yield. Moreover, expression levels of *taxadiene synthase (TS)* and *DBAT* genes were increased by MJ in plants. Importantly, transgenic cells could not produce significant taxol levels without MJ treatment (Hoa et al., 2005). Exposito et al. (2010) found that the *Taxus x media* cell lines containing *TS* and *rol* genes of *A. rhizogenes* resulted in higher taxane contents when compared to the untransformed cells (Exposito et al., 2010). In plants, 9-*cis*-epoxycarotenoid dioxygenase (NCED) is an important enzyme in the biosynthesis pathway of phytohormone abscisic acid (ABA). Overexpression of a *NCED* gene from *T. chinensis (TcNCED1)* in transgenic *T. chinensis* cells displayed a 2.7-fold increase in paclitaxel production in comparison to the untransformed cells (Li et al., 2012).

9.4 Conclusions

In summary, the technical advancements of recombinant DNA technology greatly enabled and fine-tuned the metabolic engineering strategies for scaling up of various economic and pharmaceutically significant natural plant products. Several researchers have explored synthetic biology and metabolic engineering with gene assemblies to improve the paclitaxel intermediates in recent years. Furthermore, engineering the production strategies with promoter manipulation, regulation of transcription factors, overexpression of enzymes, stimulation of the cryptic genes and addition of elicitors or precursors greatly enhanced the paclitaxel production at industrial scale. Nevertheless, many challenges for the paclitaxel production pathway needs to be resolved and the prospective application of gene editing tools like CRISPR/Cas9 might provide an alluring solution for the engineering and has an enormous scope for optimizing the intricate paclitaxel biosynthetic pathway.

References

Abdallah, I.I., Xue, D., Pramastya, H., van Merkerk, R., Setroikromo, R., Quax, W.J., 2020. A regulated synthetic operon facilitates stable overexpression of multigene terpenoid pathway in *Bacillus subtilis*. J. Ind. Microbiol. Biotechnol. 47, 243–249.

Ajikumar, P.K., Xiao, W.H., Tyo, K.E.J., Wang, Y., Simeon, F., Leonard, E., Mucha, O., Phon, T.H., Pfeifer, B., Stephanopoulos, G., 2010. Isoprenoid pathway optimization for taxol precursor overproduction in *Escherichia coli*. Science 330, 70–74.

Amini, S.A., Shabani, L., Afghani, L., Jalalpour, Z., Sharifi-tehrani, M., 2014. Squalestatin-induced production of taxol and baccatin in cell suspension culture of yew (*Taxus baccata* L.). Turk. J. Biol. 38, 528–536.

Appendino, G., 2002. The story of taxol. In: Goodman, J., Walsh, V. (Eds.), Nature and Politics in the Pursuit of an Anti-Cancer Drug. Cambridge University Press, Cambridge, UK, p. 282.

Bailey, J.E., 1991. Toward a science of metabolic engineering. Science 252, 1668–1675.

Bemani, E., Ghanati, F., Boroujeni, L.Y., Khatami, F., 2012. Antioxidant activity, total phenolics and taxol contents response of hazel (*Corylus avellana* L.) cells to benzoic acid and cinnamic acid. Not. Bot. Horti Agrobo. 40, 69–73.

Bender, C.L., Alarcon-Chaidez, F., Gross, D.C., 1999. *Pseudomonas syringae* phytotoxins: mode of action, regulation, and biosynthesis by peptide and polyketide synthetases. Microbiol. Mol. Biol. Rev. 63, 266–292.

Bentebibel, S., Moyano, E., Palazon, J., Cusido, R.M., Bonfill, M., Eibl, R., Pinol, M.T., 2005. Effects of immobilization by entrapment in alginate and scale-up on paclitaxel and baccatin III production in cell suspension cultures of *Taxus baccata*. Biotechnol. Bioeng. 89, 647–655.

Besumbes, Ó., Sauret-Güeto, S., Phillips, M.A., Imperial, S., Rodríguez-Concepción, M., Boronat, A., 2004. Metabolic engineering of isoprenoid biosynthesis in *Arabidopsis* for the production of taxadiene, the first committed precursor of Taxol. Biotechnol. Bioeng. 88, 168–175.

Bian, G., Yuan, Y., Tao, H., Shi, X., Zhong, X., Han, Y., Fu, S., Fang, C., Deng, Z., Liu, T., 2017. Production of taxadiene by engineering of mevalonate pathway in *Escherichia coli* and endophytic fungus *Alternaria alternata* TPF6. Biotechnol. J. 12 (4), 1600697.

Biggs, B.W., Lim, C.G., Sagliani, K., Shankar, S., Stephanopoulos, G., De Mey, M., Ajikumar, P.K., 2016. Overcoming heterologous protein interdependency to optimize P450-mediated Taxol precursor synthesis in *Escherichia coli*. Proc. Natl. Acad. Sci. 113, 3209–32114.

Broun, P., 2004. Transcription factors as tools for metabolic engineering in plants. Curr. Opin. Plant Biol. 7, 202–209.

Chatzivasileiou, A.O., Ward, V., Edgar, S.M., Stephanopoulos, G., 2019. Two-step pathway for isoprenoid synthesis. Proc. Natl. Acad. Sci. U. S. A. 116, 506–511.

Coelho, M.O.C., Monteyne, A.J., Dunlop, M.V., Harris, H.C., Morrison, D.J., Stephens, F.B., Wall, B.T., 2020. Mycoprotein as a possible alternative source of dietary protein to support muscle and metabolic health. Nutr. Rev. 78, 486–497.

Croteau, R., Ketchum, R.E., Long, R.M., Kaspera, R., Wildung, M.R.J.P.R., 2006. Taxol biosynthesis and molecular genetics. Phytochem. Rev. 5, 75–97.

Cusidó, R.M., Palazón, J., Navia-Osorio, A., Mallol, A., Bonfill, M., Morales, C., Piñol, M.T., 1999. Production of Taxol® and baccatin III by a selected *Taxus baccata* callus line and its derived cell suspension culture. Plant Sci. 146, 101–107.

Cusidó, R.M., Palazón, J., Bonfill, M., Navia-Osorio, A., Morales, C., Piñol, M.T., 2002. Improved paclitaxel and baccatin III production in suspension cultures of *Taxus media*. Biotech. Progr. 18, 418–423.

DeJong, J.M., Liu, Y., Bollon, A.P., Long, R.M., Jennewein, S., Williams, D., Croteau, R.B., 2006. Genetic engineering of taxol biosynthetic genes in *Saccharomyces cerevisiae*. Biotechnol. Bioeng. 93, 212–224.

Dörnenburg, H., Knorr, D., 1995. Strategies for the improvement of secondary metabolite production in plant cell cultures. Enzyme Microbial Tech. 17, 674–684.

El-Bialy, H.A., El-Bastawisy, H.S., 2020. Elicitors stimulate paclitaxel production by endophytic fungi isolated from ecologically altered *Taxus baccata*. J. Radiat. Res. Appl. Sci. 13, 79–87.

Engels, B., Dahm, P., Jennewein, S., 2008. Metabolic engineering of taxadiene biosynthesis in yeast as a first step towards Taxol (Paclitaxel) production. Metab. Eng. 10 (3–4), 201–206.

Exposito, O., Syklowska-Baranek, K., Moyano, E., Onrubia, M., Bonfill, M., Palazón, J., Cusidó, R.M., 2010. Metabolic responses of *Taxus media* transformed cell cultures to the addition of methyl jasmonate. Biotechnol. Prog. 26, 1145–1153.

Farhadi, S., Moieni, A., Safaie, N., Sabet, M.S., Salehi, M., 2020. Fungal cell wall and methyl-beta-cyclodextrin synergistically enhance paclitaxel biosynthesis and secretion in *Corylus avellana* cell suspension culture. Sci. Rep. 10, 5427–5436.

Fett-Neto, A.G., Melanson, S.J., Nicholson, S.A., Pennington, J.J., Dicosmo, F., 1994. Improved taxol yield by aromatic carboxylic acid and amino acid feeding to cell cultures of *Taxus cuspidata*. Biotechnol. Bioeng. 44, 967–971.

Fonseca, S., Chico, J.M., Solano, R., 2009. The jasmonate pathway: the ligand, the receptor and the core signalling module. Curr. Opin. Plant Biol. 12, 539–547.

Goodman, J., Walsh, V., 2001. The Story of Taxol: Nature and Politics in the Pursuit of an Anti-Cancer Drug. Cambridge University Press.

Guo, B.H., Kai, G.Y., Hb, J., Tang, K.X., 2006. Taxol synthesis. Afr. J. Biotechnol. 5, 15–20.

Hoa, C.K., Chang, S.H., Lung, J., Tsaib, C.J., Chen, K.P., 2005. The strategies to increase taxol production by using *Taxus mairei* cells transformed with TS and DBAT genes. Int. J. Appl. Sci. Eng. 3, 179–185.

Huang, Q., Roessner, C.A., Croteau, R., Scott, A.I., 2001. Engineering *Escherichia coli* for the synthesis of taxadiene, a key intermediate in the biosynthesis of taxol. Bioorg. Med. Chem. 9, 2237–2242.

Jennewein, S., Wildung, M.R., Chau, M., Walker, K., Croteau, R., 2004. Random sequencing of an induced *Taxus* cell cDNA library for identification of clones involved in Taxol biosynthesis. Proc. Natl. Acad. Sci. 101, 9149–9154.

Kashani, K., Jalali Javaran, M., Sabet, M.S., Moieni, A., 2018. Identification of rate-limiting enzymes involved in paclitaxel biosynthesis pathway affected by coronatine and methyl-beta-cyclodextrin in *Taxus baccata* L. cell suspension cultures. DARU 26, 129–142.

Ketchum, R.E.B., Gibson, D.M., Croteau, R.B., Shuler, M.L., 1999. The kinetics of taxoid accumulation in cell suspension cultures of *Taxus* following elicitation with methyl jasmonate. Biotechnol. Bioeng. 62, 97–105.

Kundu, S., Jha, S., Ghosh, B.J.T., 2017. In: Jha, S. (Ed.), Transgenesis and Secondary Metabolism. Reference Series in Phytochemistry. Springer, Cham, pp. 463–484, https://doi.org/10.1007/978-3-319-28669-3_29.

Lenka, S.K., Boutaoui, N., Paulose, B., Vongpaseuth, K., Normanly, J., Roberts, S.C., Walker, E.L., 2012. Identification and expression analysis of methyl jasmonate responsive ESTs in paclitaxel producing *Taxus cuspidata* suspension culture cells. BMC Genomics 13, 1–10.

Lenka, S.K., Nims, N.E., Vongpaseuth, K., Boshar, R.A., Roberts, S.C., Walker, E.L., 2015. Jasmonate-responsive expression of paclitaxel biosynthesis genes in *Taxus cuspidata* cultured cells is negatively regulated by the bHLH transcription factors TcJAMYC1, TcJAMYC2, and TcJAMYC4. Front. Plant Sci. 6, 115.

Li, Y.C., Tao, W.Y., Cheng, L., 2009. Paclitaxel production using co-culture of *Taxus* suspension cells and paclitaxel-producing endophytic fungi in a co-bioreactor. Appl. Microbiol. Biotechnol. 83, 233–239.

Li, S.T., Fu, C.H., Zhang, M., Zhang, Y., Xie, S., Yu, L.J., 2012. Enhancing taxol biosynthesis by overexpressing a 9-cis-epoxycarotenoid dioxygenase gene in transgenic cell lines of *Taxus chinensis*. Plant Mol. Biol. Rep. 30, 1125–1130.

Li, M., Jiang, F., Yu, X., Miao, Z., 2015. Engineering isoprenoid biosynthesis in *Artemisia annua* L. for the production of taxadiene: a key intermediate of taxol. Biomed Res. Int. 2015, 504932.

Li, J., Mutanda, I., Wang, K., Yang, L., Wang, J., Wang, Y., 2019. Chloroplastic metabolic engineering coupled with isoprenoid pool enhancement for committed taxanes biosynthesis in *Nicotiana benthamiana*. Nat. Commun. 10, 1–12.

Linden, J.C., Phisalaphong, M., 2000. Oligosaccharides potentiate methyl jasmonate-induced production of paclitaxel in *Taxus canadensis*. Plant Sci. 158, 41–51.

Liu, W.C., Gong, T., Zhu, P., 2016. Advances in exploring alternative Taxol sources. RSC Adv. 6, 48800–48809.

Lu, X., Tang, K., Li, P., 2016. Plant metabolic engineering strategies for the production of pharmaceutical terpenoids. Front. Plant Sci. 7, 1647.

Malik, S., Cusido, R.M., Mirjalili, M.H., Moyano, E., Palazon, J., Bonfill, M., 2011. Production of the anticancer drug taxol in *Taxus baccata* suspension cultures: a review. Process Biochem. 46, 23–34.

Miele, M., Mumot, A.M., Zappa, A., Romano, P., Ottaggio, L., 2012. Hazel and other sources of paclitaxel and related compounds. Phytochem. Rev. 11, 211–225.

Naik, P.M., Al-Khayri, J.M., 2016. Impact of abiotic elicitors on in vitro production of plant secondary metabolites: a review. J. Adv. Res. Biotechnol. 1, 1–7.

Newman, M.A., Sundelin, T., Nielsen, J.T., Erbs, G., 2013. Mamp (microbe-associated molecular pattern) triggered immunity in plants. Front. Plant Sci. 4, 139–153.

Ojima, I., Lichtenthal, B., Lee, S., Wang, C., Wang, X., 2016. Taxane anticancer agents: a patent perspective. Expert Opin. Ther. Pat. 26, 1–20.

Onrubia, M., Cusidó, R.M., Ramirez, K., Hernández-Vázquez, L., Moyano, E., Bonfill, M., Palazon, J., 2013a. Bioprocessing of plant in vitro systems for the mass production of pharmaceutically important metabolites: paclitaxel and its derivatives. Curr. Med. Chem. 20, 880–891.

Onrubia, M., Moyano, E., Bonfill, M., Cusido, R.M., Goossens, A., Palazon, J., 2013b. Coronatine, a more powerful elicitor for inducing taxane biosynthesis in *Taxus media* cell cultures than methyl jasmonate. J. Plant Physiol. 170, 211–219.

Parsaeimehr, A., Sargsyan, E., Vardanyan, A., 2011. Expression of secondary metabolites in plants and their useful perspective in animal health. ABAH Bioflux 3, 115–124.

Perdue Jr., R.E., Hartwell, J.L., 1969. The search for plant sources of anticancer drugs. Morris Arbor. Bull 20, 35–53.

Qiao, W., Ling, F., Yu, L., Huang, Y., Wang, T., 2017. Enhancing taxol production in a novel endophytic fungus, *Aspergillus aculeatinus* Tax-6, isolated from *Taxus chinensis* var. *mairei*. Fungal Biol. 121, 1037–1044.

Rahpeyma, S.A., Moieni, A., Jalali Javaran, M., 2015. Paclitaxel production is enhanced in suspension-cultured hazel (*Corylus avellana* L.) cells by using a combination of sugar, precursor, and elicitor. Eng. Life Sci. 15, 234–242.

Ramirez-Estrada, K., Vidal-Limon, H., Hidalgo, D., Moyano, E., Golenioswki, M., Cusidó, R.M., Palazon, J.J.M., 2016. Elicitation, an effective strategy for the biotechnological production of bioactive high-added value compounds in plant cell factories. Molecules 21, 182. https://doi.org/10.3390/molecules21020182.

Rontein, D., Onillon, S., Herbette, G., Lesot, A., Werck-Reichhart, D., Sallaud, C., Tissier, A., 2008. CYP725A4 from yew catalyzes complex structural rearrangement of taxa-4(5),11(12)-diene into the cyclic ether 5(12)-oxa-3(11)-cyclotaxane. J. Biol. Chem. 283, 6067–6075.

Sabater-Jara, A.B., Onrubia, M., Moyano, E., Bonfill, M., Palazon, J., Pedreno, M.A., Cusido, R.M., 2014. Synergistic effect of cyclodextrins and methyl jasmonate on taxane production in *Taxus x media* cell cultures. Plant Biotechnol. J. 12, 1075–1084.

Sabzehzari, M., Zeinali, M., Naghavi, M.R., 2020. Alternative sources and metabolic engineering of Taxol: advances and future perspectives. Biotechnol. Adv. 43, 107569–107850.

Salehi, M., Moieni, A., Safaie, N., Farhadi, S., 2018. Elicitors derived from endophytic fungi *Chaetomium globosum* and *Paraconiothyrium brasiliense* enhance paclitaxel production in *Corylus avellana* cell suspension culture. Plant Cell Tissue Organ Cult. 136, 161–171.

Salehi, M., Moieni, A., Safaie, N., Farhadi, S., 2019. New synergistic co-culture of *Corylus avellana* cells and *Epicoccum nigrum* for paclitaxel production. J. Ind. Microbiol. Biotechnol. 46, 613–623.

Salehi, M., Moieni, A., Safaie, N., Farhadi, S., 2020. Whole fungal elicitors boost paclitaxel biosynthesis induction in *Corylus avellana* cell culture. PLoS One 15, e0236191.

Sharma, M., Koul, A., Sharma, D., Kaul, S., Swamy, M.K., Dhar, M.K., 2019. Metabolic engineering strategies for enhancing the production of bio-active compounds from medicinal plants. In: Akhtar, M.S., Swamy, M.K. (Eds.), Natural Bio-Active Compounds. Springer, Singapore, pp. 287–316.

Soliman, S.S., Raizada, M.N., 2013. Interactions between Co-Habitating fungi elicit synthesis of Taxol from an endophytic fungus in host *Taxus* plants. Front. Microbiol. 4, 3–17.

Somjaipeng, S., Medina, A., Magan, N., 2016. Environmental stress and elicitors enhance taxol production by endophytic strains of *Paraconiothyrium variabile* and *Epicoccum nigrum*. Enzyme Microb. Technol. 90, 69–75.

Strobel, G.A., Stierle, A., Van Kuijk, F.J.G.M., 1992. Factors influencing the in vitro production of radiolabeled taxol by Pacific yew, *Taxus brevifolia*. Plant Sci. 84, 65–74.

Subban, K., Subramani, R., Srinivasan, V.P.M., Johnpaul, M., Chelliah, J., 2019. Salicylic acid as an effective elicitor for improved taxol production in endophytic fungus *Pestalotiopsis microspora*. PLoS One 14, e0212736.

Sykłowska-Baranek, K., Grech-Baran, M., Naliwajski, M.R., Bonfill, M., Pietrosiuk, A., 2015. Paclitaxel production and PAL activity in hairy root cultures of *Taxus x media* var. *hicksii* carrying a taxadiene synthase transgene elicited with nitric oxide and methyl jasmonate. Acta Physiol. Plant. 37, 1–9.

Vidal-Limon, H., Sanchez Muñoz, R., Khojasteh, A., Moyano, E., Cusido, R.M., Palazon, J., 2018. Taxus cell cultures: an effective biotechnological tool to enhance and gain new biosynthetic insights into taxane production. In: Pavlov, A., Bley, T. (Eds.), Bioprocessing of Plant in Vitro Systems. Springer, pp. 295–316.

Walker, K., Croteau, R., 2001. Taxol biosynthetic genes. Phytochemistry 58, 1–7.

Wall, M.E., Wani, M.C., 1995. Camptothecin and taxol: discovery to clinic—thirteenth Bruce F. Cain Memorial Award Lecture. Cancer Res. 55, 753–760.

Walsh, V., Goodman, J., 2002. From taxol to taxol®: the changing identities and ownership of an anti-cancer drug. Med. Anthropol. 21, 307–336.

Wang, Y.D., Wu, J.C., Yuan, Y.J., 2007. Salicylic acid-induced taxol production and isopentenyl pyrophosphate biosynthesis in suspension cultures of *Taxus chinensis* var. *mairei*. Cell Biol. Int. 31, 1179–1183.

Wang, T., Chen, Y., Zhuang, W., Zhang, F., Shu, X., Wang, Z., Yang, Q., 2019. Transcriptome sequencing reveals regulatory mechanisms of taxol synthesis in *Taxus wallichiana* var. *mairei*. Int. J. Genomics 2019, 1596895.

Wani, M.C., Taylor, H.L., Wall, M.E., Coggon, P., McPhail, A.T., 1971. Plant antitumor agents. VI. The isolation and structure of taxol, a novel antileukemic and antitumor agent from *Taxus brevifolia*. J. Am. Chem. Soc. 93, 2325–2327.

Waugh, W.N., Trissel, L.A., Stella, V.J.J.A.J.O.H.P., 1991. Stability, compatibility, and plasticizer extraction of taxol (NSC-125973) injection diluted in infusion solutions and stored in various containers. Am. J. Hosp. Pharm. 48, 1520–1524.

Weaver, B.A., 2014. How Taxol/paclitaxel kills cancer cells. Mol. Biol. Cell 25, 2677–2681.

Woo, J.W., Kim, J., Kwon, S.I., Corvalán, C., Cho, S.W., Kim, H., Kim, S.G., Kim, S.T., Choe, S., Kim, J.S., 2015. DNA-free genome editing in plants with preassembled CRISPR-Cas9 ribonucleoproteins. Nat. Biotechnol. 33, 1162–1164.

Ye, V.M., Bhatia, S.K.J.B.J., 2012. Metabolic engineering for the production of clinically important molecules: Omega-3 fatty acids, artemisinin, and taxol. Biotechnol. J. 7, 20–33.

Yuan, Y.J., Li, C., Hu, Z.D., Wu, J.C., Zeng, A.P., 2002. Fungal elicitor-induced cell apoptosis in suspension cultures of *Taxus chinensis* var. mairei for taxol production. Process Biochem. 38, 193–198.

Yukimune, Y., Tabata, H., Higashi, H., Hara, Y., 1996. Methyl jasmonate-induced overproduction of paclitaxel and baccatin III in *Taxus* cell suspension cultures. Nat. Biotechnol. 14, 1129–1132.

Zamir, L.O., Nedea, M.E., Belair, S., Sauricol, F., Mamer, O., Jacqmain, E., Jean, F.I., Garneau, F.X., 1992. Taxanes isolated from *Taxus canadensis*. Tetrahedron Lett. 33, 5235–5236.

Zhang, C.H., Mei, X.G., Liu, L., Yu, L.J., 2000. Enhanced paclitaxel production induced by the combination of elicitors in cell suspension cultures of *Taxus chinensis*. Biotechnol. Lett. 22, 1561–1564.

Zhang, M., Li, S., Nie, L., Chen, Q., Xu, X., Yu, L., Fu, C., 2015. Two jasmonate-responsive factors, TcERF12 and TcERF15, respectively act as repressor and activator of tasy gene of taxol biosynthesis in *Taxus chinensis*. Plant Mol. Biol. 89, 463–473.

Zhao, Y., Thilmony, R., Bender, C.L., Schaller, A., He, S.Y., Howe, G.A., 2003. Virulence systems of *Pseudomonas syringae pv.* tomato promote bacterial speck disease in tomato by targeting the jasmonate signaling pathway. Plant J. 36, 485–499.

Zhu, L.Y., Chen, L.Q., 2019. Progress in research on paclitaxel and tumor immunotherapy. Cell. Mol. Biol. Lett. 24, 1–11.

10

Paclitaxel and chemoresistance

Zhuo-Xun Wu[a], Jing-Quan Wang[a], Qingbin Cui[a,b],
Xiang-Xi Xu[c], and Zhe-Sheng Chen[a]

[a]College of Pharmacy and Health Sciences, St. John's University, Queens, NY, United States, [b]Department of Cancer Biology, University of Toledo College of Medicine and Life Sciences, Toledo, OH, United States, [c]Department of Radiation Oncology, Sylvester Comprehensive Cancer Center, University of Miami Miller School of Medicine, Miami, FL, United States

10.1 Introduction

Paclitaxel (Taxol), firstly isolated from Pacific yew tree, is a broad-spectrum anticancer drug. It was approved for clinical use by US Food and Drug Administration (FDA) in 1992 for ovarian cancer treatment. Since then, paclitaxel has become a well-known and widely used chemotherapeutic drug for ovarian cancer, breast cancer, lung cancer, and other types of cancer (Yusuf et al., 2003). Paclitaxel is a microtubule stabilizing agent and the major mechanism of paclitaxel in chemotherapy is the selective binding to the subunit β of tubulin proteins (Bernabeu et al., 2017). Tubulin proteins and other microtubules play crucial roles in the mitotic apparatus for eukaryotic cells. Bound with paclitaxel will cause the polymerization and assembly of β-tubulin which will consequently stabilize the formation of microtubules. As a result, mitotic spindle are formed with dysfunction to cause mitotic arrest at G2/M phase, which eventually leads to cell apoptosis (Snyder et al., 2001). Besides the direct binding to β-tubulin, paclitaxel is also reported to inhibit tumor angiogenesis and cell growth as well as inducing cell apoptosis by modulating the expression of cytokines (Taghian et al., 2005). Therefore, paclitaxel exerts anticancer activity via the combination of cytotoxicity and antiproliferative effect.

In this chapter, we summarized the mechanisms and clinical markers of paclitaxel resistance. In addition, the current strategies to tackle paclitaxel resistance are included and discussed as well.

Paclitaxel. https://doi.org/10.1016/B978-0-323-90951-8.00002-3
Copyright © 2022 Elsevier Inc. All rights reserved.

10.2 Mechanisms of chemoresistance

10.2.1 Alteration of microtubule dynamic

Microtubules, composed of α- and β-heterodimers, are dynamic polymers. The growth or shortening of microtubules are due to addition or deletion of tubulin at their ends (Goodson and Jonasson, 2018). Given that paclitaxel targets tubulin, the altered microtubule system potentially led to paclitaxel resistance. For example, βIII-tubulin isotype has been associated with paclitaxel resistance various types of cancers (Kamath et al., 2005). An in vitro study involving Chinese hamster ovary cells expressing βI- or βIII-tublin showed the dynamic instability of microtubules was suppressed to a lesser extent in cells overexpressing βIII-tubulin than βI-tubulin (Kamath et al., 2005). Such results indicate that microtubule dynamics potentially differentiate among tubulin isotypes. Additionally, paclitaxel regulated the expression level of tubulin isotypes to different extents. For example, in paclitaxel-selected human prostate carcinoma DU145 cells, the expression level of βIII-tubulin increased by 9-fold while βIV-tubulin increased by 5-fold. Moreover, the expression level of β-tubulin isotypes showed positive correlation with the fold-resistance to paclitaxel (Ranganathan et al., 1998). Similar results were also reported in epithelial ovarian tumors, where paclitaxel-resistant tumor cells showed significant increase in certain β-tubulin isotypes such as classes I, III, and IV with 3.6- to 7.6-fold resistance (Kavallaris et al., 1997). βIII-tubulin-related paclitaxel resistance was also observed in ovarian cancer patients: significant mRNA and protein expressionup-regulation of βIII-tubulin were observed in resistant groups (Dewanjee et al., 2017; Mozzetti et al., 2005).

10.2.2 ABC transporters mediated MDR

Cancer multidrug resistance (MDR) mediated by ABC transporters has been extensively studied in the past decades. ABC transporters belong to a superfamily of membrane protein composed of seven subfamilies, namely ABCA to ABCG. Among all ABC transporters, P-glycoprotein, or ABCB1, is one of the most well-studied (Dean, 2001). Like other ABC transporters, ABCB1 shares the consensus structural features such as transmembrane domains (TMDs) and nucleotide binding domains (NBDs). Some ABCB1 substrates, including paclitaxel, have been confirmed by cryo-EM structures to bind at the pockets buried in TMDs (Alam et al., 2019). Upon binding, paclitaxel could induce an occluded conformation and a closure of NBDs, which will initialize the conformational change from inward-facing to outward-facing and translocate paclitaxel out of the cell (Alam et al., 2019). As a result, ABCB1 is responsible for mediating the chemoresistance to paclitaxel due to the decreased intracellular accumulation and thus deteriorated therapeutic effects (Feng et al., 2020b). Studies on overcoming paclitaxel resistance mediated by ABCB1 are mostly focusing on inhibiting the transportation via small molecule inhibitors, regulating the protein expression level or silencing the ABCB1 gene expression (Daood et al., 2008; Lage, 2016).

Another member of ABC transporter MRP7 or ABCC10, was also reported to mediate the drug resistance to paclitaxel, which was validated in transfected HEK293 cells and mice xenograft models (Hopper-Borge et al., 2011). Moreover, the resistance could be antagonized by inhibitors such as tariquidar and cepharanthine (Sun et al., 2013; Zhou et al., 2009). It is worth noting that both tariquidar and cepharanthine also exhibited inhibitory effects on ABCB1 thus ABCB1-mediated paclitaxel-resistance. Additionally, synthetic compounds displayed reversal effects in paclitaxel-resistant-ABCC10-overexpressing cells. For example, a synthetic 1, 2, 3-triazole-pyrimidine hybrid was reported to antagonize paclitaxel resistance in transfected HEK/ABCC10 cells (Wang et al., 2020).

Besides ABCB1 and ABCC10, other ABC transporters were also reported to confer resistance to paclitaxel in cancer cells. For example, Overbeck et al. (2013) discovered ABCA3 expressed in non-small-cell lung cancer cells significantly affects the susceptibility to paclitaxel and cisplatin (Overbeck et al., 2013). ABCC2, also named MRP2, was found to transport paclitaxel in transfected MDCKII-MRP2 cells. Moreover, co-administration of probenecid stimulates the activity of ABCC2 due to unknown drug-drug interactions (Huisman et al., 2005). A recent study in 2017 revealed that the forkhead box molecule 1 (FOXM1) along with ABCC5 contributes to paclitaxel resistance in nasopharyngeal carcinoma cells (Hou et al., 2017).

10.2.3 Genetic factor-related paclitaxel resistance

A number of genes have been implicated in modulating paclitaxel resistance, including tumor-suppressor genes (TSGs) and Bcl family. Xu et al. (2016) identified a set of TSGs that are highly related to paclitaxel resistance. These genes, including BRCA1, TP53, PTEN, APC, can affect the therapeutic response of certain cancers. BRCA1 is mostly associated with breast and ovarian cancers (Narod and Foulkes, 2004). Several studies suggested that BRCA1 mutation or expression down-regulation may confer drug resistance to paclitaxel-based chemotherapy (Chabalier et al., 2006; Gilmore et al., 2004). TP53 is another major TSG that regulates a broad range of cellular processes, such as cell cycle progression, apoptosis, and DNA repair (Meek, 2015). In vitro studies have confirmed that the increased intracellular p53 level may sensitize cancer cells to paclitaxel (Guntur et al., 2010). In vivo study also suggested that TP53 mutation may correlate with response of paclitaxel-based chemotherapy (Lu and El-Deiry, 2009). In addition, Seagle et al. (2015) demonstrated that in ovarian cancer, TP53 hotspot mutations may further upregulate ABCB1 expression level, thereby conferring resistance to paclitaxel (Seagle et al., 2015). While the relationship between these genes and paclitaxel resistance has been extensively studied, more studies are needed to decipher the network.

Considerable evidence has suggested that Bcl-2 family of proteins may participate in the development of paclitaxel resistance. The anti-apoptotic Bcl-2 and Bcl-XL proteins were found to confer resistance to paclitaxel. Chun and Lee (2004) reported that the efficacy of paclitaxel in hepatocellular cancer cell line SNU-398 was attenuated due to the high levels of Bcl-2 and Bcl-XL. In another study, Shi et al. (2017) demonstrated that the downregulation of Bcl-2 and Bcl-XL can significantly increase the cytotoxicity of paclitaxel and reverse the drug resistance in paclitaxel-resistant esophageal cancer cell line EC109/PTX (Shi et al., 2017). Similar results were also observed in other studies, confirming that Bcl-2 family proteins are important regulators of paclitaxel resistance (Akyol et al., 2016; Ferlini et al., 2003; Tabuchi et al., 2009).

10.2.4 Other potential mechanisms of resistance

Besides all the aforementioned mechanisms of paclitaxel resistance, several minor reasons have also been confirmed to cause paclitaxel resistance. As well-evolved cells with indomitable properties, cancer cells develop varied defensive systems to combat toxic effects of any forms induced by exocentric agents, including paclitaxel.

10.2.4.1 PI3K/AKT activation

Phosphatidylinositol 3′-kinase (PI3K) plays a key role in multiple biological and pathophysiological processes, particularly in carcinogenesis and drug resistance in cancers. Growing evidence has suggested that the activation of PI3K/AKT pathway contributes to the acquired resistance of paclitaxel (Liu et al., 2020, 2015, 2019; Wu et al., 2014). PI3K/AKT may cause paclitaxel resistance via several mechanisms. First, it may enhance the expression of membrane transporters ABCB1 and ABCC1, both of which can transport paclitaxel out of cancer cells, leading to drug resistance (Tazzari et al., 2007). In paclitaxel-resistant lung cancer A549 cells, the activated PI3K/AKT pathway was found to be accompanied with upregulated expression of ABCB1 as compared to parent A549 cells (Li et al., 2020), similar results were also observed in paclitaxel-resistant human ovarian cancer A2780/Taxol cells (Huang et al., 2017). Second, PI3K/ AKT may activate other critical survival pathways that may combat the apoptosis induced by paclitaxel (Liu et al., 2020). For instance, the activation of PI3K/AKT can activate antiapoptotic protein, such as Bcl-2 family (Asnaghi et al., 2004; MacKeigan et al., 2002), members of Inhibitor of Apoptosis proteins (IAP) (Gagnon et al., 2008), etc. These activated antiapoptotic proteins may also contribute in paclitaxel resistance in cancers.

10.2.4.2 MAPK

Mitogen-activated protein kinase (MAPK) family, composed with extracellular signal-regulated kinase (ERK), c-Jun NH2-terminal kinase (JNK), and the p38 kinases, etc., are in charge of responding to a diverse range of stimuli in cell biology, including growth factors, hormones, neurotransmitters, and cellular stress, controlling fundamental cellular processes such as growth, proliferation, differentiation, migration, and apoptosis (Dhillon et al., 2007). Research showed that members of MAPK family proteins were activated in paclitaxel-treated H460/R and 226B/R lung cancer cells (Park et al., 2016), human ovarian carcinoma A2780 cells (Lu et al., 2007), HeLa and Chinese hamster ovary cells (McDaid and Horwitz, 2001), suggesting their role in inducing paclitaxel resistance. In addition, the selective inhibition of phospho-ERK by PD98059 and U0126 in HeLa cells, or p38MAPK inhibitor SB203580 in A2780, H460/R and 226B/R cells could remarkably enhance paclitaxel's efficacies (McDaid and Horwitz, 2001; Lu et al., 2007; Park et al., 2016).

10.2.4.3 Antioxidative systems

Paclitaxel-treated cancer cells, which show resistant profile, were found to possess higher level of ROS via enhancing the activity of NADPH oxidase (NOX) associated with plasma membranes (Alexandre et al., 2007). The activation/upregulation of antioxidative systems may lead to paclitaxel resistance as well (Ramanathan et al., 2005; Cui et al., 2018). Antioxidant proteins such as catalase, glutathione peroxidase, cystathionine β-synthase, glutathione and glutathione-synthesis associated proteins, etc., were found to be upregulated in paclitaxel-resistant lung cancer A549 cells (Datta et al., 2017), breast cancer MCF-7 cells (Liebmann et al., 1993), ovarian cancer cells (Nunes and Serpa, 2018; Feng et al., 2020a), as well as many other resistant cancer cells. The overexpressed antioxidant proteins may either eliminate ROS induced by paclitaxel, or mediate certain cell pro-apoptotic pathway, both of which may counteract the cytotoxicity of paclitaxel and lead to its resistance (Xue et al., 2020). Consequently, the inhibition of associated antioxidant proteins could sensitize paclitaxel to resistant cancer cells, further verifying their role in inducing resistance (Sugiyama et al., 2020; Cui et al., 2018). While it's worth mentioning that, there are other paclitaxel-resistant cancer cells that show lower level of certain antioxidant proteins (Bi et al., 2014), suggesting that the antioxidative systems-mediated resistant mechanism is cancer type/cell dependent, requiring further study.

10.2.4.4 IRF9

Interferon (IFN) signaling plays a critical role in the immune response and regulates pathways involved not only in antiviral defense, but also in cancer cells proliferation, apoptosis, and metastasis. IFN regulatory factors (IRFs), which modulated signal transduced by IFNs, are often dysregulated in certain resistant cancer cells (Alsamman and El-Masry, 2018; Yi et al., 2013). The overexpression of interferon regulatory factor 9 (IRF9) may confer resistance to antimicrotubule agents in breast, colon, and ovarian cancer cells (Kolosenko et al., 2015). Luker et al. (2001) reported that in paclitaxel-resistant clonal sublines of MCF-7 breast adenocarcinoma cells and in the tissues of pretreatment patient samples of breast and uterine tumors, the levels of signal transducer and activator of transcription 1, 2 (STAT1, 2), and IRF9 were all significantly upregulated, while only IRF9 was found to play the major role in the resistance profile of paclitaxel. The authors showed that the overexpression of IRF9 by transient transfection caused paclitaxel resistance (Luker et al., 2001); meanwhile, functional lacking of IRF9 sensitized a fibrosarcoma cell line to several chemotherapeutic drugs including paclitaxel (Weihua et al., 1997), further verifying IRF9's critical role in inducing paclitaxel resistance.

10.2.4.5 Other key factors

Other minor mechanisms that involved in the resistance of paclitaxel include plasma membrane calcium ATPase 2 (PMCA2) (Baggott et al., 2012), sorcin, another calcium-associated protein (Parekh et al., 2002), antiapoptotic survivin (Zaffaroni et al., 2002) and its interplay with BRAC1 (Promkan et al., 2009, 2011), intracellular mitochondrial protein MM-TRAG (also named as MGC4175) (Duan et al., 2004), as well as epidermal growth factor receptor (EGFR) (Cheng et al., 2011).

The above information may provide hints of therapeutic implications via combinational therapy to overcome paclitaxel resistance.

10.3 Clinical markers of paclitaxel resistance

The dramatic progression in clinical diagnosis, screening, and genetic sequencing can provide the prediction of paclitaxel responses among cancer patients, achieving better medication programs. Apart from the above discussed major and minor mechanisms that can serve as response and prognosis indicators, there are other key factors that can also predict the paclitaxel response in cancer patients.

Among breast cancer patients, Feng et al. (2019) screened and identified about 90 genes in regulating apoptosis, cell cycle, and DNA damage and repair, which could serve as the prediction parameter of nonresponding individuals ER + and ER − breast cancer patients. Another study conducted among HER2-positive breast cancer patients, Koo et al. (2015) identified that the high level of Tau-protein, a microtubule-associated protein, and low level of PTEN, a tumor suppressor, were correlated with poor response to paclitaxel plus trastuzumab (a monoclonal antibody targeting HER2. In 2012, a phase II clinical trial enrolled 46 patients with stage III (plus stage II with tumor ≥ 5 cm) breast cancer revealed that the markers such as estrogen receptor (ER) and ki-67 (a key protein regulating cells proliferation), IAP member survivin and MAPK member pERK, can predict the response of a combinational therapy composed with three chemotherapeutics including gemcitabine, doxorubicin, and paclitaxel (Sanchez-Rovira et al., 2012). Recently, Rodrigues-Ferreira et al. (2019) reported that *MTUS1* gene, and its product, ATIP3 whose low level can predict the response among breast cancer patients toward paclitaxel (Rodrigues-Ferreira et al., 2019). In addition, a previous in vitro study in the same group also showed that the depletion of ATIP3 combined with paclitaxel could improve the treatment outcomes in metastatic breast tumors (Molina et al., 2013). A study analyzing the paclitaxel response rate among 140 women with locally advanced breast cancer (LABC) showed that the levels of four single nucleotide polymorphisms (SNPs) including *LPHN2*, *ROBO1*, *SNTG1*, and *GRIK1* genes negative correlated with paclitaxel response rate, providing key information for personal medicine for patients of LABC (Perez-Ortiz et al., 2017). In patients of triple-negative breast cancer (TNBC), the *OTUD7B* transcripts, AT-rich interaction domain 1A (ARID1A), positively correlated with paclitaxel response and the poor prognosis as shown in Chiu et al.' studies (Chiu et al., 2018; Lin et al., 2018). Other biomarker in advanced breast cancer includes class III β-tubulin (Wang et al., 2013) which also could serve as a prognostic biomarker for poor overall survival in ovarian cancer patients (Roque et al., 2014) and lung cancer patients (Zhang et al., 2012) upon paclitaxel treatment.

Among ovarian cancer patients, CA-125 is a well-known marker for paclitaxel response (Paulsen et al., 2000; Meyer and Rustin, 2000). Komatsu et al. (2006) identified and developed formulae composed with three genes that can predict the response of paclitaxel combined with platinum-base drugs, including *TNFSF13B*, *IFIT3* and *BTN3A2*.

RASSF1A gene (Kassler et al., 2012), markers of cancer stem cells (CSCs) including ALDH, CD133, and bone morphogenetic protein 2 (BMP2) (van Zyl et al., 2018), etc.

Among pancreatic cancer patients, high level of ADAM12, a member of the A Disintegrin And Metalloproteases (ADAM) protein family that regulate cell adhesion, was found to be a predicting marker for paclitaxel response, as confirmed in the blood samples of 60 patients, as well as in the patient-derived xenograft model (Veenstra et al., 2018). Other potential predicting marker includes CA19-9 (Chiorean et al., 2016).

10.4 Strategies to overcome paclitaxel resistance

10.4.1 Overcoming paclitaxel resistance with novel tubulin inhibitors

Another well-established paclitaxel resistance mechanism is the alteration in the expression patterns of tubulin. A major strategy to tackle paclitaxel resistance mediated by tubulin isotypes is by developing novel microtubule-stabilizing agents. Examples include epothilones A and B, natural products with minimal structural similarity to paclitaxel. Epothilones A and B showed similar antiproliferative effects in sensitive human cell lines as paclitaxel, while in paclitaxel-resistant colon carcinoma and ovarian cancer cell lines epothilones A and B remained effective (Kowalski et al., 1997). In order to overcome paclitaxel resistance due to overexpressed βIII-tubulin, a seco-taxane IDN5390 was developed as a promising concomitant treatment with paclitaxel against resistant cancer. Specifically, IDB5390 exhibited selective activity in βIII-tubulin-overexpressing resistant cells. Paclitaxel/IDN5390 co-treatment showed strong synergism, which made them candidates for diverse spectrum drugs against tubulin isotypes (Ferlini et al., 2005).

Given that ABCB1-mediated MDR is one of the major factors causing paclitaxel resistance, novel tubulin inhibitors are designed to overcome MDR. For example, BMS-247550 (also known as ixabepilone) is a semisynthetic analog of epothilone with similar mechanism of action as paclitaxel (Lee et al., 2001). Preclinical studies showed that ixabepilone has potent anticancer efficacy against paclitaxel-resistant cancer cells with high ABCB1 expression or β-tubulin mutation. Subsequently, ixabepilone was approved by FDA for breast cancer treatment. Other novel inhibitors are still under development such as IMB5046 (Zheng et al., 2016), and DJ95 (Arnst et al., 2019).

10.4.2 Inhibiting the function of ABCB1 transporter

Overexpression of ABCB1 transporter has been characterized as one of the major mechanisms of paclitaxel resistance. Functional inhibitors of ABCB1 have been shown to reverse drug resistance and increase the efficacy of ABCB1 substrate drugs. To effectively overcome ABCB1-mediated MDR, researchers have developed several approaches, such as combining paclitaxel with ABCB1 reversal agents or using different drug carriers.

In the past decades, extensive studies have been performed on developing small molecule drugs targeting ABCB1 as well as repurposing preclinical and clinical drugs as ABCB1 inhibitors. A wide range of small molecule drugs have been validated in vitro and in vivo to exhibit ABCB1 inhibitory effects including TKIs such as gefitinib, poziotinib (Zhang et al., 2020; Wu et al., 2019), as well as natural products such as curcumin, cepharanthine, piperine (Dewanjee et al., 2017). These identified inhibitors can block the drug efflux process mediated by ABCB1 transporter, thereby increasing the intracellular accumulation of substrate drugs in MDR cancer cells (Wu et al., 2011). Although most of the small molecule ABCB1 inhibitors, including specific inhibitor, have sensitizing effects in animal models when combined with paclitaxel, none of them have been approved by FDA specifically for treating paclitaxel resistance due to increased toxicity, since ABCB1 also plays important role in protecting cells from xeno-toxins (Daood et al., 2008).

Another effective method to overcome ABCB1-mediated paclitaxel resistance is by utilizing new drug delivery system. Different types of drug carriers including nanoparticles, liposomes, micelles have been developed to overcome paclitaxel resistance. Xie et al. (2018) suggested that paclitaxel loaded tetrahedral DNA nanostructures can be a promising candidate to overcome ABCB1-mediated paclitaxel resistance. Zou et al. (2017) established a paclitaxel and borneol co-loaded PEG-PAMAM nanoparticle method that can effectively deliver paclitaxel into ABCB1-overexpressing cancer cells.

10.4.3 Inhibiting signal transduction pathways

As aforementioned, signal transduction pathways are also involved in the development of paclitaxel resistance. Therefore, research is conducted to identify effective inhibitors which targeting the signaling pathways such as NF-κB, PI3K/AKT, and EGFR, etc. Kawiak et al. found that a natural product plumbagin is able to overcome paclitaxel resistance in breast cancer cells (Kawiak et al., 2019). The reversal effect was achieved by downregulating ERK expression, thereby reducing p-ERK level and increasing apoptosis in resistant breast cancer cells. In another study conducted by Zhang et al. (2015) they reported that inhibiting the SET/PI3K/Akt pathway can effectively reverse paclitaxel resistance in a breast cancer MCF7/PTX cell line. Interestingly, downregulating the activity of SET/PI3K/AKT pathway contributes to the decreased expression level of ABCB1, which sensitizes the resistant cells to paclitaxel. Similar phenomenon was observed when inhibiting the TLR4, NF-κB (Sun et al., 2018), STAT3 (Seo et al., 2017), and MAPK (Mei et al., 2014) signaling pathways. As the major reversal mechanism, the inhibitions of these pathways commonly result in the downregulation of ABCB1, confirming that ABCB1 plays a crucial role in mediating paclitaxel resistance.

10.5 Summary

Since its approval in 1992, paclitaxel serves as one of the most successful and widely used chemotherapeutic drugs. Through binding to the β-tubulin, paclitaxel stabilizes the formation of microtubules, leading to apoptosis of cancer cells. However, drug resistance has been a major issue for chemotherapy and currently there is still no effective ways to completely overcome the drug resistance. Substantial studies have facilitated the understanding of resistance mechanism of paclitaxel. Generally, overexpression of ABCB1 transporter, alteration of β-tubulin dynamic as well as variation in genetic factors may confer cancer cells resistance to paclitaxel. To overcome the drug resistance, several strategies are being considered. These include the development of ABCB1 inhibitors, renovated drug delivery methods, or novel tubulin inhibitors to antagonize drug resistance mediated by ABCB1 or β-tubulin isotype/mutation. In clinical settings, paclitaxel resistance may be comprised of multiple mechanisms instead of single factor, making it difficult to develop a universal strategy to combat drug resistance. However, there is no doubt that overcoming paclitaxel resistance would be of great beneficial to cancer patients.

References

Akyol, Z., Çoker-Gürkan, A., Arisan, E.D., Obakan-Yerlikaya, P., Palavan-Ünsal, N., 2016. DENSpm overcame Bcl-2 mediated resistance against paclitaxel treatment in MCF-7 breast cancer cells via activating polyamine catabolic machinery. Biomed. Pharmacother. 84, 2029–2041.

Alam, A., Kowal, J., Broude, E., Roninson, I., Locher, K.P., 2019. Structural insight into substrate and inhibitor discrimination by human P-glycoprotein. Science 363, 753–756.

Alexandre, J., Hu, Y., Lu, W., Pelicano, H., Huang, P., 2007. Novel action of paclitaxel against cancer cells: bystander effect mediated by reactive oxygen species. Cancer Res. 67, 3512–3517.

Alsamman, K., El-Masry, O.S., 2018. Interferon regulatory factor 1 inactivation in human cancer. Biosci. Rep. 38 (3). BSR20171672.

Arnst, K.E., Wang, Y., Lei, Z.N., Hwang, D.J., Kumar, G., Ma, D., Parke, D.N., Chen, Q., Yang, J., White, S.W., Seagroves, T.N., Chen, Z.S., Miller, D.D., Li, W., 2019. Colchicine binding site agent DJ95 overcomes drug resistance and exhibits antitumor efficacy. Mol. Pharmacol. 96, 73–89.

Asnaghi, L., Calastretti, A., Bevilacqua, A., D'Agnano, I., Gatti, G., Canti, G., Delia, D., Capaccioli, S., Nicolin, A., 2004. Bcl-2 phosphorylation and apoptosis activated by damaged microtubules require mTOR and are regulated by Akt. Oncogene 23, 5781–5791.

Baggott, R.R., Mohamed, T.M., Oceandy, D., Holton, M., Blanc, M.C., Roux-Soro, S.C., Brown, S., Brown, J.E., Cartwright, E.J., Wang, W., Neyses, L., Armesilla, A.L., 2012. Disruption of the interaction between PMCA2 and calcineurin triggers apoptosis and enhances paclitaxel-induced cytotoxicity in breast cancer cells. Carcinogenesis 33, 2362–2368.

Bernabeu, E., Cagel, M., Lagomarsino, E., Moretton, M., Chiappetta, D.A., 2017. Paclitaxel: what has been done and the challenges remain ahead. Int. J. Pharm. 526, 474–495.

Bi, W., Wang, Y., Sun, G., Zhang, X., Wei, Y., Li, L., Wang, X., 2014. Paclitaxel-resistant HeLa cells have up-regulated levels of reactive oxygen species and increased expression of taxol resistance gene 1. Pak. J. Pharm. Sci. 27, 871–878.

Chabalier, C., Lamare, C., Racca, C., Privat, M., Valette, A., Larminat, F., 2006. BRCA1 downregulation leads to premature inactivation of spindle checkpoint and confers paclitaxel resistance. Cell Cycle 5, 1001–1007.

Cheng, H., An, S.J., Dong, S., Zhang, Y.F., Zhang, X.C., Chen, Z.H., Jian, S., Wu, Y.L., 2011. Molecular mechanism of the schedule-dependent synergistic interaction in EGFR-mutant non-small cell lung cancer cell lines treated with paclitaxel and gefitinib. J. Hematol. Oncol. 4, 5.

Chiorean, E.G., Von Hoff, D.D., Reni, M., Arena, F.P., Infante, J.R., Bathini, V.G., Wood, T.E., Mainwaring, P.N., Muldoon, R.T., Clingan, P.R., Kunzmann, V., Ramanathan, R.K., Tabernero, J., Goldstein, D., McGovern, D., Lu, B., Ko, A., 2016. CA19-9 decrease at 8 weeks as a predictor of overall survival in a randomized phase III trial (MPACT) of weekly nab-paclitaxel plus gemcitabine versus gemcitabine alone in patients with metastatic pancreatic cancer. Ann. Oncol. 27, 654–660.

Chiu, H.W., Lin, H.Y., Tseng, I.J., Lin, Y.F., 2018. OTUD7B upregulation predicts a poor response to paclitaxel in patients with triple-negative breast cancer. Oncotarget 9, 553–565.

Chun, E., Lee, K.Y., 2004. Bcl-2 and Bcl-xL are important for the induction of paclitaxel resistance in human hepatocellular carcinoma cells. Biochem. Biophys. Res. Commun. 315, 771–779.

Cui, Q., Wang, J.Q., Assaraf, Y.G., Ren, L., Gupta, P., Wei, L., Ashby Jr., C.R., Yang, D.H., Chen, Z.S., 2018. Modulating ROS to overcome multidrug resistance in cancer. Drug Resist. Updat. 41, 1–25.

Daood, M., Tsai, C., Ahdab-Barmada, M., Watchko, J., 2008. ABC transporter (P-gp/ABCB1, MRP1/ABCC1, BCRP/ABCG2) expression in the developing human CNS. Neuropediatrics 39, 211–218.

Datta, S., Choudhury, D., Das, A., Das Mukherjee, D., Das, N., Roy, S.S., Chakrabarti, G., 2017. Paclitaxel resistance development is associated with biphasic changes in reactive oxygen species, mitochondrial membrane potential and autophagy with elevated energy production capacity in lung cancer cells: a chronological study. Tumour Biol. 39. 1010428317694314.

Dean, M., 2001. The human ATP-binding cassette (ABC) transporter superfamily. Genome Res. 11, 1156–1166.

Dewanjee, S., Dua, T.K., Bhattacharjee, N., Das, A., Gangopadhyay, M., Khanra, R., Joardar, S., Riaz, M., Feo, V., Zia-ul-Haq, M., 2017. Natural products as alternative choices for P-glycoprotein (P-gp) inhibition. Molecules 22 (6), 871.

Dhillon, A.S., Hagan, S., Rath, O., Kolch, W., 2007. MAP kinase signalling pathways in cancer. Oncogene 26, 3279–3290.

Duan, Z., Brakora, K.A., Seiden, M.V., 2004. MM-TRAG (MGC4175), a novel intracellular mitochondrial protein, is associated with the taxol- and doxorubicin-resistant phenotype in human cancer cell lines. Gene 340, 53–59.

Feng, X., Wang, E., Cui, Q., 2019. Gene expression-based predictive markers for paclitaxel treatment in ER + and ER- breast cancer. Front. Genet. 10, 156.

Feng, Q., Li, X., Sun, W., Sun, M., Li, Z., Sheng, H., Xie, F., Zhang, S., Shan, C., 2020a. Targeting G6PD reverses paclitaxel resistance in ovarian cancer by suppressing GSTP1. Biochem. Pharmacol. 178, 114092.

Feng, W., Zhang, M., Wu, Z.-X., Wang, J.-Q., Dong, X.-D., Yang, Y., Teng, Q.-X., Chen, X.-Y., Cui, Q., Yang, D.-H., 2020b. Erdafitinib antagonizes ABCB1-mediated multidrug resistance in cancer cells. Front. Oncol. 10, 955.

Ferlini, C., Raspaglio, G., Mozzetti, S., Distefano, M., Filippetti, F., Martinelli, E., Ferrandina, G., Gallo, D., Ranelletti, F.O., Scambia, G., 2003. Bcl-2 down-regulation is a novel mechanism of paclitaxel resistance. Mol. Pharmacol. 64, 51–58.

Ferlini, C., Raspaglio, G., Mozzetti, S., Cicchillitti, L., Filippetti, F., Gallo, D., Fattorusso, C., Campiani, G., Scambia, G., 2005. The seco-taxane IDN5390 is able to target class III beta-tubulin and to overcome paclitaxel resistance. Cancer Res. 65, 2397–2405.

Gagnon, V., Van Themsche, C., Turner, S., Leblanc, V., Asselin, E., 2008. Akt and XIAP regulate the sensitivity of human uterine cancer cells to cisplatin, doxorubicin and taxol. Apoptosis 13, 259–271.

Gilmore, P.M., Mccabe, N., Quinn, J.E., Kennedy, R.D., Gorski, J.J., Andrews, H.N., Mcwilliams, S., Carty, M., Mullan, P.B., Duprex, W.P., Liu, E.T., Johnston, P.G., Harkin, D.P., 2004. BRCA1 interacts with and is required for paclitaxel-induced activation of mitogen-activated protein kinase kinase kinase 3. Cancer Res. 64, 4148–4154.

Goodson, H.V., Jonasson, E.M., 2018. Microtubules and microtubule-associated proteins. Cold Spring Harb. Perspect. Biol. 10 (6), a022608.

Guntur, V.P., Waldrep, J.C., Guo, J.J., Selting, K., Dhand, R., 2010. Increasing p53 protein sensitizes non-small cell lung cancer to paclitaxel and cisplatin in vitro. Anticancer Res. 30, 3557–3564.

Hopper-Borge, E.A., Churchill, T., Paulose, C., Nicolas, E., Jacobs, J.D., Ngo, O., Kuang, Y., Grinberg, A., Westphal, H., Chen, Z.-S., Klein-Szanto, A.J., Belinsky, M.G., Kruh, G.D., 2011. Contribution of Abcc10 (Mrp7) to in vivo paclitaxel resistance as assessed in *Abcc10* −/− mice. Cancer Res. 71, 3649–3657.

Hou, Y., Zhu, Q., Li, Z., Peng, Y., Yu, X., Yuan, B., Liu, Y., Liu, Y., Yin, L., Peng, Y., Jiang, Z., Li, J., Xie, B., Duan, Y., Tan, G., Gulina, K., Gong, Z., Sun, L., Fan, X., Li, X., 2017. The FOXM1-ABCC5 axis contributes to paclitaxel resistance in nasopharyngeal carcinoma cells. Cell Death Dis. 8, e2659.

Huang, C.Z., Wang, Y.F., Zhang, Y., Peng, Y.M., Liu, Y.X., Ma, F., Jiang, J.H., Wang, Q.D., 2017. Cepharanthine hydrochloride reverses Pglycoprotein-mediated multidrug resistance in human ovarian carcinoma A2780/taxol cells by inhibiting the PI3K/Akt signaling pathway. Oncol. Rep. 38, 2558–2564.

Huisman, M.T., Chhatta, A.A., Van Tellingen, O., Beijnen, J.H., Schinkel, A.H., 2005. MRP2 (ABCC2) transports taxanes and confers paclitaxel resistance and both processes are stimulated by probenecid. Int. J. Cancer 116, 824–829.

Kamath, K., Wilson, L., Cabral, F., Jordan, M.A., 2005. βIII-tubulin induces paclitaxel resistance in association with reduced effects on microtubule dynamic instability. J. Biol. Chem. 280, 12902–12907.

Kassler, S., Donninger, H., Birrer, M.J., Clark, G.J., 2012. RASSF1A and the taxol response in ovarian cancer. Mol. Biol. Int. 2012, 263267.

Kavallaris, M., Kuo, D.Y., Burkhart, C.A., Regl, D.L., Norris, M.D., Haber, M., Horwitz, S.B., 1997. Taxol-resistant epithelial ovarian tumors are associated with altered expression of specific beta-tubulin isotypes. J. Clin. Invest. 100, 1282–1293.

Kawiak, A., Domachowska, A., Lojkowska, E., 2019. Plumbagin increases paclitaxel-induced cell death and overcomes paclitaxel resistance in breast cancer cells through ERK-mediated apoptosis induction. J. Nat. Prod. 82, 878–885.

Kolosenko, I., Fryknas, M., Forsberg, S., Johnsson, P., Cheon, H., Holvey-Bates, E.G., Edsbacker, E., Pellegrini, P., Rassoolzadeh, H., Brnjic, S., Larsson, R., Stark, G.R., Grander, D., Linder, S., Tamm, K.P., De Milito, A., 2015. Cell crowding induces interferon regulatory factor 9, which confers resistance to chemotherapeutic drugs. Int. J. Cancer 136, E51–E61.

Komatsu, M., Hiyama, K., Tanimoto, K., Yunokawa, M., Otani, K., Ohtaki, M., Hiyama, E., Kigawa, J., Ohwada, M., Suzuki, M., Nagai, N., Kudo, Y., Nishiyama, M., 2006. Prediction of individual response to platinum/paclitaxel combination using novel marker genes in ovarian cancers. Mol. Cancer Ther. 5, 767–775.

Koo, D.H., Lee, H.J., Ahn, J.H., Yoon, D.H., Kim, S.B., Gong, G., Son, B.H., Ahn, S.H., Jung, K.H., 2015. Tau and PTEN status as predictive markers for response to trastuzumab and paclitaxel in patients with HER2-positive breast cancer. Tumour Biol. 36, 5865–5871.

Kowalski, R.J., Giannakakou, P., Hamel, E., 1997. Activities of the microtubule-stabilizing agents epothilones A and B with purified tubulin and in cells resistant to paclitaxel (Taxol(R)). J. Biol. Chem. 272, 2534–2541.

Lage, H., 2016. Gene therapeutic approaches to overcome ABCB1-mediated drug resistance. Recent Results in Cancer Res. 209, 87–94.

Lee, F.Y.F., Borzilleri, R., Fairchild, C.R., Kim, S.-H., Long, B.H., Reventos-Suarez, C., Vite, G.D., Rose, W.C., Kramer, R.A., 2001. BMS-247550. A novel epothilone analog with a mode of action similar to paclitaxel but possessing superior antitumor efficacy. Clin. Cancer Res. 7, 1429–1437.

Li, J., Zheng, L., Yan, M., Wu, J., Liu, Y., Tian, X., Jiang, W., Zhang, L., Wang, R., 2020. Activity and mechanism of flavokawain A in inhibiting P-glycoprotein expression in paclitaxel resistance of lung cancer. Oncol. Lett. 19, 379–387.

Liebmann, J.E., Hahn, S.M., Cook, J.A., Lipschultz, C., Mitchell, J.B., Kaufman, D.C., 1993. Glutathione depletion by L-buthionine sulfoximine antagonizes taxol cytotoxicity. Cancer Res. 53, 2066–2070.

Lin, Y.F., Tseng, I.J., Kuo, C.J., Lin, H.Y., Chiu, I.J., Chiu, H.W., 2018. High-level expression of ARID1A predicts a favourable outcome in triple-negative breast cancer patients receiving paclitaxel-based chemotherapy. J. Cell. Mol. Med. 22, 2458–2468.

Liu, Z., Zhu, G.J., Getzenberg, R.H., Veltri, R.W., 2015. The upregulation of PI3K/Akt and MAP kinase pathways is associated with resistance of microtubule-targeting drugs in prostate cancer. J. Cell. Biochem. 116, 1341–1349.

Liu, J.J., Ho, J.Y., Lee, H.W., Baik, M.W., Kim, O., Choi, Y.J., Hur, S.Y., 2019. Inhibition of phosphatidylinositol 3-kinase (pi3k) signaling synergistically potentiates antitumor efficacy of paclitaxel and overcomes paclitaxel-mediated resistance in cervical cancer. Int. J. Mol. Sci. 20 (14), 3383.

Liu, R., Chen, Y., Liu, G., Li, C., Song, Y., Cao, Z., Li, W., Hu, J., Lu, C., Liu, Y., 2020. PI3K/AKT pathway as a key link modulates the multidrug resistance of cancers. Cell Death Dis. 11, 797.

Lu, C., El-Deiry, W.S., 2009. Targeting p53 for enhanced radio- and chemo-sensitivity. Apoptosis 14, 597–606.

Lu, M., Xiao, L., Li, Z., 2007. The relationship between p38MAPK and apoptosis during paclitaxel resistance of ovarian cancer cells. J. Huazhong Univ. Sci. Technolog. Med. Sci. 27, 725–728.

Luker, K.E., Pica, C.M., Schreiber, R.D., Piwnica-Worms, D., 2001. Overexpression of IRF9 confers resistance to antimicrotubule agents in breast cancer cells. Cancer Res. 61, 6540–6547.

MacKeigan, J.P., Taxman, D.J., Hunter, D., Earp, H.S., Graves, L.M., Ting, J.P.Y., 2002. Inactivation of the antiapoptotic phosphatidylinositol 3-kinase-Akt pathway by the combined treatment of taxol and mitogen-activated protein kinase kinase inhibition. Clin. Cancer Res. 8, 2091–2099.

McDaid, H.M., Horwitz, S.B., 2001. Selective potentiation of paclitaxel (taxol)-induced cell death by mitogen-activated protein kinase kinase inhibition in human cancer cell lines. Mol. Pharmacol. 60, 290–301.

Meek, D.W., 2015. Regulation of the p53 response and its relationship to cancer. Biochem. J. 469, 325–346.

Mei, M., Xie, D., Zhang, Y., Jin, J., You, F., Li, Y., Dai, J., Chen, X., 2014. A new 2α,5α,10β,14β-tetraacetoxy-4(20),11-taxadiene (SIA) derivative overcomes paclitaxel resistance by inhibiting MAPK signaling and increasing paclitaxel accumulation in breast cancer cells. PLoS One 9. e104317.

Meyer, T., Rustin, G.J., 2000. Role of tumour markers in monitoring epithelial ovarian cancer. Br. J. Cancer 82, 1535–1538.

Molina, A., Velot, L., Ghouinem, L., Abdelkarim, M., Bouchet, B.P., Luissint, A.C., Bouhlel, I., Morel, M., Sapharikas, E., Di Tommaso, A., Honore, S., Braguer, D., Gruel, N., Vincent-Salomon, A., Delattre, O., Sigal-Zafrani, B., Andre, F., Terris, B., Akhmanova, A., Di Benedetto, M., Nahmias, C., Rodrigues-Ferreira, S., 2013. ATIP3, a novel prognostic marker of breast cancer patient survival, limits cancer cell migration and slows metastatic progression by regulating microtubule dynamics. Cancer Res. 73, 2905–2915.

Mozzetti, S., Ferlini, C., Concolino, P., Filippetti, F., Raspaglio, G., Prislei, S., Gallo, D., Martinelli, E., Ranelletti, F.O., Ferrandina, G., Scambia, G., 2005. Class III beta-tubulin overexpression is a prominent mechanism of paclitaxel resistance in ovarian cancer patients. Clin. Cancer Res. 11, 298–305.

Narod, S.A., Foulkes, W.D., 2004. BRCA1 and BRCA2: 1994 and beyond. Nat. Rev. Cancer 4, 665–676.

Nunes, S.C., Serpa, J., 2018. Glutathione in ovarian cancer: a double-edged sword. Int. J. Mol. Sci. 19, 1882.

Overbeck, T.R., Hupfeld, T., Krause, D., Waldmann-Beushausen, R., Chapuy, B., Güldenzoph, B., Aung, T., Inagaki, N., Schöndube, F.A., Danner, B.C., Truemper, L., Wulf, G.G., 2013. Intracellular ATP-binding cassette transporter A3 is expressed in lung cancer cells and modulates susceptibility to cisplatin and paclitaxel. Oncology 84, 362–370.

Parekh, H.K., Deng, H.B., Choudhary, K., Houser, S.R., Simpkins, H., 2002. Overexpression of sorcin, a calcium-binding protein, induces a low level of paclitaxel resistance in human ovarian and breast cancer cells. Biochem. Pharmacol. 63, 1149–1158.

Park, S.H., Seong, M.A., Lee, H.Y., 2016. p38 MAPK-induced MDM2 degradation confers paclitaxel resistance through p53-mediated regulation of EGFR in human lung cancer cells. Oncotarget 7, 8184–8199.

Paulsen, T., Marth, C., Kaern, J., Nustad, K., Kristensen, G.B., Trope, C., 2000. Effects of paclitaxel on CA-125 serum levels in ovarian cancer patients. Gynecol. Oncol. 76, 326–330.

Perez-Ortiz, A.C., Ramirez, I., Cruz-Lopez, J.C., Villarreal-Garza, C., Luna-Angulo, A., Lira-Romero, E., Jimenez-Chaidez, S., Diaz-Chavez, J., Matus-Santos, J.A., Sanchez-Chapul, L., Mendoza-Lorenzo, P., Estrada-Mena, F.J., 2017. Pharmacogenetics of response to neoadjuvant paclitaxel treatment for locally advanced breast cancer. Oncotarget 8, 106454–106467.

Promkan, M., Liu, G., Patmasiriwat, P., Chakrabarty, S., 2009. BRCA1 modulates malignant cell behavior, the expression of survivin and chemosensitivity in human breast cancer cells. Int. J. Cancer 125, 2820–2828.

Promkan, M., Liu, G., Patmasiriwat, P., Chakrabarty, S., 2011. BRCA1 suppresses the expression of survivin and promotes sensitivity to paclitaxel through the calcium sensing receptor (CaSR) in human breast cancer cells. Cell Calcium 49, 79–88.

Ramanathan, B., Jan, K.Y., Chen, C.H., Hour, T.C., Yu, H.J., Pu, Y.S., 2005. Resistance to paclitaxel is proportional to cellular total antioxidant capacity. Cancer Res. 65, 8455–8460.

Ranganathan, S., Benetatos, C.A., Colarusso, P.J., Dexter, D.W., Hudes, G.R., 1998. Altered beta-tubulin isotype expression in paclitaxel-resistant human prostate carcinoma cells. Br. J. Cancer 77, 562–566.

Rodrigues-Ferreira, S., Nehlig, A., Moindjie, H., Monchecourt, C., Seiler, C., Marangoni, E., Chateau-Joubert, S., Dujaric, M.E., Servant, N., Asselain, B., De Cremoux, P., Lacroix-Triki, M., Arnedos, M., Pierga, J.Y., Andre, F., Nahmias, C., 2019. Improving breast cancer sensitivity to paclitaxel by increasing aneuploidy. Proc. Natl. Acad. Sci. USA 116, 23691–23697.

Roque, D.M., Buza, N., Glasgow, M., Bellone, S., Bortolomai, I., Gasparrini, S., Cocco, E., Ratner, E., Silasi, D.A., Azodi, M., Rutherford, T.J., Schwartz, P.E., Santin, A.D., 2014. Class III beta-tubulin overexpression within the tumor microenvironment is a prognostic biomarker for poor overall survival in ovarian cancer patients treated with neoadjuvant carboplatin/paclitaxel. Clin. Exp. Metastasis 31, 101–110.

Sanchez-Rovira, P., Anton, A., Barnadas, A., Velasco, A., Lomas, M., Rodriguez-Pinilla, M., Ramirez, J.L., Ramirez, C., Rios, M.J., Castella, E., Garcia-Andrade, C., San Antonio, B., Carrasco, E., Palacios, J.L., 2012. Classical markers like ER and ki-67, but also survivin and pERK, could be involved in the pathological response to gemcitabine, adriamycin and paclitaxel (GAT) in locally advanced breast cancer patients: results from the GEICAM/2002-01 phase II study. Clin. Transl. Oncol. 14, 430–436.

Seagle, B.L., Yang, C.P., Eng, K.H., Dandapani, M., Odunsi-Akanji, O., Goldberg, G.L., Odunsi, K., Horwitz, S.B., Shahabi, S., 2015. TP53 hot spot mutations in ovarian cancer: selective resistance to microtubule stabilizers in vitro and differential survival outcomes from the cancer genome atlas. Gynecol. Oncol. 138, 159–164.

Seo, H.-S., Ku, J.M., Lee, H.-J., Woo, J.-K., Cheon, C., Kim, M., Jang, B.-H., Shin, Y.C., Ko, S.-G., 2017. SH003 reverses drug resistance by blocking signal transducer and activator of transcription 3 (STAT3) signaling in breast cancer cells. Biosci. Rep. 37. https://doi.org/10.1042/BSR20170125.

Shi, X., Dou, Y., Zhou, K., Huo, J., Yang, T., Qin, T., Liu, W., Wang, S., Yang, D., Chang, L., Wang, C., 2017. Targeting the Bcl-2 family and P-glycoprotein reverses paclitaxel resistance in human esophageal carcinoma cell line. Biomed. Pharmacother. 90, 897–905.

Snyder, J.P., Nettles, J.H., Cornett, B., Downing, K.H., Nogales, E., 2001. The binding conformation of taxol in beta-tubulin: a model based on electron crystallographic density. Proc. Natl. Acad. Sci. USA 98, 5312–5316.

Sugiyama, A., Ohta, T., Obata, M., Takahashi, K., Seino, M., Nagase, S., 2020. xCT inhibitor sulfasalazine depletes paclitaxel-resistant tumor cells through ferroptosis in uterine serous carcinoma. Oncol. Lett. 20, 2689–2700.

Sun, Y.-L., Chen, J.-J., Kumar, P., Chen, K., Sodani, K., Patel, A., Chen, Y.-L., Chen, S.-D., Jiang, W.-Q., Chen, Z.-S., 2013. Reversal of MRP7 (ABCC10)-mediated multidrug resistance by tariquidar. PLoS One 8. e55576.

Sun, N.K., Huang, S.L., Chang, T.C., Chao, C.C., 2018. TLR4 and NFκB signaling is critical for taxol resistance in ovarian carcinoma cells. J. Cell. Physiol. 233, 2489–2501.

Tabuchi, Y., Matsuoka, J., Gunduz, M., Imada, T., Ono, R., Ito, M., Motoki, T., Yamatsuji, T., Shirakawa, Y., Takaoka, M., Haisa, M., Tanaka, N., Kurebayashi, J., Jordan, V.C., Naomoto, Y., 2009. Resistance to paclitaxel therapy is related with Bcl-2 expression through an estrogen receptor mediated pathway in breast cancer. Int. J. Oncol. 34, 313–319.

Taghian, A.G., Abi-Raad, R., Assaad, S.I., Casty, A., Ancukiewicz, M., Yeh, E., Molokhia, P., Attia, K., Sullivan, T., Kuter, I., Boucher, Y., Powell, S.N., 2005. Paclitaxel decreases the interstitial fluid pressure and improves oxygenation in breast cancers in patients treated with neoadjuvant chemotherapy: clinical implications. J. Clin. Oncol. 23, 1951–1961.

Tazzari, P.L., Cappellini, A., Ricci, F., Evangelisti, C., Papa, V., Grafone, T., Martinelli, G., Conte, R., Cocco, L., Mccubrey, J.A., Martelli, A.M., 2007. Multidrug resistance-associated protein 1 expression is under the control of the phosphoinositide 3 kinase/Akt signal transduction network in human acute myelogenous leukemia blasts. Leukemia 21, 427–438.

Van Zyl, B., Tang, D., Bowden, N.A., 2018. Biomarkers of platinum resistance in ovarian cancer: what can we use to improve treatment. Endocr. Relat. Cancer 25, R303–R318.

Veenstra, V.L., Damhofer, H., Waasdorp, C., Van Rijssen, L.B., Van De Vijver, M.J., Dijk, F., Wilmink, H.W., Besselink, M.G., Busch, O.R., Chang, D.K., Bailey, P.J., Biankin, A.V., Kocher, H.M., Medema, J.P., Li, J.S., Jiang, R., Pierce, D.W., Van Laarhoven, H.W.M., Bijlsma, M.F., 2018. ADAM12 is a circulating marker for stromal activation in pancreatic cancer and predicts response to chemotherapy. Oncogene 7, 87.

Wang, Y., Sparano, J.A., Fineberg, S., Stead, L., Sunkara, J., Horwitz, S.B., Mcdaid, H.M., 2013. High expression of class III beta-tubulin predicts good response to neoadjuvant taxane and doxorubicin/cyclophosphamide-based chemotherapy in estrogen receptor-negative breast cancer. Clin. Breast Cancer 13, 103–108.

Wang, J.-Q., Lei, Z.-N., Teng, Q.-X., Wang, B., Ma, L.-Y., Liu, H.-M., Chen, Z.-S., 2020. Abstract 2983: a synthetic derivative of 1,2,3-triazole-pyrimidine hybrid reverses multidrug resistance mediated by MRP7. In: Proc. AACR Annual Meeting 80, p. 2983.

Weihua, X., Lindner, D.J., Kalvakolanu, D.V., 1997. The interferon-inducible murine p48 (ISGF3gamma) gene is regulated by protooncogene c-myc. Proc. Natl. Acad. Sci. USA 94, 7227–7232.

Wu, C.P., Ohnuma, S., Ambudkar, S.V., 2011. Discovering natural product modulators to overcome multidrug resistance in cancer chemotherapy. Curr. Pharm. Biotechnol. 12, 609–620.

Wu, G., Qin, X.Q., Guo, J.J., Li, T.Y., Chen, J.H., 2014. AKT/ERK activation is associated with gastric cancer cell resistance to paclitaxel. Int. J. Clin. Exp. Pathol. 7, 1449–1458.

Wu, Z.X., Teng, Q.X., Cai, C.Y., Wang, J.Q., Lei, Z.N., Yang, Y., Fan, Y.F., Zhang, J.Y., Li, J., Chen, Z.S., 2019. Tepotinib reverses ABCB1-mediated multidrug resistance in cancer cells. Biochem. Pharmacol. 166, 120–127.

Xie, X., Shao, X., Ma, W., Zhao, D., Shi, S., Li, Q., Lin, Y., 2018. Overcoming drug-resistant lung cancer by paclitaxel loaded tetrahedral DNA nanostructures. Nanoscale 10, 5457–5465.

Xu, J.H., Hu, S.L., Shen, G.D., Shen, G., 2016. Tumor suppressor genes and their underlying interactions in paclitaxel resistance in cancer therapy. Cancer Cell Int. 16, 13.

Xue, D., Zhou, X., Qiu, J., 2020. Emerging role of NRF2 in ROS-mediated tumor chemoresistance. Biomed. Pharmacother. 131, 110676.

Yi, Y., Wu, H., Gao, Q., He, H.W., Li, Y.W., Cai, X.Y., Wang, J.X., Zhou, J., Cheng, Y.F., Jin, J.J., Fan, J., Qiu, S.J., 2013. Interferon regulatory factor (IRF)-1 and IRF-2 are associated with prognosis and tumor invasion in HCC. Ann. Surg. Oncol. 20, 267–276.

Yusuf, R.Z., Duan, Z., Lamendola, D.E., Penson, R.T., Seiden, M.V., 2003. Paclitaxel resistance: molecular mechanisms and pharmacologic manipulation. Curr. Cancer Drug Targets 3, 1–19.

Zaffaroni, N., Pennati, M., Colella, G., Perego, P., Supino, R., Gatti, L., Pilotti, S., Zunino, F., Daidone, M.G., 2002. Expression of the anti-apoptotic gene survivin correlates with taxol resistance in human ovarian cancer. Cell. Mol. Life Sci. 59, 1406–1412.

Zhang, H.L., Ruan, L., Zheng, L.M., Whyte, D., Tzeng, C.M., Zhou, X.W., 2012. Association between class III beta-tubulin expression and response to paclitaxel/vinorebine-based chemotherapy for non-small cell lung cancer: a meta-analysis. Lung Cancer 77, 9–15.

Zhang, W., Cai, J., Chen, S., Zheng, X., Hu, S., Dong, W., Lu, J., Xing, J., Dong, Y., 2015. Paclitaxel resistance in MCF-7/PTX cells is reversed by paeonol through suppression of the SET/phosphatidylinositol 3-kinase/Akt pathway. Mol. Med. Rep. 12, 1506–1514.

Zhang, Y., Wu, Z.-X., Yang, Y., Wang, J.-Q., Li, J., Sun, Z., Teng, Q.-X., Ashby, C.R., Yang, D.-H., 2020. Poziotinib inhibits the efflux activity of the ABCB1 and ABCG2 transporters and the expression of the ABCG2 transporter protein in multidrug resistant colon cancer cells. Cancer 12, 3249.

Zheng, Y.-B., Gong, J.-H., Liu, X.-J., Wu, S.-Y., Li, Y., Xu, X.-D., Shang, B.-Y., Zhou, J.-M., Zhu, Z.-L., Si, S.-Y., Zhen, Y.-S., 2016. A novel nitrobenzoate microtubule inhibitor that overcomes multidrug resistance exhibits antitumor activity. Sci. Rep. 6, 31472.

Zhou, Y., Hopper-Borge, E., Shen, T., Huang, X.-C., Shi, Z., Kuang, Y.-H., Furukawa, T., Akiyama, S.-I., Peng, X.-X., Ashby, C.R., Chen, X., Kruh, G.D., Chen, Z.-S., 2009. Cepharanthine is a potent reversal agent for MRP7(ABCC10)-mediated multidrug resistance. Biochem. Pharmacol. 77, 993–1001.

Zou, L., Wang, D., Hu, Y., Fu, C., Li, W., Dai, L., Yang, L., Zhang, J., 2017. Drug resistance reversal in ovarian cancer cells of paclitaxel and borneol combination therapy mediated by PEG-PAMAM nanoparticles. Oncotarget 8, 60453–60468.

11

Paclitaxel and cancer treatment: Non-mitotic mechanisms of paclitaxel action in cancer therapy

Elizabeth R. Smith[a], Zhe-Sheng Chen[b], and Xiang-Xi Xu[a]

[a]Department of Radiation Oncology, Sylvester Comprehensive Cancer Center, University of Miami Miller School of Medicine, Miami, FL, United States, [b]College of Pharmacy and Health Sciences, St. John's University, Queens, NY, United States

11.1 Introduction

11.1.1 Taxol/paclitaxel is a key agent in the management of solid tumors

The class of taxane drugs, including paclitaxel (trade name-Taxol) and docetaxel (trade name-Taxotere), is among the most effective anticancer agents commonly used in clinics today to treat several major cancers, such as breast, ovarian, prostatic, lung, pancreatic, and cervical cancers (Bates and Eastman, 2017; Bookman, 2016; Lemstrova et al., 2016; Joshi et al., 2014; Yang and Horwitz, 2017). Currently, a cisplatin (or carboplatin)/paclitaxel regimen following debulking surgery is a standard frontline chemotherapy for ovarian cancer (Bookman, 2016; Runowicz et al., 1993; Ozols et al., 2004), and a dose intensive regimen of paclitaxel is also used in salvage treatment following recurrence (Baird et al., 2010; Baker, 2003; Jain et al., 2011).

Despite the impressive clinical success of paclitaxel as a frontline and salvage cancer therapy (Gallego-Jara et al., 2020; Jain et al., 2011; Rowinsky and Donehower, 1995), a major challenge is the development of drug resistance in recurrent cancer (Mosca et al., 2021; Blagosklonny and Fojo, 1999; Fojo and Menefee, 2007; Kavallaris, 2010; Visconti and Grieco, 2017). Extensive investigations led to the proposal of a list of possible mechanisms for the important clinical question of paclitaxel resistance (Maloney et al., 2020; Yang and

Paclitaxel. https://doi.org/10.1016/B978-0-323-90951-8.00005-9
Copyright © 2022 Elsevier Inc. All rights reserved.

Horwitz, 2017). Alterations in tubulin (Horwitz et al., 1993; Orr et al., 2003) and overexpression of ABC transporters (Fojo and Menefee, 2007) have been investigated to be mechanisms of paclitaxel resistance. Although tubulin mutations were found in laboratory cell models, they were not observed in human cancer (Blagosklonny and Fojo, 1999; Orr et al., 2003). Expression of tubulin isoforms, such as beta-3 tubulin is another possible mechanism (English et al., 2013; Mozzetti et al., 2005). However, the common capability of cancer cells to acquire taxane resistance indicates that additional major mechanism(s) have not yet been uncovered (Blagosklonny and Fojo, 1999; Orr et al., 2003; Visconti and Grieco, 2017; Yu et al., 2015). The cytoskeletal protein tubulin is the molecular target of paclitaxel, and paclitaxel binds strongly to tubulin, altering the dynamic turnover and assembly, and thereby stabilizing microtubule filaments (Jordan, 2002; Jordan and Wilson, 2004; Schiff et al., 1979; Schiff and Horwitz, 1980; Yang and Horwitz, 2017). Generally, paclitaxel was thought to kill cells by inducing mitotic arrest, leading to apoptosis in cancer cells (Horwitz, 1994; Jordan, 2002). However, recent evidence suggests that paclitaxel operates by a not-yet-defined mechanism rather than as a mitotic inhibitor (Blagosklonny and Fojo, 1999; Florian and Mitchison, 2016; Fürst and Vollmar, 2013; Komlodi-Pasztor et al., 2011; Mitchison, 2012; Field et al., 2014). Moreover, the effort to develop additional anti-mitotic agents, inspired by the achievement of paclitaxel, has not proven to be successful (Komlodi-Pasztor et al., 2012; Olziersky and Labidi-Galy, 2017), reinforcing doubt that paclitaxel acts solely through mitotic mechanisms. Since investigation of new microtubule-stabilizing agents, such as epithelones (ixabepilone), laulimalide, and discodermolide, is under development (Altaha et al., 2002; Zhao et al., 2016), our understanding of the mechanism of taxanes and other microtubule-stabilizing drugs is important and may have a significant clinical implication in the years to come (Florian and Mitchison, 2016; Joshi et al., 2014).

11.2 Microtubule stabilization and anti-mitotic mechanisms

The discovery in the 1980s that paclitaxel stabilizes microtubules (Schiff et al., 1979; Schiff and Horwitz, 1980; Jordan and Wilson, 2004) and inhibits mitosis (Jordan, 2002; Jordan et al., 1993) in culture cells propelled the development of the compound into a common anti-cancer drug (Gallego-Jara et al., 2020). Cell culture studies provided clear evidence that paclitaxel inhibited mitosis, and the mechanism that paclitaxel acts as a mitotic inhibitor quickly gained widespread acceptance and is now considered dogma (Yang and Horwitz, 2017).

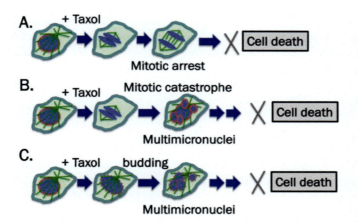

Fig. 11.1 Mechanisms of paclitaxel in instigating cancer cell death by stabilizing microtubules. (A) The generally accepted mechanism is that paclitaxel binds microtubules and interfering with its function in chromosome segregation in mitotic phase. Following a prolonged arrest at mitosis, the cells undergo apoptosis, for which the signaling and triggering of apoptosis are not yet well understood. (B) A relatively new idea is that the cells escape mitotic arrest and undergo mitotic catastrophe and aberrant chromosome segregation. The resulted multi-nucleated and lobulated cells die, with not yet well-defined mechanisms. (C) A new proposal is that the rigid microtubule filaments induced by paclitaxel can promotes massive formation of micronuclei and nuclear multiple micronucleation by nuclear budding of cells at interphases. The resulted multi-nucleated and lobulated cells die, with not yet well-defined mechanisms.

Generally, paclitaxel was thought to induce mitotic arrest and subsequently apoptosis in cancer cells (Bhalla, 2003; Wang et al., 2000) (Fig. 11.1A). The idea seems reasonable and self-evident, as cancer cells exhibit uncontrolled growth and are usually more proliferative, thus the mitotic targeting provides a specificity of paclitaxel for neoplastic compared to normal cells. Indeed, paclitaxel can cause significant off-target effects in normal, non-neoplastic cells that divide rapidly, such as hemopoietic cells (Rowinsky et al., 1993) and cells of the hair follicle matrix (Purba et al., 2019), resulting in neutropenia and alopecia, respectively.

Nevertheless, paclitaxel causes cancer cell death rather than acts as a mere cytostatic agent (Komlodi-Pasztor et al., 2011; Mitchison, 2012), though how paclitaxel-induced cell growth arrest triggers death is not well understood (Bhalla, 2003; Wang et al., 2000; Gascoigne and Taylor, 2009).

11.3 Mitotic catastrophe

Additional careful studies of the effects of paclitaxel on cancer cells in culture revealed that the cells often escape mitotic arrest and undergo aberrant mitosis (Shi and Mitchison, 2017). Thus, it is recognized that the unsuccessful mitosis in the presence of paclitaxel-induced microtubule malfunction, known as mitotic catastrophe, may be a major mechanism of cell killing (Morse et al., 2005). Experiments using time-lapse video microscopy revealed that paclitaxel-treated cells become multi-nucleated, often a result of multi-polar division (Gascoigne and Taylor, 2009; Weaver, 2014; Zasadil et al., 2014; Zhu et al., 2014). Thus, aberrant mitosis that forms multiple micronuclei, or nuclear lobules, as a result of paclitaxel targeting microtubules, is believed to be the major mechanism of drug action (Weaver, 2014).

The formation of micronuclei following paclitaxel treatment was initially observed many years ago (Theodoropoulos et al., 1999; Merlin et al., 2000), though it was only followed more recently. Generally, the formation of micronuclei is thought to be the result of chromosome mis-segregation during mitosis (Weaver, 2014; Zasadil et al., 2014; Zhu et al., 2014) (Fig. 11.1B).

11.4 Non-mitotic mechanisms

Another puzzle about the mechanism of paclitaxel action is the issue with mitosis as the target (Fürst and Vollmar, 2013). Unlike cultured cells, cancer cells are much less proliferative in vivo, with a doubling time much longer than cultured cells (Shi and Mitchison, 2017; Komlodi-Pasztor et al., 2011). At any given time, only a small fraction of cancer cells is undergoing mitosis (Mitchison, 2012; Shi and Mitchison, 2017; Komlodi-Pasztor et al., 2011). Particularly, the susceptibility of cancer cells to killing by paclitaxel does not correlate with the proliferative index of the cancer (Schimming et al., 1999). This problem inspired the concept of "proliferative index paradox" (Mitchison, 2012), denoting that mitosis may not be a key target of paclitaxel or explained its efficacy as an anti-cancer agent (Komlodi-Pasztor et al., 2011, 2012; Shi and Mitchison, 2017; Yan et al., 2020).

Additionally, efforts to develop additional specific anti-mitotic agents inspired by the success of Taxol have not been successful (Olziersky and Labidi-Galy, 2017; Shi and Mitchison, 2017; Yan et al., 2020), leading to skepticism about the rationale for targeting mitosis (Komlodi-Pasztor et al., 2011, 2012). Indeed, recent studies underscore the idea that mitosis may not be essential for cell killing by the microtubule targeting agents (Smith et al., 2021).

A few studies investigated and proposed non-mitotic mechanisms for paclitaxel in causing cancer cell cytotoxicity, including that paclitaxel influences bcl-2 phosphorylation (Pucci et al., 1999); paclitaxel targets microtubules involved in cellular transport (Giannakakou et al., 2002); and the drug impacts nuclear pores and transport (Theodoropoulos et al., 1999). Nevertheless, more investigations to identify a robust and general non-mitotic function of paclitaxel in targeting cancer cells seems obligatory.

A new proposal has been postulated, that paclitaxel kills cancer cells by breaking the nuclear envelope, and is based on results of ongoing work (Smith et al., 2021). These studies showed that paclitaxel and other microtubule-stabilizing agents induce rigid microtubules that cause the breakage/fragmentation of the malleable nucleus of cancer cells, but not the sturdier nucleus present in normal cells (Fig. 11.1C). Thus, a malleable nuclear envelope (caused by a reduction in Lamin

A/C and perhaps other nuclear envelope structural proteins) underlies the specificity of microtubule stabilizing drugs such as paclitaxel in killing malignant cells.

The proposed mechanism provides a possible alternative explanation for the well-established dogma that paclitaxel targets mitosis in cancer therapy; rather, paclitaxel likely aims at the weakened nuclear envelope of malignant cells. Thus, paclitaxel can be predicted to be effective to treat cancer that shows a deformed nuclear envelope, such as in the case of the cervical cancer cells that can be detected by a PAP test (Papanicolaou, 1942; Smith et al., 2018a). The idea holds a new realization that paclitaxel can induce the generation of micronuclei in S phase by a non-mitotic mechanism (Smith et al., 2021).

11.5 Importance of micronucleation

Although the cancer killing mechanism of paclitaxel is not well understood, the formation of micronuclei induced by paclitaxel likely may be important (Mitchison et al., 2017). Here, the micronuclei generated are speculated to undergo rupture and trigger innate immune activation and inflammatory processes that are capable of killing cancer cells (Mitchison et al., 2017) (Fig. 11.2).

The formation of multiple nuclear envelope fragments upon treatment of cancer cells with paclitaxel has been observed previously (Theodoropoulos et al., 1999; Merlin et al., 2000), though few studies have followed up the observation until recently. Generally, the formation of multiple micronuclei is thought to be a result of aberrant,

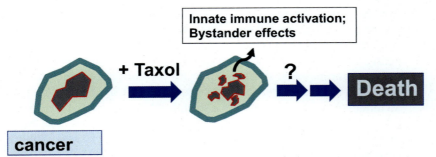

Fig. 11.2 Mechanisms of paclitaxel-induced multiple micronucleation in cancer killing efficacy. Paclitaxel induces the breaking of nuclei of neoplastic cells and the formation of multiple micronuclei. The micronuclei are defective in membrane structure and have high propensity for rupture and release of chromatin materials, resulting in cell death. Additionally, the DNA released into cytoplasm triggers the activation of cGAS-Sting innate immunity pathway. The expression and release of cytokines triggers immune responses of surrounding cells leading to additional cell killing. This phenomenon is known as the bystander effect.

multipolar mitosis (Zasadil et al., 2014; Zhu et al., 2014). Observations also lead to the proposal that nuclear budding of non-mitotic cells may be an important mechanism in producing micronuclei (Capo-Chichi et al., 2011a, 2011b, 2016; Smith et al., 2018b), as microtubules associating with nuclear envelope physically pull and distort the structure (Tapley and Starr, 2013; Tariq et al., 2017).

However, the proposal of physical force exerted by paclitaxel-induced rigid microtubule filaments in breaking meleable cancer nuclei provides a non-mitotic mechanism (Smith et al., 2021). How the formation of micronuclei leads to cell death is not obvious. It has been observed that micronuclei often undergo catastrophic rupture (Hatch et al., 2013; Capo-Chichi et al., 2011a, 2011b, 2016), which may lead to aneuploidy and cell death. Another notion is that the micronuclei formed may trigger the cellular DNA sensing innate immune pathway, which then contributes to cancer killing activity (Mitchison et al., 2017).

11.6 Innate immunity leading to the bystander effect

The innate immune system is critical for the early detection of invading pathogens and for initiating cellular host defense counter measures, which include the production of type I interferon (IFN) (Takaoka and Taniguchi, 2008; Kuriakose and Kanneganti, 2018; Platanias and Fish, 1999). Immune dysfunction develops in patients with many cancer types and may contribute to tumor progression and failure of immunotherapy (Critchley-Thorne et al., 2009; Gajewski and Corrales, 2015; Herbert, 2019; Kwon and Bakhoum, 2020; Motz and Coukos, 2013; Rabinovich et al., 2007; Woo et al., 2015). In mammalian cells, the cytosolic DNA sensing and signaling pathways have been recently discovered (Barber, 2014; Cai et al., 2014). Sting was identified to be a central player of the cellular innate immunity to defend the cells from infection by viruses and microbes, and Sting is essential for all innate responses triggered by cytosolic DNA species (Barber, 2014). Sting is activated by binding to 3'-5' cyclic dinucleotides (cdiGMP), which is a molecule produced by bacteria (Cai et al., 2014). Among several cytosolic DNA sensors identified, cyclic GMP-AMP synthase (cGAS) produces non-canonical 2'-3' cyclic dinucleotides (cGAMP) from ATP and GTP. cGAMP activates the immune response through Sting (Cai et al., 2014). Upon recognition and detection of the cytosolic DNA species, signaling pathways are activated to trigger the induction of interferons (IFNs), and interferon subsequently activates an arsenal of genes to control viral and microbial infection by shutting down cellular transcription and translation machineries, inducing cell

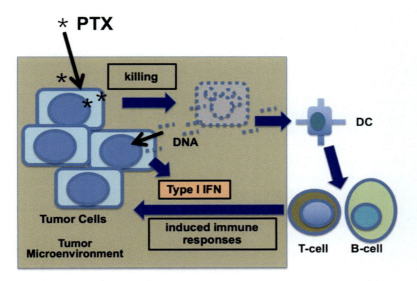

Fig. 11.3 Bystander effects of paclitaxel action. We speculate that paclitaxel eliminates tumor by two phases. In phase I, paclitaxel kills selectively proliferating cancer cells. The dying cells release DNA fragments and trigger innate and induced immunity in neighboring tumor cells. In Phase II, the DNA released from the dead tumor cells stimulate T- and B-cell through dendritic cells to activate immunity that help to clear additional tumor cells.

death, and promoting the adapted immune system of B and T cells to clear microbe infections (Takaoka and Taniguchi, 2008; Barber, 2014; Cai et al., 2014). DNA released from paclitaxel-induced multiple micronuclei is considered to be the source triggering the activation of the cGAS-Sting pathway (Lohard et al., 2020; Mitchison et al., 2017; Zierhut et al., 2019).

Thus, one suggested model puts forward that paclitaxel eliminates tumors by two phases. In phase I, paclitaxel kills selectively proliferating cancer cells. The dying cells release DNA fragments and trigger innate and induced immunity in neighboring tumor cells. In Phase II, the DNA released from the dead tumor cells stimulate T- and B-cells through dendritic cells to activate immunity that helps to clear additional tumor cells. The ability of paclitaxel to kill cancer cell indirectly is referred as a bystander effect (Fig. 11.3) (Mitchison et al., 2017).

11.7 Cellular retention of paclitaxel

Paclitaxel has high binding affinity to beta-tubulin located in microtubule filaments (Manfredi et al., 1982), and the binding can approach 1-to-1 ratio (Jordan and Wilson, 2004; Holmfeldt et al., 2009). In a cell culture study, a short-term exposure of cancer cells to paclitaxel produces a long-term, persistent inhibition of cell proliferation and induction of cell death (Michalakis et al., 2007). In vivo, although paclitaxel is rapidly cleared from the circulation following infusion, the drug and activity are retained in cells and persist for several days (Koshiba et al., 2009; Michalakis et al., 2007; Mori et al., 2006)

Fig. 11.4 Retention of paclitaxel enables efficient killing of tumor cells but also causes side effects such as peripheral neuron toxicity and hair follicle damage. Paclitaxel (PTX) is administrated to patients over 3–6 h, and taxane concentration reaches a peak level in plasma by the end of drug infusion. Over the next 6 h, paclitaxel level declines rapidly, and the drug is concentrated in cells (partly by binding to microtubules) several hundred times over the blood level (illustrated by *yellow* color). 3. Paclitaxel is present in high level inside cells for next 2–3 days (illustrated by *brown dots*), and the drug triggers death of cancer cells (*) over the next 2–3 days, but also causes damage of hair follicles and toxicity in peripheral neurons (*).

(Fig. 11.4). Presumably, the high concentration of paclitaxel within cells interferes with microtubule-dependent cellular functions several days after drug administration. The retention of paclitaxel within cancer cells likely is important for killing of cancer cells, but the persistent presence of paclitaxel in peripheral neurons and hair follciles also causes the well-known side effects of paclitaxel, such as peripheral neuropathy (Mielke et al., 2006) and alopecia (Paus et al., 2013) (Fig. 11.4).

Microtubules are polymers of alpha- and beta-tubulin heterodimers (Desai and Mitchison, 1997; Holmfeldt et al., 2009), and play multiple roles in cellular functions (Cassimeris et al., 2012; Desai and Mitchison, 1997). Microtubules are dynamic: the filaments are constantly extending and shortening, with a balance between the cellular pool of alpha- and beta-tubulin dimers and microtubule polymers, which are about half and half under normal conditions (Cassimeris et al., 2012; Jordan and Wilson, 2004; Zhai and Borisy, 1994). Paclitaxel promotes 90%–100% of tubulin monomers into polymerized forms (Jordan and Wilson, 2004; Holmfeldt et al., 2009; Diaz and Andreu, 1993; Zhai and Borisy, 1994). Because of the importance of microtubules in cellular function, the homeostasis of tubulins is tightly regulated (Caron et al., 1985; Gasic et al., 2019; Lin et al., 2020). Tubulins control their synthesis by autoregulation at the level of mRNA stability (Gasic et al., 2019; Lin et al., 2020). Thus, addition of paclitaxel to eliminate alpha- and beta-tubulin dimers (into polymers) stimulate production of new tubulins. Production of new tubulins will further sequester paclitaxel, until all available paclitaxel molecules are eliminated.

Tubulins are relatively stable, and the tubulin protein is removed by proteasome- (but not lysosome-) mediated degradation (Huff et al., 2010) and via degradation by cathepsin D (Johnson et al., 1991). Cells take up, sequester, and concentrate paclitaxel at several hundreds of times the concentration found in the extracellular space (Jordan and Wilson, 2004). Indeed, intracellular paclitaxel is not washed out, but rather retained over several days after exposure, including in human scalp hair follicles (Purba et al., 2019), during which time the rigid microtubules persist (Jordan and Wilson, 2004; Michalakis et al., 2007;

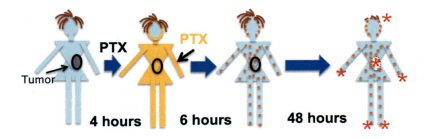

Mori et al., 2006). The ability of cells to uptake and concentrate paclitaxel results in part from paclitaxel sequestration by binding to abundant microtubules and tubulins (in the range of 10–20 μM inside cells) (Cassimeris et al., 2012; Jordan and Wilson, 2004; Zhai and Borisy, 1994).

Thus, a special feature of the pharmacokinetics of paclitaxel is the long retention of the drug inside cells from sequestration by binding to the ample cellular microtubules, despite rapid clearance of the molecules in circulation (Jordan and Wilson, 2004; Michalakis et al., 2007; Mori et al., 2006). We speculate that the long retention is likely a key factor contributing to the success of paclitaxel efficacy over non-microtubular targeting mitotic inhibitors and other anti-neoplastic cytotoxic agents.

11.8 Combination therapy

11.8.1 Drug combination with paclitaxel: Carboplatin, FTI, CDK4/6, and PD1/PDL-1

Paclitaxel is found to be active when used alone as a dose intensive schedule for treatment of recurrent ovarian, lung, and metastatic breast cancer (Rowinsky and Donehower, 1995). The current use of paclitaxel is generally in a combination regime with a platinum drug, either cisplatin or carboplatin (Bookman, 2016; Ozols et al., 2004). Paclitaxel was found early to have additive or synergistic effects with platinum agent in pre-clinical as well as clinical studies (Ozols et al., 2004). Also, the schedule and sequences of drug administration appears to influence efficacy (Ozols et al., 2004). Currently, a cisplatin (or carboplatin)/paclitaxel regimen following debulking surgery is a standard frontline chemotherapy for ovarian cancer (Bookman, 2016; Ozols et al., 2004), and a dose intensive regimen of paclitaxel is also used in salvage treatment as a second line drug following recurrence (Baird et al., 2010; Baker, 2003; Jain et al., 2011). Paclitaxel is also used in combination with other agent(s), such as with doxorubicin (anthracycline) in metastatic breast cancer (Holmes, 1996; Friedrich et al., 2004), and with Bevacizumab in lung cancer (Behera et al., 2015). The rationale combination of paclitaxel with other agents is an issue that oncologists must consider to increase the treatment response and efficacy (Herbst and Khuri, 2003; Barbuti and Chen, 2015).

The efforts to test a combination of paclitaxel with FTI (Farnesyl Transferase Inhibitor) are specially discussed here. FTI was first developed to target Ras with a rationale that the enzymatic activity in post-translation modification of the Ras protein is required for oncogenic Ras localization to plasma membrane to achieve its signaling function (Cox et al., 2015). However, FTI was found to have anti-cancer activity in vitro, independent of a Ras mutation that was the original rationale for the development of the drug (Basso et al., 2005; Rose et al., 2001). Despite strong preclinical demonstration of the activity of the addition of FTI to paclitaxel in targeting cancer cells and tumor models in multiple systems and experiments by several groups (Basso et al., 2005; Marcus et al., 2005, 2006; Rose et al., 2001; Taylor et al., 2008), the single agent FTI and combination paclitaxel trials failed to demonstrate the benefits in several cases including ovarian cancer (Meier et al., 2012) and breast cancer (Andreopoulou et al., 2013), though some increased anti-cancer activity was found for non-small cell lung cancer patients (Kim et al., 2005). In all these trials, FTI was generally given after paclitaxel administration. In patients, the plasma concentration of paclitaxel falls rapidly following infusion (Wiernik et al., 1987), although a fraction of the compound that enters the tumors will produce cytotoxicity and killing in the next few days (Jordan and Wilson, 2004; Michalakis et al., 2007; Mori et al., 2006), as described above. However, little overlap between paclitaxel and FTI in the circulation existed with the protocol design of the trials. In the combination trials using FTI (lonafarnib), paclitaxel was given by infusion over 3–6 h, followed by FTI 3 days later, given twice a day (Meier et al., 2012). After six cycles of chemotherapy, FTI was again given for up to 6 months, though paclitaxel plasma levels would be long depleted following infusion (Wiernik et al., 1987). Thus, in contrast to preclinical studies in mouse xenograft tumor models, FTI and paclitaxel have little overlap for potential drug interaction and synergy in clinical trials (Andreopoulou et al., 2013; Kim et al., 2005; Meier et al., 2012). Possibly, modifications or adjustments in the sequences of drug administration and scheduling may contribute to very different outcome in future combination trials of paclitaxel and FTI.

As immune therapy becomes a new strategy to increase cancer therapeutic efficacy, the combination of anti-PD-1/PDL-1 immunomodulators with paclitaxel in cancer treatment has been investigated (Hu et al., 2017). So far, it appears that paclitaxel has no addititive or positive interaction with immune therapy (Hanna, 2015). Possibly, paclitaxel may suppress the activation of immune cells. However, paclitaxel and immune therapy may be still used for treatment, providing careful consideration of scheduling to avoid negative drug interactions.

Cell cycle kinase drugs, the CDK4/6 inhibitors, are new agents found to have activity in cancer treatment (Harbeck et al., 2021; Hu et al., 2017). At first thought, since CDK4/6 inhibitors block cell cycle progression and mitosis, they would not be predicted to have a positive interaction with paclitaxel. Indeed, an in vitro study showed that CDK4/6 inhibitors rescued killing of proliferative hair follicle matrix cells by paclitaxel (Purba et al., 2019). Nevertheless, preclinical studies indicate that CDK4/6 inhibitors do have additive results with paclitaxel in suppressing neoplastic growth (Cretella et al., 2019; Salvador-Barbero et al., 2020). This likely can be explained by the importance of non-mitotic mechanisms of paclitaxel in causing cancer cell death.

As paclitaxel is a cornerstone of current cancer therapy for many solid tumors, any new agents developed will likely be considered and investigated for combination with paclitaxel (Herbst and Khuri, 2003; Barbuti and Chen, 2015; Hu et al., 2017). The sequences and schedules of drug administration can influence efficacy, implicating the mechanisms of drug interaction.

11.9 Prospective: New formulation of paclitaxel and additional microtubule stabilizing drugs

Investigated in the 1970–1980s and entered into clinical use in the early 1990s (Rowinsky and Donehower, 1995; Gallego-Jara et al., 2020; Mosca et al., 2021), taxane/paclitaxel is still the most commonly used cancer drug today after treating millions of patients over the last 40 + years (Gallego-Jara et al., 2020; Ojima et al., 2016). The development of taxanes for cancer therapy has been a celebrated success story (Pabla and Sparreboom, 2015; Ojima et al., 2016; Wani and Horwitz, 2014), and new drugs with similar mechanism of actions as microtubule stabilization agents have a promising future in cancer treatment (Hunt, 2009; Zhao et al., 2016).

Additional formulations such as Abraxane and liposomal taxane provide improvement on the delivery of paclitaxel and reduction of the hypersensitivity side effects (Kundranda and Niu, 2015; Sofias et al., 2017). Although the mechanism of paclitaxel drug action is still under study to gain a better understanding, microtubule stabilizing activity seems to be a key mechanism driving cancer killing activity (Blagosklonny and Fojo, 1999; Florian and Mitchison, 2016). Thus, a class of additional non-taxane microtubule-stabilizing agents, such as epothilones (ixabepilone), laulimalide, and discodermolide, isolated from microbiomes, sponges, and corals, is undergoing clinical development and testing in trials (Altaha et al., 2002; Hunt, 2009; Zhao et al., 2016; Cao et al., 2018). These new paclitaxel-like microtubule stabilizing agents may be useful for cancer that develops resistance to taxanes, and also has potential ability to be orally administrated, and have higher water solubility. Thus, continuing research and understanding of the microtubule stabilizing agents for their mechanism in efficient cancer cell killing will have a significant clinical implication in the years to come (Gallego-Jara et al., 2020; Barbuti and Chen, 2015).

11.10 Conclusions

Taxanes (paclitaxel, decetaxol) and additional microtubule stabilizing agents appear still to be important drugs for many types of major solid tumors. The development of new molecules and delivery methods will definitely improve and expand their application in cancer treatment. Surprisingly, the mechanism of action for the class of microtubule stabilizing drugs is not yet well understood, and especially the non-mitotic mechanism(s) discussed here is a timely addition to our appreciation of the taxanes. The continuous study and careful analyses of paclitaxel cellular mechanisms to achieve a new understanding will likely lead to better use of the taxanes, and the understanding may also give us a new strategy (such as the breaking of neoplastic nuclei) in future cancer therapy.

References

Altaha, R., Fojo, T., Reed, E., 2002. Abraham Epothilones: a novel class of non-taxane microtubule-stabilizing agents. J. Curr. Pharm. Des. 8 (19), 1707–1712.

Andreopoulou, E., Vigoda, I.S., Valero, V., Hershman, D.L., Raptis, G., Vahdat, L.T., Han, H.S., Wright, J.J., Pellegrino, C.M., Cristofanilli, M., Alvarez, R.H., Fehn, K., Fineberg, S., Sparano, J.A., 2013. Phase I-II study of the farnesyl transferase inhibitor tipifarnib plus sequential weekly paclitaxel and doxorubicin-cyclophosphamide in HER2/neu-negative inflammatory carcinoma and non-inflammatory estrogen receptor-positive breast carcinoma. Breast Cancer Res. Treat. 141 (3), 429–435.

Baird, R.D., Tan, D.S., Kaye, S.B., 2010. Weekly paclitaxel in the treatment of recurrent ovarian cancer. Nat. Rev. Clin. Oncol. 7 (10), 575–582.

Baker, V.V., 2003. Salvage therapy for recurrent epithelial ovarian cancer. Hematol. Oncol. Clin. North Am. 17 (4), 977–988.

Barber, G.N., 2014. STING-dependent cytosolic DNA sensing pathways. Trends Immunol. 35.

Barbuti, A.M., Chen, Z.S., 2015. Paclitaxel through the ages of anticancer therapy: exploring its role in chemoresistance and radiation therapy. Cancers (Basel) 7 (4), 2360–2371.

Basso, A.D., Mirza, A., Liu, G., Long, B.J., Bishop, W.R., Kirschmeier, P., 2005. The farnesyl transferase inhibitor (FTI) SCH66336 (lonafarnib) inhibits Rheb farnesylation and mTOR signaling. Role in FTI enhancement of taxane and tamoxifen anti-tumor activity. J. Biol. Chem. 280 (35), 31101–31108.

Bates, D., Eastman, A., 2017. Microtubule destabilising agents: far more than just antimitotic anticancer drugs. Br. J. Clin. Pharmacol. 83, 255–268.

Behera, M., Pillai, R.N., Owonikoko, T.K., Kim, S., Steuer, C., Chen, Z., Saba, N.F., Belani, C.P., Khuri, F.R., Ramalingam, S.S., 2015. Bevacizumab in combination with taxane versus non-taxane containing regimens for advanced/metastatic nonsquamous non-small-cell lung cancer: a systematic review. J. Thorac. Oncol. 10 (8), 1142–1147.

Bhalla, K.N., 2003. Microtubule-targeted anticancer agents and apoptosis. Oncogene 22, 9075–9086.

Blagosklonny, M.V., Fojo, T., 1999. Molecular effects of paclitaxel: myths and reality (a critical review). Int. J. Cancer 83 (2), 151–156.

Bookman, M.A., 2016. Optimal primary therapy of ovarian cancer. Ann. Oncol. 27 (Suppl 1), i58–i62.

Cai, X., Chiu, Y.H., Chen, Z.J., 2014. The cGAS-cGAMP-STING pathway of cytosolic DNA sensing and signaling. Mol. Cell 54, 289–296.

Cao, Y.N., Zheng, L.L., Wang, D., Liang, X.X., Gao, F., Zhou, X.L., 2018. Recent advances in microtubule-stabilizing agents. Eur. J. Med. Chem. 143, 806–828.

Capo-Chichi, C.D., Cai, K.Q., Smedberg, J., Ganjei-Azar, P., Godwin, A.K., Xu, X.X., 2011a. Loss of A-type lamin expression compromises nuclear envelope integrity in breast cancer. Chin. J. Cancer 30, 415–425.

Capo-Chichi, C.D., Cai, K.Q., Simpkins, F., Ganjei-Azar, P., Godwin, A.K., Xu, X.X., 2011b. Nuclear envelope structural defects cause chromosomal numerical instability and aneuploidy in ovarian cancer. BMC Med. 9, 28.

Capo-Chichi, C.D., Yeasky, T.M., Smith, E.R., Xu, X.X., 2016. Nuclear envelope structural defect underlies the main cause of aneuploidy in ovarian carcinogenesis. BMC Cell Biol. 17 (1), 37.

Caron, J.M., Jones, A.L., Kirschner, M.W., 1985. Autoregulation of tubulin synthesis in hepatocytes and fibroblasts. J. Cell Biol. 101, 1763–1772.

Cassimeris, L., Silva, V.C., Miller, E., Ton, Q., Molnar, C., Fong, J., 2012. Fueled by microtubules: does tubulin dimer/polymer partitioning regulate intracellular metabolism? Cytoskeleton 69, 133–143.

Cox, A.D., Der, C.J., Philips, M.R., 2015. Targeting RAS membrane association: back to the future for anti-RAS drug discovery? Clin. Cancer Res. 21 (8), 1819–1827.

Cretella, D., Fumarola, C., Bonelli, M., Alfieri, R., La Monica, S., Digiacomo, G., Cavazzoni, A., Galetti, M., Generali, D., Petronini, P.G., 2019. Pre-treatment with the CDK4/6 inhibitor palbociclib improves the efficacy of paclitaxel in TNBC cells. Sci. Rep. 9 (1), 13014.

Critchley-Thorne, R.J., Simons, D.L., Yan, N., Miyahira, A.K., Dirbas, F.M., Johnson, D.L., Swetter, S.M., Carlson, R.W., Fisher, G.A., Koong, A., Holmes, S., Lee, P.P., 2009. Impaired interferon signaling is a common immune defect in human cancer. Proc. Natl. Acad. Sci. U. S. A. 106, 9010–9015.

Desai, A., Mitchison, T.J., 1997. Microtubule polymerization dynamics. Annu. Rev. Cell Dev. Biol. 13, 83–117.

Diaz, J.F., Andreu, J.M., 1993. Assembly of purified GDP-tubulin into microtubules induced by taxol and taxotere: reversibility, ligand stoichiometry, and competition. Biochemistry 32, 2747–2755.

English, D.P., Roque, D.M., Santin, A.D., 2013. Class III b-tubulin overexpression in gynecologic tumors: implications for the choice of microtubule targeted agents? Expert. Rev. Anticancer. Ther. 13 (1), 63–74.

Field, J.J., Kanakkanthara, A., Miller, J.H., 2014. Microtubule-targeting agents are clinically successful due to both mitotic and interphase impairment of microtubule function. Bioorg. Med. Chem. 22 (18), 5050–5059.

Florian, S., Mitchison, T.J., 2016. Anti-microtubule drugs. Methods Mol. Biol. 1413, 403–421.

Fojo, T., Menefee, M., 2007. Mechanisms of multidrug resistance: the potential role of microtubule-stabilizing agents. Ann. Oncol. 18 (Suppl 5), v3–v8.

Friedrich, M., Diesing, D., Villena-Heinsen, C., Felberbaum, R., Kolberg, H.C., Diedrich, K., 2004. Taxanes in the first-line chemotherapy of metastatic breast cancer: review. Eur. J. Gynaecol. Oncol. 25 (1), 66–70.

Fürst, R., Vollmar, A.M., 2013. A new perspective on old drugs: non-mitotic actions of tubulin-binding drugs play a major role in cancer treatment. Pharmazie 68 (7), 478–483.

Gajewski, T.F., Corrales, L., 2015. New perspectives on type I IFNs in cancer. Cytokine Growth Factor Rev. 26, 175–178.

Gallego-Jara, J., Lozano-Terol, G., Sola-Martínez, R.A., Cánovas-Díaz, M., de Diego Puente, T., 2020. A compressive review about taxol: history and future challenges. Molecules 25 (24), 5986.

Gascoigne, K., Taylor, S.S., 2009. How do anti-mitotic drugs kill cancer cells. J. Cell Sci. 122, 2579–2585.

Gasic, I., Boswell, S.A., Mitchison, T.J., 2019. Tubulin mRNA stability is sensitive to change in microtubule dynamics caused by multiple physiological and toxic cues. PLoS Biol. 17 (4), e3000225.

Giannakakou, P., Nakano, M., Nicolaou, K.C., O'Brate, A., Yu, J., Blagosklonny, M.V., Greber, U.F., Fojo, T., 2002. Enhanced microtubule-dependent trafficking and p53 nuclear accumulation by suppression of microtubule dynamics. Proc. Natl. Acad. Sci. U. S. A. 99 (16), 10855–10860.

Hanna, N., 2015. Current standards and clinical trials in systemic therapy for stage III lung cancer: what is new? Am. Soc. Clin. Oncol. Educ. Book., e442–e447.

Harbeck, N., Bartlett, M., Spurden, D., Hooper, B., Zhan, L., Rosta, E., Cameron, C., Mitra, D., Zhou, A., 2021. CDK4/6 inhibitors in HR+/HER2- advanced/metastatic breast cancer: a systematic literature review of real-world evidence studies. Future Oncol. 2021. https://doi.org/10.2217/fon-2020-1264. Online ahead of print.

Hatch, E.M., Fischer, A.H., Deerinck, T.J., Hetzer, M.W., 2013. Catastrophic nuclear envelope collapse in cancer cell micronuclei. Cell 154 (1), 47–60.

Herbert, A., 2019. ADAR and immune silencing in cancer. Trends Cancer. 5 (5), 272–282.

Herbst, R.S., Khuri, F.R., 2003. Mode of action of docetaxel—a basis for combination with novel anticancer agents. Cancer Treat. Rev. 29 (5), 407–415.

Holmes, F.A., 1996. Paclitaxel combination therapy in the treatment of metastatic breast cancer: a review. Semin. Oncol. 23 (5 Suppl 11), 46–56.

Holmfeldt, P., Sellin, M.E., Gullberg, M., 2009. Predominant regulators of tubulin monomer-polymer partitioning and their implication for cell polarization. Cell. Mol. Life Sci. 66, 3263–3276.

Horwitz, S.B., 1994. Taxol (paclitaxel): mechanisms of action. Ann. Oncol. 5 (Suppl 6), S3–S6.

Horwitz, S.B., Cohen, D., Rao, S., Ringel, I., Shen, H.J., Yang, C.P., 1993. Taxol: mechanisms of action and resistance. J. Natl. Cancer Inst. Monogr. (15), 55–61.

Hu, X., Huang, W., Fan, M., 2017. Emerging therapies for breast cancer. J. Hematol. Oncol. 10 (1), 98.

Huff, L.M., Sackett, D.L., Poruchynsky, M.S., Fojo, T., 2010. Microtubule-disrupting chemotherapeutics result in enhanced proteasome-mediated degradation and disappearance of tubulin in neural cells. Cancer Res. 70 (14), 5870–5879.

Hunt, J.T., 2009. Discovery of ixabepilone. Mol. Cancer Ther. 8 (2), 275–281.

Jain, A., Dubashi, B., Reddy, K.S., Jain, P., 2011. Weekly paclitaxel in ovarian cancer-the latest success story. Curr. Oncol. 18 (1), 16–17.

Johnson, G.V., Litersky, J.M., Whitaker, J.N., 1991. Proteolysis of microtubule-associated protein 2 and tubulin by cathepsin D. J. Neurochem. 57 (5), 1577–1583.

Jordan, M.A., 2002. Mechanism of action of antitumor drugs that interact with microtubules and tubulin. Curr. Med. Chem. Anticancer Agents 2 (1), 1–17.

Jordan, M.A., Wilson, L., 2004. Microtubules as a target for anticancer drugs. Nat. Rev. Cancer 4, 253–265.

Jordan, M.A., Toso, R.J., Thrower, D., Wilson, L., 1993. Mechanism of mitotic block and inhibition of cell proliferation by taxol at low concentrations. Proc. Natl. Acad. Sci. U. S. A. 90, 9552–9556.

Joshi, M., Liu, X., Belani, C.P., 2014. Taxanes, past, present, and future impact on non-small cell lung cancer. Anti-Cancer Drugs 25 (5), 571–583.

Kavallaris, M., 2010. Microtubules and resistance to tubulin-binding agents. Nat. Rev. Cancer 10 (3), 194–204.

Kim, E.S., Kies, M.S., Fossella, F.V., Glisson, B.S., Zaknoen, S., Statkevich, P., Munden, R.F., Summey, C., Pisters, K.M., Papadimitrakopoulou, V., Tighiouart, M., Rogatko, A., Khuri, F.R., 2005. Phase II study of the farnesyltransferase inhibitor lonafarnib with paclitaxel in patients with taxane-refractory/resistant nonsmall cell lung carcinoma. Cancer 104 (3), 561–569.

Komlodi-Pasztor, E., Sackett, D., Wilkerson, J., Fojo, T., 2011. Mitosis is not a key target of microtubule agents in patient tumors. Nat. Rev. Clin. Oncol. 8 (4), 244–250.

Komlodi-Pasztor, E., Sackett, D.L., Fojo, A.T., 2012. Inhibitors targeting mitosis: tales of how great drugs against a promising target were brought down by a flawed rationale. Clin. Cancer Res. 18 (1), 51–63.

Koshiba, H., Hosokawa, K., Mori, T., Kubo, A., Watanabe, A., Honjo, H., 2009. Intravenous paclitaxel is specifically retained in human gynecologic carcinoma tissues in vivo. Int. J. Gynecol. Cancer 19 (4), 484–4488.

Kundranda, M.N., Niu, J., 2015. Albumin-bound paclitaxel in solid tumors: clinical development and future directions. Drug Des. Devel. Ther. 9, 3767–3777.

Kuriakose, T., Kanneganti, T.D., 2018. ZBP1: innate sensor regulating cell death and inflammation. Trends Immunol. 39 (2), 123–134.

Kwon, J., Bakhoum, S.F., 2020. The cytosolic DNA-sensing cGAS–STING pathway in cancer. Cancer Discov. 10 (1), 26–39.

Lemstrova, R., Melichar, B., Mohelnikova-Duchonova, B., 2016. Therapeutic potential of taxanes in the treatment of metastatic pancreatic cancer. Cancer Chemother. Pharmacol. 78 (6), 1101–1111.

Lin, Z., Gasic, I., Chandrasekaran, V., Peters, N., Shao, S., Mitchison, T.J., Hegde, R.S., 2020. TTC5 mediates autoregulation of tubulin via mRNA degradation. Science 367 (6473), 100–104.

Lohard, S., Bourgeois, N., Maillet, L., Gautier, F., Fétiveau, A., Lasla, H., Nguyen, F., Vuillier, C., Dumont, A., Moreau-Aubry, A., Frapin, M., David, L., Loussouarn, D., Kerdraon, O., Campone, M., Jézéquel, P., Juin, P.P., Barillé-Nion, S., 2020. STING-dependent paracriny shapes apoptotic priming of breast tumors in response to anti-mitotic treatment. Nat. Commun. 11 (1), 259.

Maloney, S.M., Hoover, C.A., Morejon-Lasso, L.V., Prosperi, J.R., 2020. Mechanisms of taxane resistance. Cancers (Basel) 12 (11), 3323.

Manfredi, J.J., Parness, J., Horwitz, S.B., 1982. Taxol binds to cellular microtubules. J. Cell Biol. 94 (3), 688–696.

Marcus, A.I., Zhou, J., O'Brate, A., Hamel, E., Wong, J., Nivens, M., El-Naggar, A., Yao, T.P., Khuri, F.R., Giannakakou, P., 2005. The synergistic combination of the farnesyl transferase inhibitor lonafarnib and paclitaxel enhances tubulin acetylation and requires a functional tubulin deacetylase. Cancer Res. 65 (9), 3883–3893.

Marcus, A.I., O'Brate, A.M., Buey, R.M., Zhou, J., Thomas, S., Khuri, F.R., Andreu, J.M., Díaz, F., Giannakakou, P., 2006. Farnesyltransferase inhibitors reverse taxane resistance. Cancer Res. 66 (17), 8838–8846.

Meier, W., du Bois, A., Rau, J., Gropp-Meier, M., Baumann, K., Huober, J., Wollschlaeger, K., Kreienberg, R., Canzler, U., Schmalfeldt, B., Wimberger, P., Richter, B., Schröder, W., Belau, A., Stähle, A., Burges, A., Sehouli, J., 2012. Randomized phase II trial of carboplatin and paclitaxel with or without lonafarnib in first-line treatment of epithelial ovarian cancer stage IIB-IV. Gynecol. Oncol. 126 (2), 236–240.

Merlin, J.L., Bour-Dill, C., Marchal, S., Bastien, L., Gramain, M.P., 2000. Resistance to paclitaxel induces time-delayed multinucleation and DNA fragmentation into large fragments in MCF-7 human breast adenocarcinoma cells. Anti-Cancer Drugs 11, 295–302.

Michalakis, J., Georgatos, S.D., de Bree, E., Polioudaki, H., Romanos, J., Georgoulias, V., Tsiftsis, D.D., Theodoropoulos, P.A., 2007. Short-term exposure of cancer cells to micromolar doses of paclitaxel, with or without hyperthermia, induces long-term inhibition of cell proliferation and cell death in vitro. Ann. Surg. Oncol. 14 (3), 1220–1228.

Mielke, S., Sparreboom, A., Mross, K., 2006. Peripheral neuropathy: a persisting challenge in paclitaxel-based regimes. Eur. J. Cancer 42 (1), 24–30.

Mitchison, T.J., 2012. The proliferation rate paradox in antimitotic chemotherapy. Mol. Biol. Cell 23 (1), 1–6.

Mitchison, T.J., Pineda, J., Shi, J., Florian, S., 2017. Is inflammatory micronucleation the key to a successful anti-mitotic cancer drug? Open Biol. 7 (11), 170182.

Mori, T., Kinoshita, Y., Watanabe, A., Yamaguchi, T., Hosokawa, K., Honjo, H., 2006. Retention of paclitaxel in cancer cells for 1 week in vivo and in vitro. Cancer Chemother. Pharmacol. 58 (5), 665–6672.

Morse, D.L., Gray, H., Payne, C.M., Gillies, R.J., 2005. Docetaxel induces cell death through mitotic catastrophe in human breast cancer cells. Mol. Cancer Ther. 4 (10), 1495–1504.

Mosca, L., Ilari, A., Fazi, F., Assaraf, Y.G., Colotti, G., 2021. Taxanes in cancer treatment: activity, chemoresistance and its overcoming. Drug Resist. Updat. 54, 100742.

Motz, G.T., Coukos, G., 2013. Deciphering and reversing tumor immune suppression. Immunity 39, 61–73.

Mozzetti, S., Ferlini, C., Concolino, P., Filippetti, F., Raspaglio, G., Prislei, S., Gallo, D., Martinelli, E., Ranelletti, F.O., Ferrandina, G., Scambia, G., 2005. Class III beta-tubulin overexpression is a prominent mechanism of paclitaxel resistance in ovarian cancer patients. Clin. Cancer Res. 11 (1), 298–305.

Ojima, I., Lichtenthal, B., Lee, S., Wang, C., Wang, X., 2016. Taxane anticancer agents: a patent perspective. Expert Opin. Ther. Pat. 26 (1), 1–20.

Olziersky, A.M., Labidi-Galy, S.I., 2017. Clinical development of anti-mitotic drugs in cancer. Adv. Exp. Med. Biol. 1002, 125–152.

Orr, G.A., Verdier-Pinard, P., McDaid, H., Horwitz, S.B., 2003. Mechanisms of taxol resistance related to microtubules. Oncogene 22 (47), 7280–7295.

Ozols, R.F., Bookman, M.A., Connolly, D.C., Daly, M.B., Godwin, A.K., Schilder, R.J., Xu, X.X., Hamilton, T.C., 2004. Focus on epithelial ovarian cancer. Cancer Cell 5, 19–24.

Pabla, N., Sparreboom, A., 2015. CCR 20th anniversary commentary: BMS-247550—microtubule stabilization as successful targeted therapy. Clin. Cancer Res. 21 (6), 1237–1239.

Papanicolaou, G.N., 1942. A new procedure for staining vaginal smears. Science 95, 438–439.

Paus, R., Haslam, I.S., Sharov, A.A., Botchkarev, V.A., 2013. Pathobiology of chemotherapy-induced hair loss. Lancet Oncol. 14 (2), e50–e59.

Platanias, L.C., Fish, E.N., 1999. Signaling pathways activated by interferons. Exp. Hematol. 27, 1583–1592.

Pucci, B., Bellincampi, L., Tafani, M., Masciullo, V., Melino, G., Giordano, A., 1999. Paclitaxel induces apoptosis in Saos-2 cells with CD95L upregulation and Bcl-2 phosphorylation. Exp. Cell Res. 252 (1), 134–143.

Purba, T.S., Ng'andu, K., Brunken, L., Smart, E., Mitchell, E., Hassan, N., O'Brien, A., Mellor, C., Jackson, J., Shahmalak, A., Paus, R., 2019. CDK4/6 inhibition mitigates stem cell damage in a novel model for taxane-induced alopecia. EMBO Mol. Med. 11 (10), e11031.

Rabinovich, G.A., Gabrilovich, D., Sotomayor, E.M., 2007. Immunosuppressive strategies that are mediated by tumor cells. Annu. Rev. Immunol. 25, 267–296.

Rose, W.C., Lee, F.Y., Fairchild, C.R., Lynch, M., Monticello, T., Kramer, R.A., Manne, V., 2001. Preclinical antitumor activity of BMS-214662, a highly apoptotic and novel farnesyltransferase inhibitor. Cancer Res. 61 (20), 7507–7517.

Rowinsky, E.K., Donehower, R.C., 1995. Paclitaxel (taxol). N. Engl. J. Med. 332 (15), 1004–1014.

Rowinsky, E.K., Eisenhauer, E.A., Chaudhry, V., Arbuck, S.G., Donehower, R.C., 1993. Clinical toxicities encountered with paclitaxel (Taxol). Semin. Oncol. 20 (4 Suppl 3), 1–15.

Runowicz, C.D., Wiernik, P.H., Einzig, A.I., Goldberg, G.L., Horwitz, S.B., 1993. Taxol in ovarian cancer. Cancer 71 (4 Suppl), 1591–1596.

Salvador-Barbero, B., Álvarez-Fernández, M., Zapatero-Solana, E., El Bakkali, A., Menéndez, M.D.C., López-Casas, P.P., Di Domenico, T., Xie, T., VanArsdale, T., Shields, D.J., Hidalgo, M., Malumbres, M., 2020. CDK4/6 inhibitors impair recovery from cytotoxic chemotherapy in pancreatic adenocarcinoma. Cancer Cell 37 (3), 340–353. e6.

Schiff, P.B., Horwitz, S.B., 1980. Taxol stabilizes microtubules in mouse fibroblast cells. Proc. Natl. Acad. Sci. U. S. A. 77 (3), 1561–1565.

Schiff, P.B., Fant, J., Horwitz, S.B., 1979. Promotion of microtubule assembly in vitro by taxol. Nature 277 (5698), 665–667.

Schimming, R., Mason, K.A., Hunter, N., Weil, M., Kishi, K., Milas, L., 1999. Lack of correlation between mitotic arrest or apoptosis and antitumor effect of docetaxel. Cancer Chemother. Pharmacol. 43 (2), 165–172.

Shi, J., Mitchison, T.J., 2017. Cell death response to anti-mitotic drug treatment in cell culture, mouse tumor model and the clinic. Endocr. Relat. Cancer 24 (9), T83–T96.

Smith, E.R., George, S.H., Kobetz, E., Xu, X.X., 2018a. New biological research and understanding of Papanicolaou's test. Diagn. Cytopathol. 46 (6), 507–515.

Smith, E.R., Capo-Chichi, C.D., Xu, X.X., 2018b. Defective nuclear lamina in aneuploidy and carcinogenesis. Front. Oncol. 8, 529.

Smith, E.R., Leal, J., Amaya, C., Li, B., Xu, X.X., 2021. Nuclear Lamin A/C expression is a key determinant of paclitaxel sensitivity. Mol. Cell. Biol. 41 (7), e0064820. https://doi.org/10.1128/MCB.00648-20. Epub 2021 Jun 23. PMID: 33972393.

Sofias, A.M., Dunne, M., Storm, G., Allen, C., 2017. The battle of "nano" paclitaxel. Adv. Drug Deliv. Rev. 122, 20–30.

Takaoka, A., Taniguchi, T., 2008. Cytosolic DNA recognition for triggering innate immune responses. Adv. Drug Deliv. Rev. 60, 847–857.

Tapley, E.C., Starr, D.A., 2013. Connecting the nucleus to the cytoskeleton by SUN-KASH bridges across the nuclear envelope. Curr. Opin. Cell Biol. 25 (1), 57–62.

Tariq, Z., Zhang, H., Chia-Liu, A., Shen, Y., Gete, Y., Xiong, Z.M., Tocheny, C., Campanello, L., Wu, D., Losert, W., Cao, K., 2017. Lamin A and microtubules collaborate to maintain nuclear morphology. Nucleus 8 (4), 433–446.

Taylor, S.A., Marrinan, C.H., Liu, G., Nale, L., Bishop, W.R., Kirschmeier, P., Liu, M., Long, B.J., 2008. Combining the farnesyltransferase inhibitor lonafarnib with paclitaxel results in enhanced growth inhibitory effects on human ovarian cancer models in vitro and in vivo. Gynecol. Oncol. 109 (1), 97–106.

Theodoropoulos, P.A., Polioudaki, H., Kostaki, O., Derdas, S.P., Georgoulias, V., Dargemont, C., Georgatos, S.D., 1999. Taxol affects nuclear lamina and pore complex organization and inhibits import of karyophilic proteins into the cell nucleus. Cancer Res. 59 (18), 4625–4633.

Visconti, R., Grieco, D., 2017. Fighting tubulin-targeting anticancer drug toxicity and resistance. Endocr. Relat. Cancer 24 (9), T107–T117.

Wang, T.H., Wang, H.S., Soong, Y.K., 2000. Paclitaxel-induced cell death: where the cell cycle and apoptosis come together. Cancer 88 (11), 2619–2628.

Wani, M.C., Horwitz, S.B., 2014. Nature as a remarkable chemist: a personal story of the discovery and development of Taxol. Anti-Cancer Drugs 25 (5), 482–487.

Weaver, B.A., 2014. How Taxol/paclitaxel kills cancer cells. Mol. Biol. Cell 25 (18), 2677–2681.

Wiernik, P.H., Schwartz, E.L., Strauman, J.J., Dutcher, J.P., Lipton, R.B., Paietta, E., 1987. Phase I clinical and pharmacokinetic study of taxol. Cancer Res. 47 (9), 2486–2493.

Woo, S.R., Corrales, L., Gajewski, T.F., 2015. Innate immune recognition of cancer. Annu. Rev. Immunol. 33, 445–474.

Yan, V.C., Butterfield, H.E., Poral, A.H., Yan, M.J., Yang, K.L., Pham, C.D., Muller, F.L., 2020. Why great mitotic inhibitors make poor cancer drugs. Trends Cancer. 6, 924–941.

Yang, C.H., Horwitz, S.B., 2017. Taxol®: the first microtubule stabilizing agent. Int. J. Mol. Sci. 18 (8), 1733.

Yu, Y., Gaillard, S., Phillip, J.M., Huang, T.C., Pinto, S.M., Tessarollo, N.G., Zhang, Z., Pandey, A., Wirtz, D., Ayhan, A., Davidson, B., Wang, T.L., Shih, I.M., 2015. Inhibition of spleen tyrosine kinase potentiates paclitaxel-induced cytotoxicity in ovarian cancer cells by stabilizing microtubules. Cancer Cell 28, 82–96.

Zasadil, L.M., Andersen, K.A., Yeum, D., Rocque, G.B., Wilke, L.G., Tevaarwerk, A.J., Raines, R.T., Burkard, M.E., Weaver, B.A., 2014. Cytotoxicity of paclitaxel in breast cancer is due to chromosome missegregation on multipolar spindles. Sci. Transl. Med. 6 (229), 229ra43.

Zhai, Y., Borisy, G.G.Q., 1994. Quantitative determination of the proportion of microtubule polymer present during the mitosis-interphase transition. J. Cell Sci. 107 (Pt 4), 881–890.

Zhao, Y., Mu, X., Du, G., 2016. Microtubule-stabilizing agents: new drug discovery and cancer therapy. Pharmacol. Ther. 162, 134–143.

Zhu, Y., Zhou, Y., Shi, J., 2014. Post-slippage multinucleation renders cytotoxic variation in anti-mitotic drugs that target the microtubules or mitotic spindle. Cell Cycle 13 (11), 1756–1764.

Zierhut, C., Norihiro Yamaguchi, N., Paredes, M., Luo, J.D., Carroll, T., Funabiki, H., 2019. The cytoplasmic DNA sensor cGAS promotes mitotic cell death. Cell 178 (2), 302–315. e23.

12

An update on paclitaxel treatment in breast cancer

Tuyelee Das[a], Samapika Nandy[a], Devendra Kumar Pandey[b], Abdel Rahman Al-Tawaha[c], Mallappa Kumara Swamy[d], Vinay Kumar[e], Potshangbam Nongdam[f], and Abhijit Dey[a]

[a]*Department of Life Sciences, Presidency University, Kolkata, India,* [b]*Department of Biotechnology, Lovely Faculty of Technology and Sciences, Lovely Professional University, Phagwara, India,* [c]*Department of Biological Sciences, Al-Hussein Bin Talal University, Maan, Jordon,* [d]*Department of Biotechnology, East West First Grade College, Bengaluru, Karnataka, India,* [e]*Department of Biotechnology, Modern College (Savitribai Phule Pune University), Pune, India,* [f]*Department of Biotechnology, Manipur University, Imphal, Manipur, India*

12.1 Introduction

Breast cancer is the second most common cause of death due to cancer in women in the U.S. Breast cancer-related death rates have increased since 2007 in women. It is estimated that in the U.S. about 43,600 women are likely to die in 2021 because of breast cancer (https://www.breastcancer.org/symptoms/understand_bc/statistics#:~:text = About%2043%2C600%20women%20in%20the, year%20from%202,013%20to%20, 2018). In black women breast cancer is more common as compared with white women under the age of 45 years. The risk of rising breast cancer in Asian, Hispanic, and Native-American women is lower. Besides female breast cancer, the rate of male breast cancer incidence is also increasing. In India, breast cancer is likely to be found more in younger women (Mannan et al., 2016). Approximately, 350 cases of male breast cancer are detected in the UK every year (Amith et al., 2015). Men developed breast cancer due to exposure to radiation for a long time from a young age or increased alcohol intake linked to breast cancer in males (Amith et al., 2015). Male breast cancer relatively gets little attention as compared to female breast cancer, however, paclitaxel plus bevacizumab or paclitaxel plus trastuzumab is effective against invasive ductal carcinoma

Paclitaxel. https://doi.org/10.1016/B978-0-323-90951-8.00013-8
Copyright © 2022 Elsevier Inc. All rights reserved.

with HER2-positive (Hayashi et al., 2009) or HER2-negative male breast cancer (Huang et al., 2014). In comparison to female, male breast cancer is detected lately. Ly et al. (2013) studied breast cancer trends in males and females between 1988 and 2002, and revealed that 1.24/100,000/year male cancer rate was detected in Israel, and 90.7/100,000/year woman cancer rate was observed in the U.S. They also concluded that male and female breast cancer occurrences shared common risk factors (Ly et al., 2013). Treatments available for breast cancer include surgery, radiation therapy, chemotherapy, biological therapy, and hormonal therapy (https://www.nationalbreastcancer.org/breast-cancer-treatment/, n.d.). Chemotherapeutic drugs such as carboplatin, cyclophosphamide, doxorubicin, gemcitabine, paclitaxel, and docetaxel have been efficiently used for breast cancer treatment. These drugs also developed toxicities in breast cancer patients including neutropenia, neuropathy, sepsis, nausea, vomiting, anemia, fever, alopecia, and diarrhea. Among the other anti-cancer drugs, Paclitaxel has been widely used for the treatment of different types of breast cancer as first-line therapy, second-line therapy, and salvage therapy, or as an adjuvant and neoadjuvant treatment. Combination therapy of paclitaxel with other active antineoplastic agents, inhibitors of receptor on the breast cancer cells, inhibitors of the interleukins, and mTOR (mechanistic target of rapamycin) (Sara et al., 2015; Schott et al., 2017), have shown efficient result for breast cancer treatment in clinical trials worldwide. Besides the benefits of paclitaxel in breast cancer treatment, severe adverse events like hypersensitive reaction, cardiotoxicities, neutropenia, and peripheral neuropathy, chemoresistance are also linked with paclitaxel treatment toxicities. Table 12.1 describes the doses and outcomes of paclitaxel treatment in breast cancer with adverse events. Based on previous clinical trials, in vitro research on breast cancer cell with paclitaxel treatment, this book chapter aims to give a concise idea about the current roles and adverse events due to paclitaxel treatment in different breast cancer.

12.2 Types of breast cancer

Breast cancer is a heterogeneous disease including different entities with different biological features and clinical behavior. Histopathologically breast cancer can be classified into biologically and clinically expressive subgroups. Histological types are differentiated by the morphological pattern of the growth of the tumors. The types of breast carcinoma broadly include non-invasive breast cancers and invasive breast cancers. Invasive breast cancers further include invasive ductal carcinoma of no special type, which is the most common type (Ellis et al., 2003), invasive lobular carcinoma (Reed et al., 2015),

Table 12.1 Doses and outcomes of paclitaxel treatment in breast cancer with adverse events.

Types of breast cancer	Design phase	Patients involved in numbers	Chemotherapy, doses	Results	Adverse events	References
Advanced breast cancer	Phase II	44	P 200 mg/m^2 IV infusion → Ci 75 mg/m^2 IV	CR (11.9%), PR (40.5%), ORR (52.4%), median response duration (10.6 months)	Grade 4 neutropenia (68%), grade 1 neurotoxicity (96%), grade 2 neurotoxicity (52%)	Wasserheit et al. (1996)
Metastatic breast cancer	Phase II	33	P 100 mg/m^2	ORR (53%), CR (10%), PRs (43%)	grade 3/4 neutropenia (12%), grade 3 neuropathy (63%)	Seidman et al. (1998)
Metastatic breast cancer	Phase III	209	P 200 mg/m^2 IV over 3 h or C - 100 mg/m^2 orally (days 1–14) + M 40 mg/m^2 IV on (1, 8) + F 600 mg/m^2 IV on (1, 8) + Pr 40 mg/m^2/d orally (days 1–14)	P: median survival (17.3 months) CMFPr: median survival (9 months)	P: neutropenia (39%), peripheral neuropathy (1%), leukopenia (6%) CMFPr: neutropenia (38%), peripheral neuropathy (0%), leukopenia (24%)	Bishop et al. (1999)
Bilateral, locally advanced, or metastatic cancer	Phase II		P 80 mg/m^2	ORR (21.5%), OS (12.8 months), median progression time (4.7 months)	Grade 3/4 hematologic toxicity (15%), grade 3 neurotoxicity (9%)	Perez et al. (2001)
Locally advanced breast cancer	Phase I/II	44	P mg/m^2 IV for 1 h twice weekly for a total of 8 to 10 weeks, + radiotherapy (45 Gy at 1.8 Gy/fraction)	CRR (91%), PCR (16%), PPR (18%)	Grade 3 skin desquamation (7%), hypersensitivity (2%), stomatitis (2%)	Formenti et al. (2003)
Metastatic breast cancer	Phase III	739	A (60 mg/m^2) or P (175 mg/m2/24 h) or A + P (50 mg/m^2 + 150 mg/m2/24 h)	A: median survival (18.9 months), ORR (36%) P: median survival (22.2 months), ORR (34%) A + P: median survival 22.0 (months) ORR (47%)	A: neurologic complications (1.6%) P: neurologic complications (3.7%) A + P: neurologic complications (10.7%)	Sledge et al. (2003)

Continued

Table 12.1 Doses and outcomes of paclitaxel treatment in breast cancer with adverse events—cont'd

Types of breast cancer	Design phase	Patients involved in numbers	Chemotherapy, doses	Results	Adverse events	References
Advanced breast cancer	Phase III	327	P 175 mg/m^2 for 3-h infusion → EPI 80 mg/m^2 or P 175 mg/m2 → Cp 6 mg*min/ml every 3 weeks	P + EPI: median survival (23.5 months) P + Cp: median survival (27.8 months)	P + EPI: severe side-effect (24%) P + Cp: severe side-effect (29%)	Fountzilas et al. (2004)
Metastatic breast cancer	Parallel-group study	–	A 60 mg/m^2 + P 200 mg/m^2 for 3-h infusion or A 60 mg/m^2 + C 600 mg/m^2 every 3 weeks	A + P: OCR (80%), DFS (91%) A + C: OCR 70%, DFS (70%)	A + P: grade 3–4 neutropenia (97%), febrile neutropenia (11%) A + C: grade 3–4 neutropenia (76%), febrile neutropenia (0%)	Diéras et al. (2004)
Unresectable locally advanced breast cancer	Phase I/II	33	P 20–30 mg/m^2/day (days 2–5) + vinorelbine 20 mg/m^2 + radiotherapy (60–70 Gy)	Median follow-up (43.8 months), DFS (33%), OS (56%)	Moist desquamation (24%), grade 3–4 neutropenia (0.09%)	Kao et al. (2005)
HER-2– over-expressing metastatic breast cancer	Phase III	196	Tr 4 mg/kg + 2 mg/kg weekly + P 175 mg/m^2 every 3 weeks + Ca, or Tr 4 mg/kg + paclitaxel 175 mg/m^2	P + Tr + Ca: PFS (10.7 months), ORR (52%) P + Tr: 7.1 months, ORR (36%)	P + Tr + Ca: neutropenia, fatigue, pain (51%) P + Tr: neutropenia, fatigue, pain 32%	Robert et al. (2006)
Advanced breast cancer overexpressing HER-2	Phase II	124	P 80 mg/m^2 or P 80 mg/m^2 + Tr 4 mg/kg	P: ORR (56.9%), CR (13.8%) P + Tr: ORR (75%), CR (21.7%)	P: grade 3 neutropenia (0.032%) P + Tr: grade 3 neutropenia (0.064%)	Gasparini et al. (2007)

Metastatic HER-2– positive breast	Phase III	579	P 175 mg/m^2 IV 3 h on day 1, + L 1500 mg/d or paclitaxel + placebo once daily	P + L: CBR (69.4%), ORR (63.3%), TPP (36.4 weeks) P + placebo once daily: CBR (40.5%), ORR (37.8%), TPP (25.1 weeks)	P + L: neutropenia (26%), myalgia (32%), nausea (34%), diarrhea (58%) P + placebo: neutropenia (20%), myalgia (74%), nausea (36%), diarrhea (26%)	Di Leo et al. (2008)
Locally advanced breast cancer	–	105	P (30 mg/m^2 intravenously twice a week) for 10–12 weeks	pCR (34%), Median DFS (57 months), median OS (84 months)	–	Adams et al. (2010)
Inflammatory breast cancer	Phase II	42	L at 1500 mg/d for 14 days, → L at 1500 mg/d + weekly paclitaxel (80 mg/m^2) for 12 weeks	CRR 78.6%	Diarrhea (55%), rash, alopecia, nausea (> 50%)	Boussen et al. (2010)
Locally advanced breast cancer	28	131	P 80 mg/m^2/weekly for 16 weeks + weekly Ci 25 mg/m^2 (days 1, 8, 15)	tpCR 34.4%, pCR (44.3%)	Anemia, leukopenia, peripheral sensory neuropathy	Zhou et al. (2017)
Metastatic breast cancer	Phase II	74	P 80 mg/m^2 IV (days 1, 8, 15) + pelareorep 3 × 1010 TCID$_{50}$ IV (days 1, 2, 8, 9, 15, 16) every 4 weeks or P 80 mg/m^2	P + pelareorep 3 × 1010 TCID$_{50}$: PFS (3.78 months), median OS (17.4 months) P: PFS 3.38 months, median OS 17.4 months, median OS 10.4 (months)	P + pelareorep 3 × 1010 TCID$_{50}$: fatigue (16%), diarrhea (0%), anorexia (7%) P: fatigue (13%), diarrhea (8%), anorexia (3%)	Bernstein et al. (2018)

Continued

Table 12.1 Doses and outcomes of paclitaxel treatment in breast cancer with adverse events—cont'd

Types of breast cancer	Design phase	Patients involved in numbers	Chemotherapy, doses	Results	Adverse events	References
Metastatic triple-negative breast cancer	Phase II		P 80 mg/m^2 (days 1, 8, 15) + I 400 mg or P + placebo once per day (days 1–21) every 28 days	P + I: median follow-up (10·4 months), median PFS (6·2 months) P + placebo: median follow-up (10·2 months), median PFS (3·6–5·4 months)	P + I: grade 3 diarrhea (23%), neutrophil count decreased (8%), neutropenia (10%) P + placebo: grade 3 diarrhea (0%), neutrophil count decreased (6%), neutropenia (2%)	Kim et al. (2018)
Metastatic triple negative breast cancer	Phase II	46	P 90 mg/m^2 + Ca (days 1, 8, 15)	CR (15.2%), PR (50%), ORR (65.2%), median PFS (10.3 months), median OS (25.7 months)	Grade 3 neutropenia (28.3%), grade 4 neutropenia (13.04%)	Saloustros et al. (2018)
Metastatic triple-negative breast cancer	Phase II	140	P 90 mg/m^2 (days 1, 8, 15) + Cap (400 mg twice daily) or P + placebo (days 2–5, 9–12, 16–19)	P + Cap: median OS (19.1 months), median PFS (5.9 months) P + placebo: median OS (12.6 months), median PFS (4.2 months)	P + Cap: diarrhea (13%), infection (4%), rash (4%), fatigue (4%), neutropenia (3%) P + placebo: diarrhea (1%), infection (1%), rash (0%), fatigue (0%), neutropenia (3%)	Schmid et al. (2020)

Note: *Cp*, carboplatin; *Cap*, capivasertib; *Ci*, cisplatin; *C*, cyclophosphamide; *CR*, complete response; *CRR*, clinical response rate; *A*, doxorubicin; *DFS*, disease-free survival; *EPI*, epirubicin; *F*, fluorouracin; *I*, ipatasertib; *L*, lapatinib; *M*, methotrexate; *OCR*, objective clinical response; *OS*, overall survival; *P*, paclitaxel; *PR*, partial response; *PPR*, pathologic partial response; *PFS*, progression-free survival; *pCr*, pathological complete response; *Pr*, prednisone; *Tr*, trastuzumab; *tpCR*, total pathological complete response; →, followed by.

inflammatory breast cancer (Overmoyer and Pierce, 2014), phyllodes tumors of the breast (Tan et al., 2016), locally advanced breast cancer (Costa et al., 2018) and metastatic breast cancer. Subtypes of breast cancer are classified into HER2 positive (HER2 +), hormone receptor-positive breast cancer, and triple-negative breast cancer (TNBC). Intrinsic molecular subtypes of breast cancer have been identified as luminal A and B, HER2- enriched, basal-like and claudin- low (Prat et al., 2015). Some of the rare types of breast malignancy comprise acinic-cell carcinoma (Conlon et al., 2016), glycogen-rich clear cell carcinoma (Markopoulos et al., 2008; Toikkanen and Joensu, 1991), sebaceous carcinoma (Murakami et al., 2009), Oncocytic carcinoma (Ragazzi et al., 2011), secretory carcinoma (Amott et al., 2006), pure tubular carcinoma (Peters et al., 1981), and medullary carcinoma (Reinfuss et al., 1995).

12.3   Molecular mechanism of paclitaxel in breast cancer

Paclitaxel binds to tubulin and promotes microtubules stabilization that causes G2M cell cycle arrest. Sunters et al. (2006) reported that apoptosis of MCF-7 breast cancer cells is dependent upon JNK (c-Jun NH2-terminal kinase) activation when treated with paclitaxel. JNK enhanced FOXO3a (forkhead box O3) activity and suppressed the AKT pathway (Sunters et al., 2006). They revealed that paclitaxel treatment (1–10 nmol/L) causes nuclear translocation of FOXO3a transcription factors in several breast cancer cells. The enhanced FOXO3a level induced proapoptotic gene *Bim* (Bcl-2 interacting mediator of cell death). FOXO3a expression induced the apoptosis of breast cancer cells. FOXO3a is a class of forkhead protein that works after signaling through PI3K/AKT (phosphatidylinositol-3-kinase/protein kinase B) pathway. PI3K induced PIP2 [phosphatydylinositol (4, 5) diphosphate] phosphorylation leads to forming PIP3 [phosphatydylinositol (3–5) triphosphate]. After that, IP3 (inositol trisphosphate) binds to AKT and AKT further phosphorylates serine and threonine residues of FOXOs. Due to phosphorylation on FOXOs, it cannot bind with DNA instead, started binding with 14-3-3, which is a chaperone protein. Therefore, FOXOs and 14-3-3 complexes are then promoting cell cycle arrest and apoptosis (Dijkers et al., 2000; Van Der Heide et al., 2004). In combinational chemotherapy of coralyne with paclitaxel help in the inhibition of MCF-7 and MDA-MB-231 breast cancer cell lines. Paclitaxel and coralyne treatment reduced ki-67 (proliferation marker) expression and improve pro-apoptotic gene (*Bax*) expression with decreased anti-apoptotic factor Bcl-2 expression (Kumari et al., 2017). Shi et al. (2015), studied breast treatment efficacy by applying a combination of

BIBR1532 (2-[(E)-3-naphtalen-2-yl-but-2- enoylamino]- benzoic acid) and paclitaxel in MCF-7 cells in a dose-dependent manner. BIBR1532 induced cell cycle arrest in the G1 phase and paclitaxel-induced cell cycle arrest at the G2/M phase. Both paclitaxel and BIBR1532 combination blocked S phase cells from ingoing to the G2/M phase (Shi et al., 2015).

12.4 Paclitaxel treatment in different types of breast cancer

12.4.1 Metastatic breast cancer (MBC)

Paclitaxel is widely used for the treatment of metastatic breast cancer (MBC) as first-line chemotherapy, however, a complete cure for MBC is not yet discovered. MBC is stage IV advanced breast cancer. Symptoms of MBC are including constant back, or joint pain nipple discharge, pain, numbness, and shortness of breath. Microtubule-associated protein tau when preincubated with tubulin resulted in a decrease in paclitaxel binding and microtubule polymerization. The presence of low tau reduces microtubules exposure to paclitaxel for binding and breast cancer cells become hypersensitive to this paclitaxel (Neal and Yu, 2006). Tanaka et al. (2009), examined 35 patients and determined tau expression by immunohistochemistry. They reported 15 cases out of 35 cases were tau-negative and other cases were tau-positive. Moreover, they also reported that 85% of tau-positive expression displayed lower susceptibility after paclitaxel treatment but 60% of tau-negative expression exhibited favorable response (Tanaka et al., 2009). Paclitaxel and lapatinib treatment or placebo in HER-2–negative and HER-2–uncharacterized MBC showed no significant differences in TTP (time to progression), PFS (progression free survival), or OS (overall survival) (Di Leo et al., 2008). Weekly paclitaxel treatment is active and well-tolerated in patients with MBC (Perez et al., 2001; Seidman et al., 1998). Combination therapy of paclitaxel plus doxorubicin is effective than doxorubicin plus cyclophosphamide (Diéras et al., 2004). Paclitaxel has been compared with doxorubicin in Phase III trials (Sledge et al., 2003; Van Vreckem et al., 2000). Van Vreckem et al. (2000) administrated paclitaxel $200 \, \text{mg/m}^2$ for 3-h or intravenous bolus $75 \, \text{mg/m}^2$ doxorubicin to breast cancer patients showed better efficiency for doxorubicin single therapy (Van Vreckem et al., 2000). Sledge et al. (2003), reported equivalent results for single therapy of doxorubicin or paclitaxel. Combination therapy of doxorubicin and paclitaxel results in higher OS and median time to treatment failure but doxorubicin and paclitaxel treatment did not expand the quality of life of the patient as compared with single-agent therapy (Sledge et al., 2003). Paclitaxel has shown to be equal to cyclophosphamide, methotrexate, fluorouracil, prednisone treatment in patients with MBC. No significant difference was shown between OS, RR, and PFS (Bishop et al., 1999).

12.4.2 Inflammatory breast cancer (IBC)

Paclitaxel has been used with anthracyclines for preoperative chemotherapy. Treating options for inflammatory breast cancer (IBC) are very limited, and there is no treatment modality that can completely cure IBC. However, combined therapy, surgery, hormonal therapy, and radiotherapy can improve IBC condition. Efforts to the analysis of this multidisciplinary approach show that paclitaxel seems to be the greater efficient drug for IBC when added with anthracyclines. The addition of paclitaxel improved ORR (overall response rate) and survival with lower cardiotoxicity. Combined therapy for the management of IBC has improved both local control and survival. A dual kinase inhibitor lapatinib can reversibly inhibit EGFR (epidermal growth factor receptor) and HER2 receptor which further inhibits tumor cell growth. When paclitaxel and lapatinib therapy is given to patients for 12 weeks after 14 days lapatinib monotherapy in phase II study resulted in improved OS with limited toxicity of IBC patients with HER2-overexpressing tumors (Boussen et al., 2010). Cristofanilli et al. (2004), reported that paclitaxel addition to anthracycline-based therapy improved overall outcome in ER-negative IBC patients. Group 1 trial included 5-fluorouracil/doxorubicin/cyclophosphamide anti-cancer drugs, whereas group 2 included 5-fluorouracil/doxorubicin/cyclophosphamide followed by paclitaxel treatment. Median follow-up durations are better in the case of group 2 (148 months) than group 1(45 months). The percentage of negative estrogen receptors was more in group 2 than in group 1 (Cristofanilli et al., 2004). Paclitaxel also recovers tumor resectability in anthracycline-refractory IBC (Cristofanilli et al., 2001). Paclitaxel was given to patients those shown less clinical response than partial response after anthracycline treatment. 44% of patients out of 16 were able to undergo mastectomy (Cristofanilli et al., 2001).

12.4.3 Locally advanced breast cancer (LABC)

Neoadjuvant paclitaxel therapies for women with breast cancer proved effective in locally advanced breast cancer (LABC). Patient with LABC treated with neoadjuvant weekly paclitaxel ($30 \, \text{mg/m}^2$; intravenously) for 10–12 weeks also delivered radiotherapy to breast, axillary, and supraclavicular lymph nodes before surgery. Pathologic response rate was more in patients with HR negative tumors than in patients with HR-positive tumors (Maghous et al., 2018). Twice-weekly paclitaxel with parallel radiation is a possible key treatment for LABC with limited toxicities. Formenti et al. (2003), studied the safety and efficacy of paclitaxel and concurrent radiation in phase I/II trial. An ORR of 91% was observed in patients treated with paclitaxel 30 mg/m^2 intravenously for 1 h, followed by radical mastectomy. Among the 44 patients, pathologic complete responses occurred in 6% and pathologic partial responses in 18% (Formenti et al., 2003). Concurrent week-on/week-of radiation therapy and paclitaxel plus vinorelbine is also an effective therapy for unresectable locally advanced breast cancer with grade 3–4 neutropenia (Kao et al., 2005). Another neoadjuvant chemotherapy by paclitaxel and cisplatin combination weekly was also highly efficacious with minimum toxicities in the triple-negative and HER2-positive tumors. The phase II trial invested neoadjuvant cisplatin/paclitaxel in patients with large operable stage of locally advanced breast cancer (Zhou et al., 2017). The addition of paclitaxel to vantictumab, a monoclonal antibody that interacts to frizzled receptors and suppresses canonical WNT (Wingless and Int-1) signaling. A phase Ib trial that has generated considerable signs in the combination of the anti-frizzled antibody with paclitaxel in patients with locally advanced or metastatic HER2-negative breast cancer. In this controlled phase Ib trial, patients were treated with a combination weekly paclitaxel 90 mg/m^2 and vantictumab 3.5–14 mg/kg intravenously. The ORR was 31.3% with a clinical benefit rate of 68.8% (Diamond et al., 2020). In another study, 55 breast cancer patients during their clinical phase II–III were administered with paclitaxel (80 mg/m^2) doses, every 3 months plus vorinostat (the histone deacetylase inhibitor) (200–300 mg taken orally) and trastuzumab (for Her2/neu positive disease only) on days 1–3 of each paclitaxel dose, followed by doxorubicin (60 mg/m^2) and cyclophosphamide (600 mg/m^2) every 14 days. A pathological complete response (pCR) was observed to occur in 13 of 24 patients having Her2-positive disease. Vorinostat enhanced lysine acetylation in breast cancer by ensuing the suppression of HDAC6 (Histone deacetylase 6), and reducing Hsp90 proteins that associated with cell survival (Tu et al., 2014).

12.4.4 Molecular subtypes of breast cancer

Triple-negative breast cancer (TNBC) is a molecular subtype of cancer and it constitutes a heterogeneous group of malignancies. TNBC is defined by the absence of expression of the estrogen receptor (ER), progesterone receptor (PgR), and nonamplified human epidermal growth factor receptor 2 (HER2) expression. AKT targeted therapy for TNBC treatment has shown desired results. PI3K/AKT signaling pathway is activated in TNBC by triggering PIK3CA (phosphatidylinositol-4,5-Bisphosphate 3-Kinase Catalytic Subunit Alpha) or AKT1 mutation or disabling PTEN (phosphatase and tensin homolog) alternation. PI3K/AKT signaling pathway is associated with the growth of cancer cells. Capivasertib is an AKT inhibitor when given to the patient with paclitaxel resulted in longer PFS and OS. Paclitaxel plus placebo showed a median PFS of 4.2 months and paclitaxel plus capivasertib showed a median PFS of 5.9 months. Common adverse events were also shown higher in paclitaxel plus capivasertib therapy (Schmid et al., 2020). Ipatasertib is another AKT inhibitor that has a highly selective small molecule. Also, it has been reported that ipatasertib plus paclitaxel treatment showed a higher PFS than the placebo-treated patient (Kim et al., 2018). Naturally occurring active compound thymoquinone is found in the essential oil of *Nigella sativa* L. which has a beneficial role in cancer prevention. Thymoquinone and paclitaxel have been shown to inhibit cancer cell growth in cell culture and mice. A high dose of thymoquinone upregulated VEGF (Vascular endothelial growth factor) and EGF (Epidermal growth factor) and downregulated pro-apoptotic factors such as caspases. Paclitaxel and thymoquinone together upregulated tumor suppressor genes (Şakalar et al., 2016). In the phase III trial of patients with cT1c-cT4a-d and (HER)2 + TNBC were treated with epirubicin, paclitaxel, cyclophosphamide or weekly paclitaxel plus non-pegylated liposomal doxorubicin and trastuzumab for HER2-positive TNBC. The study found that pathological complete response was detected more frequently in patients with TNBC who received additional carboplatin (48.3% vs. 48.0%, $p = 0.005$). However, 16.4% with epirubicin, paclitaxel, cyclophosphamide and 34.1% with paclitaxel plus liposomal doxorubicin discontinued treatment due to toxic events (Schneeweiss et al., 2019). Paclitaxel with neoadjuvant carboplatin significantly rises the pathological complete response in patients with triple-negative breast cancer (Von Minckwitz et al., 2014). Robert et al. (2006) compared paclitaxel, trastuzumab, carboplatin treatment with paclitaxel and trastuzumab treatment in HER-2–overexpressing MBC. That study found a 52% ORR for paclitaxel, trastuzumab, carboplatin, and 36% ORR for paclitaxel, trastuzumab (Robert et al., 2006). The role of weekly paclitaxel was also studied in the phase II trial in metastatic TNBC. In that study, women within metastatic TNBC were treated with weekly paclitaxel plus carboplatin and bevacizumab combination. At the 5-year continuation, significant PFS and median OS improvement were detected for the metastatic TNBC (Saloustros et al., 2018).

12.5 Adverse events and resistance due to paclitaxel treatment

12.5.1 Side-effects

Neuropathic symptoms in the patient with clinical use of paclitaxel are mainly dependent upon the dose administration and infusion duration. Paclitaxel stabilizes microtubules associated with induction of toxicity within the cell. In vitro studies have demonstrated that ganglion cells are formed large abnormal microtubule arrays in treatment with paclitaxel resultant in abnormal neurite outgrowth and neuronal death (Letourneau and Ressler, 1984; Masurovsky et al., 1981, 1983; Peterson and Crain, 1982). Paclitaxel toxicity observed in the patient with recurrent breast cancers and anthracycline-resistant breast cancer when treated with $175\,mg/m^2$ and $250-300\,mg/m^2$ concentration, respectively. In recurrent breast cancer and anthracycline-resistant breast cancer patients, neuropathic symptoms were 79% and 100% respectively. The patient showed mild paresis, small-diameter abnormalities, and nerve fiber abnormalities (Postma et al., 1995).

Peripheral neuropathy harms the mood and quality of life of the patient. Paclitaxel-induced peripheral neuropathy symptoms may start after 24–72 h of high single doses of paclitaxel administration (Reyes-Gibby et al., 2009). In vitro models (dorsal root ganglion neurons, human induced pluripotent stem cells) and in vivo models' studies recommend that peripheral neuropathy could be developed by the involvement of various molecular pathways. Paclitaxel encourages calcium signaling alternation, mitochondrial dysfunction, formation of ROS, and ion channel activation (Staff et al., 2020).

Mitochondrial dysfunction has been observed in in vitro models' for various types of neuropathic pain (Kober et al., 2018). The mechanism of paclitaxel-induced peripheral neuropathy in patients with breast cancer was not fully known (Ghoreishi et al., 2018). Single-nucleotide polymorphisms (SNP) are associated with paclitaxel related sensory neuropathy (Abraham et al., 2014). Abraham et al. (2014), investigated 73 SNPs in 50 genes for their contribution to paclitaxel related sensory neuropathy risk in breast cancer. They detected paclitaxel and SNPs association with paclitaxel dose a by Cox regression analysis and resulted in moderate/severe paclitaxel related sensory neuropathy development (Abraham et al., 2014). Peripheral neuropathy induced by paclitaxel persists longer in older patients (C60). Tanabe et al. (2013), reported that 212 out of 219 patients developed peripheral neuropathy and the severity of peripheral neuropathy was significantly associated with peripheral neurotoxicity duration (Tanabe et al., 2013). In clinical studies, it was shown that age, body surface area, and PR + patients (progesterone receptor-positive) as potential risk factors were associated with the development of peripheral neurotoxicity with paclitaxel treatment (Ghoreishi et al., 2018). Administrations of paclitaxel ($80\,mg/m^2$) for 12 weeks induce symptomatic and objective neuropathy in the early days and persist later (Timmins et al., 2021). Typhlitis is an inflammation of the cecum caused by aggressive chemotherapy for hematologic and solid malignancies. It has been reported to develop typhlitis in metastatic breast cancer patients treated with a combination therapy of paclitaxel ($180\,mg/m^2$) and doxorubicin, but not with a single drug therapy (Pestalozzi et al., 1993).

Not often, Paclitaxel induces respiratory injury including pulmonary opacities, acute interstitial pneumonia, and subacute interstitial pneumonia. Pneumonitis is a rare lung inflammation associated with paclitaxel treatment. Abulkhair and El Melouk (2011) reported patients with locally advanced breast cancer when treated with weekly paclitaxel and trastuzumab resulted in drug-induced interstitial pneumonitis (Abulkhair and El Melouk, 2011). Taghian et al. (2001), reported combining paclitaxel and radiotherapy increase radiation pneumonitis in the primary treatment of breast cancer. Combination therapy resulted in 14.6% radiation pneumonitis, whereas, therapy given to the patient without paclitaxel reported the development of only 1.1% radiation pneumonitis (Taghian et al., 2001).

12.5.2 Paclitaxel resistance

The microtubule-stabilizing agent paclitaxel interfere with the spindle microtubule resulting in apoptosis. Paclitaxel has become extensively known as active chemotherapeutical drugs in the treatment of breast cancer. Paclitaxel treatment progression in breast cancer is comparatively short and the quality of life of patient suffers and after few months' reversibility of this drug occur. Mechanisms of paclitaxel resistance by cancer cells have yet to be fully elucidated. One of the recent reviews broadly described the paclitaxel resistance mechanism in cancer cell caused by MDR efflux transporters overexpression, altered expression of microtubule-associated proteins or microtubule alternation (Das et al., 2021). Chemoresistance is one of the major problems of cancer treatment denotes the important cause of mortality in breast cancer patients. Long exposure to chemotherapeutic drugs may develop acquired resistance. Aoudjit and Vuori (2001), demonstrate that both breast cancer cell lines MDA-MB-231 and MDA-MB-435 induced by paclitaxel and resulted in ligation of b1 integrins associated with inhibition of apoptosis-related to cytochrome c release inhibition. This leads to activation of the PI3K/AKT pathway and is further associated with drug resistance (Aoudjit and Vuori, 2001). Ajabnoor et al. (2012), examined clinically paclitaxel doses in MCF-7 cells and found that relevant doses of paclitaxel are associated with drug-induced cytotoxicity. They also reported that paclitaxel-resistant MCF-7TaxR cells lack caspase-mediated apoptotic cell death pathways, however, an increased autophagic response is also observed (Ajabnoor et al., 2012). Moreover, paclitaxel resistance is associated with Six1 (Sine oculis 1) (Li et al., 2013), lacking GPSM2 (G protein signaling modulator; 2) mRNA (Zhang et al., 2020), Lin28 (Lin-28 homolog A) expression (Lv et al., 2012), FTH1P3 (long non-coding RNAs ferritin heavy chain 1 pseudogene; 3) (Wang et al., 2018) and overexpression of IRAK1 (interleukin-1 receptor-associated kinase; 4) in breast cancer cell lines (Wee et al., 2015). As we know that paclitaxel targets microtubules and microtubules are made up of a-and b-tubulins. Banerjee (2002), reported that the PTX-resistant breast cancer cells MCF-7 contain a two-fold higher amount of tyrosinated a-tubulin, 2.5-fold higher amounts of bIII, and 1.5-fold higher amount of bIV-tubulin than those of the wild-type MCF-7 cells (Banerjee, 2002).

12.6 Efficiency of other anti-cancer drugs over paclitaxel

The efficiency of gemcitabine plus paclitaxel or paclitaxel single therapy in advanced breast cancer patients resulted from gemcitabine plus paclitaxel as a practical choice with limited toxicities (Albain et al., 2008). Patients treated gemcitabine plus paclitaxel showed better median survival (18.6 months), TTP (6.14 months) and RR (response rate) (41.4%) than paclitaxel single therapy whereas median survival, TTP and RR are 15.8 months, 3.98 months and 26.2%, respectively (Albain et al., 2008). Another anti-cancer drug docetaxel when compared with paclitaxel in phase III trial patients with advanced breast cancer reported that the median OS and TTP, for docetaxel were longer than for paclitaxel. However, hematologic and nonhematologic toxicities are also more in docetaxel treatment than for paclitaxel treatment (Jones et al., 2005). Paclitaxel efficacy over doxorubicin was explored by given intravenous paclitaxel or doxorubicin as a single agent therapy to a patient with advanced breast cancer. Response for doxorubicin was significantly better than for paclitaxel. Toxicity among doxorubicin treated patients reported greater than paclitaxel (Van Vreckem et al., 2000).

12.7 Conclusions

The effectiveness of paclitaxel treatment has been proved in first-line therapy, salvage therapy, and anthracycline drug-refractory patients of breast cancer. In adjuvant therapy, clinical data from randomized trials have maintained the progressive use of paclitaxel in patients who were treated prior by doxorubicin and cyclophosphamide drugs. Paclitaxel combination therapy with other anticancer drugs showed unique mechanisms in breast cancer cells with limited toxicity that made paclitaxel an efficient drug from breast cancer treatment. Combination therapy with paclitaxel and anthracyclines have proved a high response rate. However, clinical trials of paclitaxel versus doxorubicin or docetaxel or gemcitabine revealed a better response rate in patients treated with doxorubicin or docetaxel or gemcitabine. Combination therapy with trastuzumab, vinorelbine, cisplatin, lapatinib, and gemcitabine appear to be promising. The weekly paclitaxel trials with $(80\text{--}100\,mg/m^2)$ have been detected to give good response rates with limited toxicity. Treatment with high dose paclitaxel showed relatively more toxicity, so, dose reduction is the strategy to suppress neurotoxicity. Multiple clinical trials by using paclitaxel in the treatment of MBE, TNBC, IBC, and LABC have been performed and confirmed the advantage of the exploitation of paclitaxel treatment. In conclusion, paclitaxel is an important drug for breast cancer treatment. Though some disagreements linked with its use in clinical trials, vastly improvement of its resistance properties needs to be addressed in cancer therapy. To reduce its side effects in the future, research should be focused on developing new carriers for targeted cancer cells.

Acknowledgment

Authors are extremely thankful to UGC, Government of India, for financial assistance. Authors are highly grateful to Presidency University-FRPDF fund, Kolkata for providing the needed research facilities.

References

Abraham, J.E., Guo, Q., Dorling, L., Tyrer, J., Ingle, S., Hardy, R., Vallier, A.L., Hiller, L., Burns, R., Jones, L., Bowden, S.J., Dunn, J.A., Poole, C.J., Caldas, C., Pharoah, P.P.D., Earl, H.M., 2014. Replication of genetic polymorphisms reported to be associated with taxane-related sensory neuropathy in patients with early breast cancer treated with paclitaxel. Clin. Cancer Res. 20, 2466–2475. https://doi.org/10.1158/1078-0432.CCR-13-3232.

Abulkhair, O., El Melouk, W., 2011. Delayed paclitaxel-trastuzumab-induced interstitial pneumonitis in breast cancer. Case Rep. Oncol. 4, 186–191. https://doi.org/10.1159/000326063.

Adams, S., Chakravarthy, A.B., Donach, M., Spicer, D., Lymberis, S., Singh, B., Bauer, J.A., Hochman, T., Goldberg, J.D., Muggia, F., Schneider, R.J., Pietenpol, J.A., Formenti, S.C., 2010. Preoperative concurrent paclitaxel-radiation in locally advanced breast cancer: pathologic response correlates with five-year overall survival. Breast Cancer Res. Treat. 124, 723–732. https://doi.org/10.1007/s10549-010-1181-8.

Ajabnoor, G.M.A., Crook, T., Coley, H.M., 2012. Paclitaxel resistance is associated with switch from apoptotic to autophagic cell death in MCF-7 breast cancer cells. Cell Death Dis. 3, 1–9. https://doi.org/10.1038/cddis.2011.139.

Albain, K.S., Nag, S.M., Calderillo-Ruiz, G., Jordaan, J.P., Llombart, A.C., Pluzanska, A., Rolski, J., Melemed, A.S., Reyes-Vidal, J.M., Sekhon, J.S., Simms, L., O'Shaughnessy, J., 2008. Gemcitabine plus paclitaxel versus paclitaxel monotherapy in patients with metastatic breast cancer and prior anthracycline treatment. J. Clin. Oncol. 26, 3950–3957. https://doi.org/10.1200/JCO.2007.11.9362.

Amith, S.R., Wilkinson, J.M., Baksh, S., Fliegel, L., 2015. The Na+/H+ exchanger (NHE1) as a novel co-adjuvant target in paclitaxel therapy of triple-negative breast cancer cells. Oncotarget 6, 1262–1275. https://doi.org/10.18632/oncotarget.2860.

Amott, D.H., Masters, R., Moore, S., 2006. Secretory carcinoma of the breast. Breast J. 12, 183. https://doi.org/10.1111/j.1075-122X.2006.00233.x.

Aoudjit, F., Vuori, K., 2001. Integrin signaling inhibits paclitaxel-induced apoptosis in breast cancer cells. Oncogene 20, 4995–5004. https://doi.org/10.1038/sj.onc.1204554.

Banerjee, A., 2002. Increased levels of tyrosinated α-, βIII-, and βIV-tubulin isotypes in paclitaxel-resistant MCF-7 breast cancer cells. Biochem. Biophys. Res. Commun. 293, 598–601. https://doi.org/10.1016/S0006-291X(02)00269-3.

Bernstein, V., Ellard, S.L., Dent, S.F., Tu, D., Mates, M., Dhesy-Thind, S.K., Panasci, L., Gelmon, K.A., Salim, M., Song, X., Clemons, M., Ksienski, D., Verma, S., Simmons, C., Lui, H., Chi, K., Feilotter, H., Hagerman, L.J., Seymour, L., 2018. A randomized phase II study of weekly paclitaxel with or without pelareorep in patients with metastatic breast cancer: final analysis of Canadian Cancer Trials Group IND.213. Breast Cancer Res. Treat. 167, 485–493. https://doi.org/10.1007/s10549-017-4538-4.

Bishop, J.F., Dewar, J., Toner, G.C., Smith, J., Tattersall, M.H., Olver, I.N., Ackland, S., Kennedy, I., Goldstein, D., Gurney, H., Walpole, E., 1999. Initial paclitaxel improves outcome compared with cmfp combination chemotherapy as front-line therapy in untreated metastatic breast cancer. J. Clin. Oncol. 17 (8), 2355.

Boussen, H., Cristofanilli, M., Zaks, T., DeSilvio, M., Salazar, V., Spector, N., 2010. Phase II study to evaluate the efficacy and safety of neoadjuvant lapatinib plus paclitaxel in patients with inflammatory breast cancer. J. Clin. Oncol. 28, 3248–3255. https://doi.org/10.1200/JCO.2009.21.8594.

Conlon, N., Sadri, N., Corben, A.D., Tan, L.K., 2016. Acinic cell carcinoma of breast: morphologic and immunohistochemical review of a rare breast cancer subtype. Hum. Pathol. 51, 16–24.

Costa, R., Hansen, N., Gradishar, W.J., 2018. Locally advanced breast cancer. Breast Compr. Manag. Benign Malig. Dis., 819–831. e6 https://doi.org/10.1016/B978-0-323-35955-9.00063-5.

Cristofanilli, M., Buzdar, A.U., Sneige, N., Smith, T., Wasaff, B., Ibrahim, N., Booser, D., Rivera, E., Murray, J.L., Valero, V., Ueno, N., Singletary, E.S., Hunt, K., Strom, E., McNeese, M., Stelling, C., Hortobagyi, G.N., 2001. Paclitaxel in the multimodality treatment for inflammatory breast carcinoma. Cancer 92, 1775–1782. https://doi.org/10.1002/1097-0142(20011001)92:7<1775::AID-CNCR1693>3.0.CO;2-E.

Cristofanilli, M., Gonzalez-Angulo, A.M., Buzdar, A.U., Kau, S.W., Frye, D.K., Hortobagyi, G.N., 2004. Paclitaxel improves the prognosis in estrogen receptor-negative inflammatory breast cancer: the M.D. Anderson Cancer center experience. Clin. Breast Cancer 4, 415–419. https://doi.org/10.3816/CBC.2004.n.004.

Das, T., Anand, U., Pandey, S.K., Ashby Jr., C.R., Assaraf, Y.G., Chen, Z., Dey, A., 2021. Therapeutic strategies to overcome taxane resistance in cancer. Drug Resist. Updat. 100754. https://doi.org/10.1016/j.drup.2021.100754.

Di Leo, A., Gomez, H.L., Aziz, Z., Zvirbule, Z., Bines, J., Arbushites, M.C., Guerrera, S.F., Koehler, M., Oliva, C., Stein, S.H., Williams, L.S., Dering, J., Finn, R.S., Press, M.F., 2008. Phase III, double-blind, randomized study comparing lapatinib plus paclitaxel with placebo plus paclitaxel as first-line treatment for metastatic breast cancer. J. Clin. Oncol. 26, 5544–5552. https://doi.org/10.1200/JCO.2008.16.2578.

Diamond, J.R., Becerra, C., Richards, D., Mita, A., Osborne, C., O'Shaughnessy, J., Zhang, C., Henner, R., Kapoun, A.M., Xu, L., Stagg, B., Uttamsingh, S., Brachmann, R.K., Farooki, A., Mita, M., 2020. Phase Ib clinical trial of the anti-frizzled antibody vantictumab (OMP-18R5) plus paclitaxel in patients with locally advanced or metastatic HER2-negative breast cancer. Breast Cancer Res. Treat. 184, 53–62. https://doi.org/10.1007/s10549-020-05817-w.

Diéras, V., Fumoleau, P., Romieu, G., Tubiana-Hulin, M., Namer, M., Mauriac, L., Guastalla, J.P., Pujade-Lauraine, E., Kerbrat, P., Maillart, P., Pénault-Llorca, F., Buyse, M., Pouillart, P., 2004. Randomized parallel study of doxorubicin plus paclitaxel and doxorubicin plus cyclophosphamide as neoadjuvant treatment of patients with breast cancer. J. Clin. Oncol. 22, 4958–4965. https://doi.org/10.1200/JCO.2004.02.122.

Dijkers, P.F., Medema, R.H., Pals, C., Banerji, L., Thomas, N.S.B., Lam, E.W.-F., Burgering, B.M.T., Raaijmakers, J.A.M., Lammers, J.-W.J., Koenderman, L., Coffer, P.J., 2000. Forkhead transcription factor FKHR-L1 modulates cytokine-dependent transcriptional regulation of p27KIP1. Mol. Cell. Biol. 20, 9138–9148. https://doi.org/10.1128/mcb.20.24.9138-9148.2000.

Ellis, I.O., Schnitt, S.J., Sastre-Garau, X., Bussolati, G., Tavassoli, F.A., 2003. Invasive breast carcinoma. In: Tavassoli, F.A., Devilee, P. (Eds.), Pathology and Genetics of Tumours of the Breast and Female Genital Tract Organs. IARC Press, Lyon, France, pp. 18–19.

Formenti, S.C., Volm, M., Skinner, K.A., Spicer, D., Cohen, D., Perez, E., Bettini, A.C., Groshen, S., Gee, C., Florentine, B., Press, M., Danenberg, P., Muggia, F., 2003. Preoperative twice-weekly paclitaxel with concurrent radiation therapy followed by surgery and postoperative doxorubicin-based chemotherapy in locally advanced breast cancer: a phase I/II trial. J. Clin. Oncol. 21, 864–870. https://doi.org/10.1200/JCO.2003.06.132.

Fountzilas, G., Kalofonos, H.P., Dafni, U., Papadimitriou, C., Bafaloukos, D., Papakostas, P., Kalogera-Fountzila, A., Gogas, H., Aravantinos, G., Moulopoulos, L.A., Economopoulos, T., Pectasides, D., Maniadakis, N., Siafaka, V., Briasoulis, E., Christodoulou, C., Tsavdaridis, D., Makrantonakis, P., Razis, E., Kosmidis, P., Skarlos, D., Dimopoulos, M.A., 2004. Paclitaxel and epirubicin versus paclitaxel and carboplatin as first-line chemotherapy in patients with advanced breast cancer: A phase III study conducted by the Hellenic cooperative oncology group. Ann. Oncol. 15, 1517–1526. https://doi.org/10.1093/annonc/mdh395.

Gasparini, G., Gion, M., Mariani, L., Papaldo, P., Crivellari, D., Filippelli, G., Morabito, A., Silingardi, V., Torino, F., Spada, A., Zancan, M., De Sio, L., Caputo, A., Cognetti, F., Lambiase, A., Amadori, D., 2007. Randomized phase II trial of weekly paclitaxel alone versus trastuzumab plus weekly paclitaxel as first-line therapy of patients with Her-2 positive advanced breast cancer. Breast Cancer Res. Treat. 101, 355–365. https://doi.org/10.1007/s10549-006-9306-9.

Ghoreishi, Z., Keshavarz, S., Asghari Jafarabadi, M., Fathifar, Z., Goodman, K.A., Esfahani, A., 2018. Risk factors for paclitaxel-induced peripheral neuropathy in patients with breast cancer. BMC Cancer 18, 5–10. https://doi.org/10.1186/s12885-018-4869-5.

Hayashi, H., Kimura, M., Yoshimoto, N., Tsuzuki, M., Tsunoda, N., Fujita, T., Yamashita, T., Iwata, H., 2009. A case of HER2-positive male breast cancer with lung metastases showing a good response to trastuzumab and paclitaxel treatment. Breast Cancer 16, 136–140. https://doi.org/10.1007/s12282-008-0060-1.

https://www.breastcancer.org/symptoms/understand_bc/statistics#:~:text=About%2043%2C600%20women%20in%20the,year%20from%202013%20to%20, 2018.

https://www.nationalbreastcancer.org/breast-cancer-treatment/, n.d.

Huang, D.P., Ye, X.H., Jin, C., 2014. Successful use of bevacizumb and paclitaxel in a male breast cancer with liver metastases. Int. J. Clin. Exp. Med. 7, 3076–3079.

Jones, S.E., Erban, J., Overmoyer, B., Budd, G.T., Hutchins, L., Lower, E., Laufman, L., Sundaram, S., Urba, W.J., Pritchard, K.I., Mennel, R., Richards, D., Olsen, S., Meyers, M.L., Ravdin, P.M., 2005. Randomized phase III study of docetaxel compared with paclitaxel in metastatic breast cancer. J. Clin. Oncol. 23, 5542–5551. https://doi.org/10.1200/JCO.2005.02.027.

Kao, J., Conzen, S.D., Jaskowiak, N.T., Song, D.H., Recant, W., Singh, R., Masters, G.A., Fleming, G.F., Heimann, R., 2005. Concomitant radiation therapy and paclitaxel for unresectable locally advanced breast cancer: results from two consecutive phase I/II trials. Int. J. Radiat. Oncol. Biol. Phys. 61, 1045–1053. https://doi.org/10.1016/j.ijrobp.2004.07.714.

Kim, P.S., Korea, S., Dent, R., Centre, N.C., Im, S., Korea, S., Louis, H.S., Blau, S., Specialties, N.M., Tan, A.R., Isakof, S.J., Hospital, M.G., Oliveira, M., Saura, C., Wongchenko, M.J., Francisco, S.S., 2018. Ipatasertib plus paclitaxel versus placebo plus paclitaxel as first- line therapy for metastatic triple-negative breast cancer (LOTUS): a multicentre, randomised, double-blind, placebo- controlled, phase 2 trial. Lancet Oncol. 18, 1360–1372. https://doi.org/10.1016/S1470-2045(17)30450-3.Ipatasertib.

Kober, K.M., Olshen, A., Conley, Y.P., Schumacher, M., Topp, K., Smoot, B., Mazor, M., Chesney, M., Hammer, M., Paul, S.M., Levine, J.D., Miaskowski, C., 2018. Expression of mitochondrial dysfunction-related genes and pathways in paclitaxel-induced peripheral neuropathy in breast cancer survivors. Mol. Pain 14. https://doi.org/10.1177/1744806918816462.

Kumari, S., Badana, A.K., Mohan, G.M., Shailender Naik, G., Malla, R.R., 2017. Synergistic effects of coralyne and paclitaxel on cell migration and proliferation of breast cancer cells lines. Biomed. Pharmacother. 91, 436–445. https://doi.org/10.1016/j.biopha.2017.04.027.

Letourneau, P.C., Ressler, A.H., 1984. Inhibition of neurite initiation and growth by taxol. J. Cell Biol. 98, 1355–1362. https://doi.org/10.1083/jcb.98.4.1355.

Li, Z., Tian, T., Hu, X., Zhang, X., Nan, F., Chang, Y., Lv, F., Zhang, M., 2013. Six1 mediates resistance to paclitaxel in breast cancer cells. Biochem. Biophys. Res. Commun. 441, 538–543. https://doi.org/10.1016/j.bbrc.2013.10.131.

Lv, K., Liu, L., Wang, L., Yu, J., Liu, X., Cheng, Y., Dong, M., Teng, R., Wu, L., Fu, P., Deng, W., Hu, W., Teng, L., 2012. Lin28 mediates paclitaxel resistance by modulating p21, Rb and let-7a miRNA in breast cancer cells. PLoS One 7, 1–8. https://doi.org/10.1371/journal.pone.0040008.

Ly, D., Forman, D., Ferlay, J., Brinton, L.A., Cook, M.B., 2013. An international comparison of male and female breast cancer incidence rates. Int. J. Cancer 132, 1918–1926. https://doi.org/10.1002/ijc.27841.

Maghous, A., Marnouche, E.A., Zaghba, N., Andaloussi, K., Elmarjany, M., Hadadi, K., Sifat, H., Moussaoui, R.D., 2018. Neoadjuvant radiotherapy of early-stage and locally advanced breast cancer: review of the literature. J. Nucl. Med. Radiat. Ther., 14–17. https://doi.org/10.4172/2155-9619.1000357.

Mannan, A.U., Singh, J., Lakshmikeshava, R., Thota, N., Singh, S., Sowmya, T.S., Mishra, A., Sinha, A., Deshwal, S., Soni, M.R., Chandrasekar, A., Ramesh, B., Ramamurthy, B., Padhi, S., Manek, P., Ramalingam, R., Kapoor, S., Ghosh, M., Sankaran, S., Ghosh, A., Veeramachaneni, V., Ramamoorthy, P., Hariharan, R., Subramanian, K., 2016. Detection of high frequency of mutations in a breast and/or ovarian cancer cohort: implications of embracing a multi-gene panel in molecular diagnosis in India. J. Hum. Genet. 61, 515–522. https://doi.org/10.1038/jhg.2016.4.

Markopoulos, C., Mantas, D., Philipidis, T., Kouskos, E., Antonopoulou, Z., Hatzinikolaou, M.L., Gogas, H., 2008. Glycogen-rich clear cell carcinoma of the breast. World J. Surg. Oncol. 6, 4–7. https://doi.org/10.1186/1477-7819-6-44.

Masurovsky, E.B., Peterson, E.R., Crain, S.M., Horwitz, S.B., 1981. Microtubule arrays in taxol-treated mouse dorsal root ganglion-spinal cord cultures. Brain Res. 217, 392–398. https://doi.org/10.1016/0006-8993(81)90017-2.

Masurovsky, E.B., Peterson, E.R., Crain, S.M., Horwitz, S.B., 1983. Morphological alterations in dorsal root ganglion neurons and supporting cells of organotypic mouse spinal cord-ganglion cultures exposed to taxol. Neuroscience 10, 491–509. https://doi.org/10.1016/0306-4522(83)90148-3.

Murakami, A., Kawachi, K., Sasaki, T., Ishikawa, T., Nagashima, Y., Nozawa, A., 2009. Sebaceous carcinoma of the breast: case report. Pathol. Int. 59, 188–192. https://doi.org/10.1111/j.1440-1827.2009.02349.x.

Neal, C.L., Yu, D., 2006. Microtubule-associated protein tau: a marker of paclitaxel sensitivity in breast cancer. Breast Dis. 16, 374–375. https://doi.org/10.1016/S1043-321X(05)80306-6.

Overmoyer, B., Pierce, L.J., 2014. Inflammatory breast cancer. Dis. Breast Fifth Ed. 60, 351–375. https://doi.org/10.3322/caac.20082.Available.

Perez, E.A., Vogel, C.L., Irwin, D.H., Kirshner, J.J., Patel, R., 2001. Multicenter phase II trial of weekly paclitaxel in women with metastatic breast cancer. J. Clin. Oncol. 19, 4216–4223.

Pestalozzi, B.C., Sotos, G.A., Choyke, P.L., Fisherman, J.S., Cowan, K.H., O'Shaughnessy, J.A., 1993. Typhlitis resulting from treatment with taxol and doxorubicin in patients with metastatic breast cancer. Cancer 71, 1797–1800. https://doi.org/10.1002/1097-0142(19930301)71:5<1797::AID-CNCR2820710514>3.0.CO;2-B.

Peters, G.N., Wolff, M., Haagensen, C.D., 1981. Tubular carcinoma of the breast. Clinical pathologic correlations based on 100 cases. Ann. Surg. 193, 138–149. https://doi.org/10.1097/00000658-198102000-00003.

Peterson, E.R., Crain, S.M., 1982. Nerve growth factor attenuates neurotoxic effects of taxol on spinal cord-ganglion explants from fetal mice. Science (80-.) 217, 377–379. https://doi.org/10.1126/science.6124041.

Postma, T.J., Vermorken, J.B., Liefting, A.J.M., Pinedo, H.M., Heimans, J.J., 1995. Paclitaxel-induced neuropathy. Ann. Oncol. 6, 489–494.

Prat, A., Pineda, E., Adamo, B., Galván, P., Fernández, A., Gaba, L., Díez, M., Viladot, M., Arance, A., Muñoz, M., 2015. Clinical implications of the intrinsic molecular subtypes of breast cancer. Breast 24, S26–S35. https://doi.org/10.1016/j.breast.2015.07.008.

Ragazzi, M., De Biase, D., Betts, C.M., Farnedi, A., Ramadan, S.S., Tallini, G., Reis-Filho, J.S., Eusebi, V., 2011. Oncocytic carcinoma of the breast: frequency, morphology and follow-up. Hum. Pathol. 42, 166–175. https://doi.org/10.1016/j.humpath.2010.07.014.

Reed, M.E.M.C., Kutasovic, J.R., Lakhani, S.R., Simpson, P.T., 2015. Invasive lobular carcinoma of the breast: morphology, biomarkers and omics. Breast Cancer Res. 17, 1–11. https://doi.org/10.1186/s13058-015-0519-x.

Reinfuss, M., Stelmach, A., Mitus, J., Rys, J., Duda, K., 1995. Typical medullary carcinoma of the breast: a clinical and pathological analysis of 52 cases. J. Surg. Oncol. 60, 89–94. https://doi.org/10.1002/jso.2930600205.

Reyes-Gibby, C.C., Morrow, P.K., Aman Buzdar, A., Shete, S., 2009. Chemotherapy-induced peripheral neuropathy as a predictor of neuropathic pain in breast cancer patients previously treated with paclitaxel. J. Pain 10, 1146–1150. https://doi.org/10.1016/j.jpain.2009.04.006.Chemotherapy-induced.

Robert, N., Leyland-Jones, B., Asmar, L., Belt, R., Llegbodu, D., Loesch, D., Raju, R., Valentine, E., Sayre, R., Cobleigh, M., Albain, K., McCullough, C., Fuchs, L., Slamon, D., 2006. Randomized phase III study of trastuzumab, paclitaxel, and carboplatin compared with trastuzumab and paclitaxel in women with HER-2–overexpressing metastatic breast cancer. Clin. Oncol. 24, 2786–2792. https://doi.org/10.1200/JCO.2005.04.1764.

Şakalar, Ç., İzgi, K., İskender, B., Sezen, S., Aksu, H., Çakır, M., Kurt, B., Turan, A., Canatan, H., 2016. The combination of thymoquinone and paclitaxel shows anti-tumor activity through the interplay with apoptosis network in triple-negative breast cancer. Tumor Biol. 37, 4467–4477. https://doi.org/10.1007/s13277-015-4307-0.

Saloustros, E., Nikolaou, M., Kalbakis, K., Polyzos, A., Christofillakis, C., Kentepozidis, N., Pistamaltzian, N., Kourousis, C., Vamvakas, L., Georgoulias, V., Mavroudis, D., 2018. Weekly paclitaxel and carboplatin plus bevacizumab as first-line treatment of metastatic triple-negative breast cancer. A multicenter phase II trial by the Hellenic Oncology Research Group. Clin. Breast Cancer 18, 88–94. https://doi.org/10.1016/j.clbc.2017.10.013.

Sara, A., Hurvitz, F., Andre, Z., Jiang, Z., Shao, M., Mano, S., Neciosup, S.P., Tseng, L.-M., Zhang, Q., Shen, K., Liu, D., Dreosti, L.M., Sara, H., Hurvitz, A., Andre, F., Jiang, Z., Shao, Z., Mano, M.S., Neciosup, S.P., Tseng, L.-M., Zhang, Q., Shen, K., Liu, D., Lydi, H., 2015. Combination of everolimus with trastuzumab plus paclitaxel as first-line therapy for HER2+ advanced breast cancer (BOLERO-1): primary results of a phase III, randomized, double-blind, multicenter trial. Oncol. Lancet 16, 1–50.

Schmid, P., Abraham, J., Chan, S., Wheatley, D., Brunt, A.M., Nemsadze, G., Baird, R.D., Park, Y.H., Hall, P.S., Perren, T., Stein, R.C., Mangel, L., Ferrero, J.M., Phillips, M., Conibear, J., Cortes, J., Foxley, A., de Bruin, E.C., McEwen, R., Stetson, D., Dougherty, B., Sarker, S.J., Prendergast, A., McLaughlin-Callan, M., Burgess, M., Lawrence, C., Cartwright, H., Mousa, K., Turner, N.C., 2020. Capivasertib plus paclitaxel versus placebo plus paclitaxel as first-line therapy for metastatic triple-negative breast cancer: the PAKT trial. J. Clin. Oncol. 38 (5), 423–433. https://doi.org/10.1200/JCO.19.00368.

Schneeweiss, A., Möbus, V., Tesch, H., Hanusch, C., Denkert, C., Lübbe, K., Huober, J., Klare, P., Kümmel, S., Untch, M., Kast, K., Jackisch, C., Thomalla, J., Ingold-Heppner, B., Blohmer, J.U., Rezai, M., Frank, M., Engels, K., Rhiem, K., Fasching, P.A., Nekljudova, V., von Minckwitz, G., Loibl, S., 2019. Intense dose-dense epirubicin, paclitaxel, cyclophosphamide versus weekly paclitaxel, liposomal doxorubicin (plus carboplatin in triple-negative breast cancer) for neoadjuvant treatment of high-risk early breast cancer (GeparOcto—GBG 84): a randomised pha. Eur. J. Cancer 106, 181–192. https://doi.org/10.1016/j.ejca.2018.10.015.

Schott, A.F., Goldstein, L.J., Cristofanilli, M., Ruffini, P.A., McCanna, S., Reuben, J.M., Perez, R.P., Kato, G., Wicha, M., 2017. Phase Ib pilot study to evaluate reparixin in combination with weekly paclitaxel in patients with HER-2–negative metastatic breast cancer. Clin. Cancer Res. 23, 5358–5365. https://doi.org/10.1158/1078-0432.CCR-16-2748.

Seidman, A.D., Hudis, C.A., Albanell, J., Albanel, J., Tong, W., Tepler, I., Currie, V., Moynahan, M.E., Theodoulou, M., Gollub, M., Baselga, J., Norton, L., 1998. Dose-dense therapy with weekly 1-hour paclitaxel infusions in the treatment of metastatic breast cancer. J. Clin. Oncol. 16, 3353–3361. https://doi.org/10.1200/JCO.1998.16.10.3353.

Shi, Y., Sun, L., Chen, G., Zheng, D., Li, L., Wei, W., 2015. A combination of the telomerase inhibitor, BIBR1532, and paclitaxel synergistically inhibit cell proliferation in breast cancer cell lines. Target. Oncol. 10, 565–573. https://doi.org/10.1007/s11523-015-0364-y.

Sledge, G.W., Neuberg, D., Bernardo, P., Ingle, J.N., Martino, S., Rowinsky, E.K., Wood, W.C., 2003. Phase III trial of doxorubicin, paclitaxel, and the combination of doxorubicin and paclitaxel as front-line chemotherapy for metastatic breast cancer: an intergroup trial (E1193). J. Clin. Oncol. 21, 588–592. https://doi.org/10.1200/JCO.2003.08.013.

Staff, N.P., Fehrenbacher, J.C., Caillaud, M., Damaj, M.I., Segal, R.A., Rieger, S., 2020. Pathogenesis of paclitaxel-induced peripheral neuropathy: a current review of *in vitro* and *in vivo* findings using rodent and human model systems. Exp. Neurol. 324, 113121. https://doi.org/10.1016/j.expneurol.2019.113121.

Sunters, A., Madureira, P.A., Pomeranz, K.M., Aubert, M., Brosens, J.J., Cook, S.J., Burgering, B.M.T., Coombes, R.C., Lam, E.W.F., 2006. Paclitaxel-induced nuclear translocation of FOXO3a in breast cancer cells is mediated by c-Jun NH2-terminal kinase and Akt. Cancer Res. 66, 212–220. https://doi.org/10.1158/0008-5472.CAN-05-1997.

Taghian, A.G., Assaad, S.I., Niemierko, A., Kuter, I., Younger, J., Schoenthaler, R., Roche, M., Powell, S.N., 2001. Risk of pneumonitis in breast cancer patients treated with radiation therapy and combination chemotherapy with paclitaxel. J. Natl. Cancer Inst. 93, 1806–1811. https://doi.org/10.1093/jnci/93.23.1806.

Tan, B.Y., Acs, G., Apple, S.K., Badve, S., Bleiweiss, I.J., Brogi, E., Calvo, J.P., Dabbs, D.J., Ellis, I.O., Eusebi, V., Farshid, G., Fox, S.B., Ichihara, S., Lakhani, S.R., Rakha, E.A., Reis-Filho, J.S., Richardson, A.L., Sahin, A., Schmitt, F.C., Schnitt, S.J., Siziopikou, K.P., Soares, F.A., Tse, G.M., Vincent-Salomon, A., Tan, P.H., 2016. Phyllodes tumours of the breast: a consensus review. Histopathology 68, 5–21. https://doi.org/10.1111/his.12876.

Tanabe, Y., Hashimoto, K., Shimizu, C., Hirakawa, A., Harano, K., Yunokawa, M., Yonemori, K., Katsumata, N., Tamura, K., Ando, M., Kinoshita, T., Fujiwara, Y., 2013. Paclitaxel-induced peripheral neuropathy in patients receiving adjuvant chemotherapy for breast cancer. Int. J. Clin. Oncol. 18, 132–138. https://doi.org/10.1007/s10147-011-0352-x.

Tanaka, S., Nohara, T., Iwamoto, M., Sumiyoshi, K., Kimura, K., Takahashi, Y., Tanigawa, N., 2009. Tau expression and efficacy of paclitaxel treatment in metastatic breast cancer. Cancer Chemother. Pharmacol. 64, 341–346. https://doi.org/10.1007/s00280-008-0877-5.

Timmins, H.C., Li, T., Trinh, T., Kiernan, M.C., Harrison, M., Boyle, F., Friedlander, M., Goldstein, D., Park, S.B., 2021. Weekly paclitaxel-induced neurotoxicity in breast cancer: outcomes and dose response. Oncologist, 1–9. https://doi.org/10.1002/onco.13697.

Toikkanen, S., Joensu, H., 1991. Glycogen-rich clear-cell carcinoma of the breast: a clinicopathologic and flow cytometric study. Hum. Pathol. 22, 81–83.

Tu, Y., Hershman, D.L., Bhalla, K., Fiskus, W., Pellegrino, C.M., 2014. A phase I-II study of the histone deacetylase inhibitor vorinostat plus sequential weekly paclitaxel and doxorubicin- cyclophosphamide in locally advanced breast cancer. Breast Cancer Res. Treat. 1, 145–152. https://doi.org/10.1007/s10549-014-3008-5.

Van Der Heide, L.P., Hoekman, M.F.M., Smidt, M.P., 2004. The ins and outs of FoxO shuttling: mechanisms of FoxO translocation and transcriptional regulation. Biochem. J. 380, 297–309. https://doi.org/10.1042/BJ20040167.

Van Vreckem, A., Sylvester, R., Awada, A., Wildiers, J., Piccart, M., 2000. Chemotherapy for metastatic breast cancer : A European Organization for Research and Treatment of Cancer Randomized Study with cross-over. Society 18, 724–733.

Von Minckwitz, G., Schneeweiss, A., Loibl, S., Salat, C., Denkert, C., Rezai, M., Blohmer, J.U., Jackisch, C., Paepke, S., Gerber, B., Zahm, D.M., Kümmel, S., Eidtmann, H., Klare, P., Huober, J., Costa, S., Tesch, H., Hanusch, C., Hilfrich, J., Khandan, F., Fasching, P.A., Sinn, B.V., Engels, K., Mehta, K., Nekljudova, V., Untch, M., 2014. Neoadjuvant carboplatin in patients with triple-negative and HER2-positive early breast cancer (GeparSixto; GBG 66): A randomised phase 2 trial. Lancet Oncol. 15, 747–756. https://doi.org/10.1016/S1470-2045(14)70160-3.

Wang, R., Zhang, T., Yang, Z., Jiang, C., Seng, J., 2018. Long non-coding RNA FTH1P3 activates paclitaxel resistance in breast cancer through miR-206/ABCB1. J. Cell. Mol. Med. 22, 4068–4075. https://doi.org/10.1111/jcmm.13679.

Wasserheit, C., Frazein, A., Oratz, R., Sorich, J., Downey, A., Hochster, H., Chachoua, A., Wernz, J., Zeleniuch-Jacquotte, A., Blum, R., Speyer, J., 1996. Phase II trial of paclitaxel and cisplatin in women with advanced breast cancer: an active regimen with limiting neurotoxicity. J. Clin. Oncol. 14, 1993–1999. https://doi.org/10.1200/JCO.1996.14.7.1993.

Wee, Z.N., Yatim, S.M.J.M., Kohlbauer, V.K., Feng, M., Goh, J.Y., Yi, B., Lee, P.L., Zhang, S., Wang, P.P., Lim, E., Tam, W.L., Cai, Y., Ditzel, H.J., Hoon, D.S.B., Tan, E.Y., Yu, Q.,

2015. IRAK1 is a therapeutic target that drives breast cancer metastasis and resistance to paclitaxel. Nat. Commun. 6. https://doi.org/10.1038/ncomms9746.

Zhang, Z., Li, Z., Deng, M., Liu, B., Xin, X., Zhao, Z., Zhang, Y., Lv, Q., 2020. Downregulation of GPSM2 is associated with primary resistance to paclitaxel in breast cancer. Oncol. Rep. 43, 965–974. https://doi.org/10.3892/or.2020.7471.

Zhou, L., Xu, S., Yin, W., Lin, Y., Du, Y., Jiang, Y., Wang, Y., Zhang, J., Wu, Z., Lu, J., 2017. Weekly paclitaxel and cisplatin as neoadjuvant chemotherapy with locally advanced breast cancer: A prospective, single arm, phase II study. Oncotarget 8, 79305–79314. https://doi.org/10.18632/oncotarget.17954.

13

Paclitaxel conjugated magnetic carbon nanotubes induce apoptosis in breast cancer cells and breast cancer stem cells in vitro

Prachi Ghoderao[a], Sanjay Sahare[a], Anjali A. Kulkarni[b,*], and Tejashree Bhave[a,*]

[a]Department of Applied Physics, Defence Institute of Advanced Technology, Pune, India, [b]Department of Botany, Savitribai Phule Pune University (Formerly University of Pune), Pune, India

13.1 Introduction

Cancer is caused by a series of clonally selected mutations in key tumor-suppressor genes and oncogenes. A cancerous mass consists of cancer cells intermingled with a heterogeneous cell-protein matrix, blood vessels, and a dense hypoxic core. The latter three together are referred to as the tumor microenvironment (TME) of a solid tumor. TME is instrumental in cancer development and progression (Quail and Joyce, 2013).

Cancer stem cells (CSCs) are a small group of cancer cells, which show drug resistance, self-renewal, and active antiapoptotic pathways. They characteristically express a number of markers such as CD44, aldehyde dehydrogenase, CD133, etc. They may or may not be quiescent (Wicha et al., 2006; Diehn et al., 2009; Al-Hajj et al., 2003). Many studies have reported the existence of CSCs with higher resistance to conventional chemotherapy in a wide range of malignancies including acute myeloid lymphoma (AML), glioblastoma, breast, endometrial, pancreatic, prostate, lung, colon cancers, etc. (Diehn et al., 2009; Li et al., 2008c). Current radio and pharmacological interventions kill the bulk of the cancer cells but fail to eradicate the CSCs that

---

[*] Equal contribution.

Paclitaxel. https://doi.org/10.1016/B978-0-323-90951-8.00001-1
Copyright © 2022 Elsevier Inc. All rights reserved.

are protected by specific resistance/repair mechanisms, genetic and cellular adaptations, multidrug resistance (MDR), relative dormancy or slow cell cycle kinetics, efficient DNA repair mechanisms, presence of ATC binding casette (ABC) transporter proteins, metastasis, and relapse (Frosina, 2009; Dean et al., 2005). Recent research has shown that CSCs get activated from their quiescent state by mitogens and proinflammatory cytokines released by cancer cells undergoing apoptosis due to chemotherapy regime and tumor relapses (Romberg et al., 2008). Thus, ideally any new cancer therapy should target both TME and CSCs.

According to the recent GLOBOCAN 2018 database, the estimated global cancer burden is now at 18.1 million new cases and 9.6 million deaths (Bray et al., 2018). Explosive population growth, aging, skewed social and economic development all seem to contribute to the spread of cancer. Cancers of the lung, female breast, and colorectum are the top three fatal cancer types. Current cancer treatment includes surgical excision, chemotherapy, and radiotherapy, individually or in combination. However, each of these has many side effects. Due to the lack of targeted therapy and precise delivery systems, chemotherapy has lapsed into undesirable and severe side effects, low bioavailability of drugs, development of off-target toxicity as well as MDR (Chabner and Roberts, 2005). Besides, these drugs are unable to target either TME or CSCs (Li et al., 2008c).

Regardless of the current progress and new therapeutic strategies in the breast cancer treatment, most of the patients develop recurrence (Gonzalez-Angulo et al., 2007), due to presence of CSCs which repopulate the tumors that got reduced by chemo or radiotherapy regimes (Economopoulou et al., 2012; Vermeulen et al., 2012). Evidently, a smart strategy composed of a conventional chemotherapeutic drug in combination with CSC targeting agent is needed to eradicate both cancer cells and CSCs. Successful examples of such synergistic combination therapies include using parthenolide-loaded liposomes/vinorelbine-loaded liposomes (Liu et al., 2008a) and all-trans retinoic acid-loaded liposomes/vinorelbine-loaded liposomes (Li et al., 2011).

Paclitaxel (PTX), a microtubule-stabilizing agent, derived from plants of genus *Taxus*, is a highly promising drug because of its high cytotoxic activity (Wani et al., 1971; Rowinsky and Donehower, 1995) including activity against ovarian cancer, breast cancer, nonsmall-cell lung cancer, and head and neck carcinomas. It is also effective against Kaposi's sarcoma. It has shown excellent results through all stages of clinical trials and the drug is routinely prescribed. On the other hand, poor aqueous solubility of PTX (sold under trade name Taxol by Bristol-Myers Squibb, United States) has limited its therapeutic efficacy. Cremophor-E, the vehicle used for delivery of therapeutic PTX, itself leads to a lot of unpleasant side effects (Sparreboom et al., 1996,

2005). Once administered, PTX is attacked by the body's defenses, necessitating larger doses that result in complications such as joint pain, diarrhea, and an impaired ability to fend off other infections. It is thus crucial to develop efficacious strategies to improve aqueous solubility of PTX, enhance antitumor efficacy, and avoid adverse side effects by developing drug-targeting strategies.

The emergence of nanotechnology has made its affirmative impact in improving the efficacy and safe transport of chemotherapeutic drugs to the tumor site, by exploiting the enhanced retention and permeability (ERP) effects of nanoparticles (NPs) (Ruoslahti, 2002; Romberg et al., 2008). They can escape the recognition of reticuloendothelial system (RES) in healthy tissues and therefore reduce the side effects of the drug. The aqueous solubility of a number of sparingly water-soluble drugs can be greatly enhanced when conjugated with water-soluble nanomaterials (Kwok and Chan, 2014; Patel and Agrawal, 2011). Currently, carbon nanotubes (CNTs) are considered as a smart drug delivery tool (Liu et al., 2008b).

Liu et al. (2008a) reported that PTX could be conjugated with branched PEG functionalized single-walled carbon nanotubes, a more water-soluble form of CNTs. These PTX-loaded, PEG-functionalized CNTs showed excellent stability and demonstrated significantly improved antitumor activity compared with neat PTX at the dose of 5 mg/kg in a 4T1 breast cancer mouse model. Lay et al. (2010) loaded PTX on the surface of PEG-functionalized CNTs. In vitro cytotoxicity studies in HeLa and MCF-7 cells exhibited that PTX-loaded CNTs significantly increased cell death compared with neat PTX. The reported CNT-based formulations by Atyabi et al. (2011) were more antiproliferative than neat PTX in both A549 and SKOV3 cells while the blank CNTs had a negligible antiproliferative effect. In this study, the drug loading was about 38% (w/w).

Based on the prior art search and our research experience, we decided to use PTX in combination with MWCNTs prepared by a unique method in our lab (Patole et al., 2008). In order to improve the tumor directed properties of MWCNTs, further decorations were done. Among various types of nanoparticles available, biocompatible Fe_3O_4 and gold (Au) NPs are the only US Food and Drug Administration (FDA) approved nanomaterials for clinical trials as of now (Ventola, 2017; Bobo et al., 2016). Hence, we used Fe_3O_4 and Au NPs to decorate MWCNTs. Au NPs are useful as a targeted drug delivery system because they are preferentially taken up by the cancer cells (Yasun et al., 2013; Heo et al., 2012; Dreaden et al., 2012; Huang et al., 2006). Mukherjee et al. (2005) demonstrated in vitro and in vivo inhibitory effects of Au NPs on endothelial cell proliferation and VEGF-mediated angiogenesis. Further, Au NPs with their surface plasmon resonance (SPR) and surface enhanced Raman scattering, showed increased sensitivity for imaging systems (El-Sayed et al., 2005; Huang and El-Sayed, 2010). They also enhanced sensitization of irradiation at megavoltage energies due to the photoelectric effect and the Compton effect (Yang et al., 2015; Hu et al., 2015).

Magnetic NPs of Fe_3O_4 possess the capability of target specificity under external magnetic control. Superparamagnetic iron oxide NPs have been efficiently used in drug delivery (Hao et al., 2010; Qiao et al., 2009). Xu et al. (2009) achieved target-specific Cisplatin delivery using Au-Fe_3O_4 NPs as nanocarriers. The complex of cisplatin-Au-Fe_3O_4-Herceptin was successfully targeted to SK-BR3 breast cancer cells and sustained release of cisplatin under low pH conditions, inside the cancer cells was achieved. This enhanced the therapeutic efficacy of complex over free cisplatin.

In our study, we synthesized MWCNTs and functionalized them from interior to exterior to improve their cellular uptake and efficacy. MWCNTs were internally filled with superparamagnetic Fe_3O_4 NPs and embellished externally with Au NPs. These superfunctionalized MWCNTs and PTX were combined in a complex to enhance their efficacy and antiproliferative effect, specifically toward cancer cells as well as cancer stem cells (CSCs), in a theranostic way (Fig. 13.1). Various physiochemical properties were characterized by IR and UV-Vis spectroscopy, HR-TEM. We demonstrated enhanced cytotoxicity of the PTX drug complex against MCF-7 breast cancer cells in vitro as compared to neat PTX

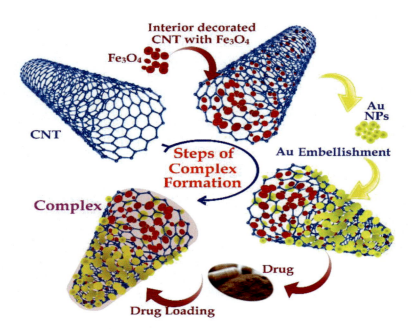

Fig. 13.1 Pictorial cartoon of PTX-Au-Mag.MWCNT complex development depicting interior filling of MWCNT with Fe_3O_4 NPs and exterior embellishment with Au NPs followed by loading of PTX onto Au-Mag. MWCNTs.

by MTT assay. The enhanced apoptosis was further studied by flow cytometry. Antiproliferative effect of the drug complex specifically on CSC population in MCF-7 breast cancer cell line in vitro was demonstrated using CD44-FITC conjugated antibody and flow cytometry.

13.2 Experimental details

13.2.1 Preparation of MWCNTs

MWCNTs (inner/outer diameter = 6–8 nm/10–15 nm) were synthesized using chemical vapor deposition technique (Patole et al., 2008). To fill interior cavities of synthesized MWCNTs, ferric nitrate (Fischer Scientific, United States) was used. MWCNTs' surface was modified using, poly (sodium 4-styrenesulfonate) (PSS, M 70,000) (Sigma-Aldrich, United States). Auric chloride (Sigma-Aldrich, United States), trisodium citrate and sodium borohydride (Fischer Scientific, United States) were used for the synthesis of gold (Au) NPs. Poly (allylamine) (PAH, M 15,000) (Sigma-Aldrich, United States) was used as a capping agent of Au NPs. 2-Propanol (Fischer Scientific, United States) was the purification agent for removal of excess PAH. Paclitaxel and dimethyl sulfoxide (DMSO) were purchased from Sigma Aldrich (United States). All other general laboratory chemicals were of analytical grade and purchased from Merck (India).

13.2.2 PTX loading onto Au-Mag.MWCNTs surface

For the synthesis of PAH capped Au NPs and magnetic MWCNTs, a previously reported method has been followed (Chen et al., 2013; Ojea-Jiménez and Campanera, 2012). For loading of PTX onto Au-Mag.MWCNTs, 0.8 mg PTX (Sigma, United States) was mixed with 10 mL Au-Mag.MWCNTs in phosphate buffered saline (PBS) and kept for 24 h in the dark at room temperature (Fig. 13.1). The complex, thus formed was collected by repeated centrifugation (1000 rpm) with PBS until the supernatant became free of PTX. The PTX-Au-Mag.MWCNT complex was re-suspended in PBS and stored at 4°C till further use. The amount of free PTX in the supernatant was determined by measuring its absorbance at 235 nm, on the basis of which the drug loading efficiency could be estimated.

The equation developed by Lay et al. (2010) has been used to determine loading of PTX onto Au-Mag.MWCNTs.

$$\text{Drug loading}(\%) = \frac{\text{Weight of drug conjugates}}{\text{Weight of conjugates}} \times 100$$

13.2.3 Characterizations

For the Fourier transform infrared spectroscopy (FT-IR) measurement, Spectrum Two IR spectrometer (L160000A, Perkin Elmer, United States) was used to record spectra in the transmission mode by KBr pellet method in between 500 and 2100 cm^{-1}.

UV-Vis absorption spectra of the PTX-Au-Mag.MWCNT complex and nondrug loaded Au-Mag.MWCNTs were recorded in the range 200–750 nm wavelengths on Specord 210 Plus (Analytic Jena, Japan).

Magnetic properties of Fe_3O_4 embellished MWCNTs were analyzed using Physical Property Measurement System (PPMS) (Quantum Design Inc., United States) equipped with a 9T superconducting magnet.

Filling of Fe_3O_4 NPs into pristine MWCNTs' interior surface and embellishment of Au NPs onto the exterior surface of MWCNT, and distribution of PTX on the Au-Mag. MWCNTs' surface were imaged using high resolution transmission electron microscopy (HR-TEM) (FEI, Tecnai-G2RT30S, Thermo Fisher Scientific, United States) at beam energy of 300 KV.

13.2.4 Cancer cell culture experiments in vitro

The human breast cancer cell line, MCF-7 was purchased from National Centre for Cell Science, Pune with passage number 18. MCF-7 cells were grown in Dulbecco's Modified Eagle Medium (DMEM, Life Technologies, Inc., United States) supplemented with 5% heat-inactivated fetal bovine serum (Sigma-Aldrich, United States), 2 mM glutamine, 1% penicillin, and streptomycin (Gibco, United States). Cells were cultured in T-25 tissue culture flasks (Corning, United States) and were kept in a CO_2 incubator (Eppendorf AG, Germany) at 37°C in a humidified atmosphere with 5% CO_2. For experimental purposes, cells in exponential growth phase (approximately 70%–80% confluency) were used. They were treated with 0.05% (w/w) trypsin whenever required (Sigma, United States). The cells were regularly observed under an inverted microscope (Olympus, Germany) to ascertain their morphology and health.

13.2.5 Detection of cytotoxicity—MTT assay

To evaluate the cytotoxic effect of neat PTX and PTX-Au-Mag. MWCNTs complex on the MCF-7 cells, MTT colorimetric assay was performed at 37°C in a 5% CO_2 atmosphere (Lay et al., 2010). Briefly, cells at a density of approximately 1×10^6 cells per well were seeded into 96-well, round-bottomed, treated tissue culture plates (Greiner M3562, United States). After 24 h, the supernatant on the cell monolayer was aspirated and 100 µL of sterile serum free medium was appended with varying log concentrations of neat PTX and complex (10, 25, 50 µg/mL) and incubated for 24 h. After further 48 h, 10 µL of 5 mg/mL MTT in PBS was added to each well and incubated for 4 h. Then the supernatant in each well was carefully removed and 100 µL of DMSO (Dimethyl sulfoxide, Sigma, United States) was added in order to dissolve the resulting formazan crystals, the amount of which could be quantified by determining absorbance at 570 nm.

13.2.6 Flow cytometry

13.2.6.1 Detection of apoptosis

Apoptosis was assessed by flow cytometry using a Dead Cell Apoptosis Kit (Cat. No: V13241, Thermo Fisher, United States), according to the manufacturer's instructions. Briefly, MCF-7 cells with a density of 1×10^6 cells/mL were seeded in each well of 96-well plates (Thermo Fisher, United States). These were allowed to grow for 24 h in a CO_2 incubator. Then the supernatant on the monolayer was aspirated and 100 µL of medium containing either neat PTX (positive control) or PTX-Au-Mag.MWCNT complex (10, 50, 100 µg/mL) was added and incubated for 24 h. A negative control was maintained by incubating cells in the absence of any drug or complex. After 24 h, the cells were harvested with a brief trypsinization treatment and then washed once with serum-containing medium followed by centrifugation at 2200 rpm for 5 min at 8°C. The supernatant was discarded and the cells were resuspended in 500 µL of 1X Annexin-binding buffer (prepared as per the Manufacturer's instructions). 5 µL of Alexa Fluor 488 annexin V (Component A) and 3 µL of 50 µg/mL PI working solution were added to each 100 µL of cell suspension. The cells were incubated at room temperature for 15 min in the dark. Then 400 µL of 1X annexin-binding buffer was added, mixed gently, and the samples were analyzed immediately by flow cytometer (Attune NxT model, Thermo Fisher Scientific, United States), measuring the fluorescence emission at 530 and 575 nm, using 488 nm excitation. The FCS files of each experimental variable were analyzed by Flow-Jo Software.

13.2.6.2 Antiproliferative studies on stem cells

To investigate the antiproliferative effect of neat PTX and synthesized PTX-Au-Mag.MWCNTs complex on cancer stem cells (CSCs), the CSCs were sorted from MCF-7 breast cancer cells population by means of a CD-44-FITC conjugated antibody (CD44 Monoclonal Antibody (IM7), APC, eBioscience, Thermo Fisher, United States) that binds exclusively to CSCs overexpressing CD-44 antigen on their surface (Chan, 2016). The freshly sorted MCF-7 cells were suspended in cold PBS and then stained with antihuman CD44-FITC on ice for

30 min. The cells were washed many times with cold PBS, suspended in 400 μL of cold PBS and analyzed by flow cytometry as mentioned earlier.

13.3 Results and discussion

13.3.1 Physical characterization of PTX-Au-Mag. MWCNT complex

FTIR spectroscopy of Au-Mag.MWCNTs, neat PTX, and PTX-Au-Mag.MWCNT complex was performed (Fig. 13.2) to comprehend the structures and chemical interactions of these materials. The wide range of wavelengths selected helped us to analyze chemical conformational status of samples with respect to the presence of PTX and MWCNTs/Fe$_3$O$_4$-Au in the conjugates.

In Au-Mag.MWCNTs, an absorption band revealing the vibrational properties of Fe–O and Au bond was observed around 515 cm^{-1}. This band could be mainly assigned to the stretching vibrations of Fe–O (Ahn et al., 2003; Li et al., 2008b). The peaks at 1655 and 1720 cm^{-1} are due to C–C stretching and C=O of the functional groups on the surface of the oxidized MWCNTs respectively (Stobinski et al., 2010). This confirmed the simultaneous presence of Fe$_3$O$_4$ nanoparticles and CNTs which is in good agreement with TEM images (Fig. 13.5). The spectra of neat PTX drug exhibited small peaks around 515–712 cm^{-1}, and these were attributed to C–H out-of-plane/C–C=O deformation. The

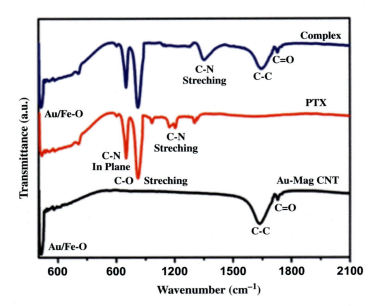

Fig. 13.2 FTIR spectra of Au-Mag.CNT *(black graph)*, PTX *(red graph)*, and PTX-Au-Mag.CNT complex *(blue graph)*.

peak at 948 cm^{-1} represented C–H in-plane deformation. The peak at 1020 cm^{-1} indicated C–O stretching available in the PTX structure. The small peaks in between 1190 and 1210 cm^{-1} were due to C–N stretching, while the peak at 1310 cm^{-1} confirmed CH$_3$ deformation (Devi and Gayathri, 2010). In the PTX-Au-Mag.MWCNT complex, the peak around 515 cm^{-1}, confirmed the simultaneous presence of Fe–O and Au in the complex. Additionally, the interaction of PTX with the Au-Mag.MWCNTs was confirmed due to the shift observed in the peaks attributed to C–N stretching toward longer wavelength, with increased intensity, probably due to the electrostatic interaction. However, peaks at 1020 cm^{-1}, and 1190–1210 cm^{-1} exhibited no spectral changes upon PTX-Au-Mag.MWCNT complex interaction. The spectrum also highlighted the presence of peaks at 1644 cm^{-1}, 1736 cm^{-1} and these were assigned to C–C stretching and C=O stretching respectively, confirming the appearance of functional groups on the surface of the oxidized MWCNTs. Thus, the FT-IR spectrum conclusively indicated successful formation of PTX-Au-Mag.MWCNT complex.

UV-Vis spectroscopy of Au-Mag.MWCNTs, neat PTX, and PTX-Au-Mag.MWCNT complex was performed (Fig. 13.3). In the neat PTX spectrum, absorption at 236 nm was due to vibrations of π–π* transitions. Spectrum of Au-Mag.MWCNTs showed a peak at 270 nm, where the red shift has been observed from 230 to 270 nm as compared with pristine CNTs (Li et al., 2008a). This red shift assured existence of carboxyl group attached onto the surface of MWCNTs, and also suggested restoration of aromatic conjugation of MWCNTs after this reduction (Liu et al., 2010). Absorption peak of MWCNTs/Fe$_3$O$_4$ was greatly influenced by the absorption of Fe$_3$O$_4$ NPs on MWCNTs, as peak at 270 nm also coincided with characteristic peak of iron oxide (Mehdipoor et al., 2011). The SPR peak of Au NPs was found at the 521 nm wavelength, confirming their presence (Ngo et al., 2015; Ojea-Jiménez et al., 2010).

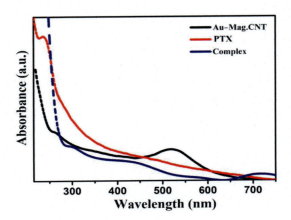

Fig. 13.3 UV-Vis spectroscopy exhibited the characteristic absorption edge of neat PTX *(red graph)* at 236 nm becoming less broader with bathochromic shift in the spectrum of PTX-Au-Mag.CNT complex *(red graph)*.

Absorbance curve of PTX-Au-Mag.MWCNT complex elucidated their probable electrostatic interaction during complex formation. The characteristic absorption peak of PTX showed a bathochromic shift at 304 cm^{-1} and the signature peak of Mag.MWCNTs was also shifted toward longer wavelength and became broader. This shift of Mag.MWCNTs absorption band could be attributed to electrostatic interactions occurring at the PTX-Au-Mag.MWCNT complex surface. Electrostatic interaction between PTX and Au-Mag.MWCNTs possibly led to the change in interfacial electron density at the surface of the complex resulting from the interaction between nitrogen-containing functionalities of PTX with Au NPs.

Magnetic properties of MWCNTs embellished with Fe$_3$O$_4$ were measured and are represented by a hysteresis curve studied by exploiting PPMS (Fig. 13.4). The magnetization measurements in between ± 10 kOe at 300 K showed that the specific saturation magnetization of 19.3 emu g^{-1} exhibited superparamagnetism, whereas pristine MWCNTs' showed saturation magnetization at 0.24 emu g^{-1} as reported elsewhere (Zhao and Gao, 2004). In addition, slow approach toward saturation and the absence of coercive field defined their superparamagnetic behavior. The saturation magnetization (M_s) values observed in nanostructured materials are usually considerably smaller than the corresponding material in bulk phase. However, fine magnetite particles show a linear correlation between M_s and particle size (Varanda et al., 2011). Highly disordered crystal orientation on the surface of CNTs emerged due to their smaller size and larger surface area resulting in remarkable decrease in their M_s values. NPs of size less than single domain (< 20 nm), show ideal superparamagnetic behavior. Also, it could be precisely characterized by slow approach toward saturation and by the absence of coercive field. An extremely faint, nonzero value of coercive field was observed that directly depicted characteristic of superparamagnetic behavior. Hence, we can state that the developed Mag.MWCNT is a superparamagnetic material.

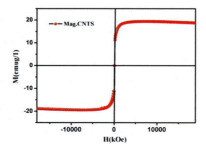

Fig. 13.4 Shows magnetic hysteresis curve of MWCNTs/Fe$_3$O$_4$. An extremely faint, nonzero value of coercive field is observed that directly corroborates superparamagnetic behavior.

HR-TEM images of pristine MWCNTs, Mag.MWCNTs, Au-Mag.MWCNTs, and PTX-Au-Mag.MWCNTs complex are shown in Fig. 13.5(i–iv) respectively. Pristine MWCNTs, showed their characteristically smooth and even outer surface (Fig. 13.5(i)). Fe_3O_4 NPs filled inside the lumen of MWCNTs could be visualized (Fig. 13.5(ii)). The outer surface of the MWCNTs remained smooth. The average size of Fe_3O_4 NPs filled in the lumen of MWCNTs was in the range of 7–8 nm. Interestingly, the image appears as if it is filled with a number of seeds in a highly ordered fashion, keeping intact the original tubular skeleton of MWCNTs after in situ filling. Vibrating sample magnetometer (VSM) studies were in agreement with HR-TEM images and proved superparamagnetic nature of NPs. TEM image of Au-Mag.MWCNTs exhibited an interesting hybrid structure, where embellishment of Au NPs on the exterior surface of Mag.MWCNTs was clearly visible as buds (Fig. 13.5(iii)). The average size of Au NPs was 20–30 nm. Fig. 13.5(iv) revealed simultaneous presence of dual particle size distribution. The grain shaped Au NPs were uniformly embellished on the surface of Mag. MWCNTs. These were densely interspersed with larger particles, the fully-grown buds of PTX drug molecules with irregular shapes, on the exterior surface of Au-Mag.MWCNTs.

13.3.2 Biological characterization of PTX-Au-Mag. MWCNT complex

13.3.2.1 Cytotoxicity and apoptosis assay

MCF-7, a breast cancer derived cell line, was chosen by us for in vitro experimentation because it is a "Luminal A" subtype of non-invasive cell line and is highly responsive to chemotherapy, especially Paclitaxel (PTX) (Holliday and Speirs, 2011; Abu Samaan et al., 2019). It also shows presence of small populations of cancer stem cells (Comşa et al., 2015).

Viability of MCF-7 cells in presence of neat PTX and complexed PTX was checked in vitro using MTT assay method and the results are depicted in Fig. 13.6. Neat PTX showed 77.9%, 68.6%, and 47.1% cell viability respectively at 10, 25, and 50 µg/mL concentration. PTX-Au-Mag.MWCNT complex with same concentrations, showed cell viability at 72.4%, 58.6%, and 44.88% respectively. Only 56% PTX was loaded onto Au-Mag.MWCNTs and yet it suppressed proliferation of the cells more prominantly.

Further, to understand the antiproliferative activity of the complex, we studied apoptotic pathways. Apoptosis is a genetically determined form of cell death in eukaryotes, playing a fundamental role during embryogenesis, in the homeostatic control of tissue integrity, tumor regression and immune response development

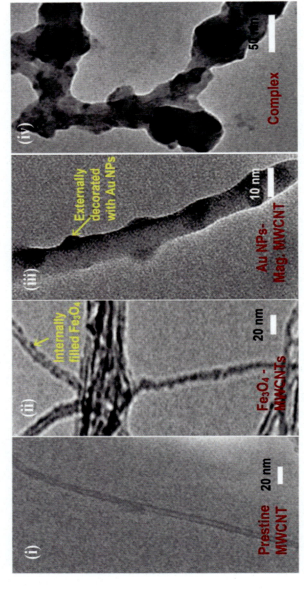

Fig. 13.5 HR-TEM images show (i) pristine MWCNTs with even and smooth surface, (ii) MWCNTs internally filled with Fe$_3$O$_4$ NPs *(yellow arrow)*, (iii) Mag.MWCNTs embellished with Au NPs on its exterior surface as buds coming out *(yellow arrow)*, (iv) PTX-Au-Mag.CNT complex exhibits concurrence of embellished Au NPs (small particles) and loaded drug molecules (bigger particles).

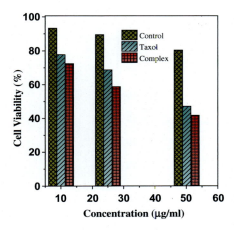

Fig. 13.6 Cell viability against MCF7 breast cancer cell lines shows that PTX complex containing only 56% of PTX has suppressed cancerous cell proliferation more efficiently.

(Wylie et al., 1997). During apoptosis, cells undergo several biochemical and morphological changes such as cytoplasmic membrane modifications with a loss of phospholipid asymmetry and exposure of phosphatidylserine (PS) on the outer leaflet, membrane blebs, cellular shrinking, nuclear condensation, DNA fragmentation, and disintegration of the dying cell into apoptotic bodies (Elmore, 2007). PS exposure, indicating membrane asymmetry, can be used to detect apoptosis by allowing FITC-labeled annexin V to bind specifically to it. PI can be used as a counter stain as it does not enter into viable cells. It will only enter in necrotic cells due to membrane disintegration. Since induction of apoptosis should be an important goal of any worthwhile cancer therapy, we wished to determine whether most of the PTX-Au-Mag.MWCNT complex-induced cell death that was observed with the MTT assay was due to apoptosis or necrosis. To differentiate between these two, MCF-7 cells treated with neat PTX and complex were double-stained with FITC-labeled annexin V and PI and observed with a flow cytometer (Figs. 13.7 and 13.8).

The flow cytometric analysis is depicted as cytograms in Fig. 13.8. The treated MCF-7 cell population got separated into four groups: (a) live cells with only a low level of fluorescence due to exclusion of PI and no binding of FITC-annexin V due to normal, healthy morphology of the plasma membrane (the lower left quadrant (Q4) of the cytograms) (FITC$^-$, PI$^-$). (b) The early apoptotic cells localized in (Q3) were FITC-annexin V positive and PI negative (FITC$^+$, PI$^-$). These cells showed green fluorescence due to FITC-labeled annexin V binding to PS on outer leaf of plasma membrane but they had cytoplasmic membrane integrity, thus excluded PI and did not show

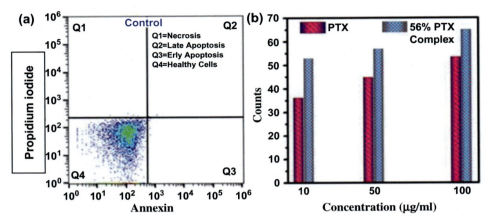

Fig. 13.7 (A) Flow cytometry of nontreated MCF-7 cells stained with FITC labeled annexin V and counter-stained with propidium iodide (PI) and (B) comparison between apoptosis induced in MCF-7 cells due to PTX and PTX-Au-Mag.MWCNTs complex at 10, 50, 100 µg/mL.

red fluorescence. The upper right quadrant (c) the late apoptotic cells localized in (Q2) showed green and red fluorescence (FITC$^+$, PI$^+$). The cells were positive for FITC-annexin V binding to PS on outer leaf of plasma membrane and showed PI uptake due to localized disruptions of cytoplasmic membrane, allowing entry of PI. (d) The upper left quadrant (Q1) represented the nonviable, necrotic cells (FITC$^-$, PI$^+$). The cells appeared negative for FITC-annexin V binding due to rupture of plasma membrane, allowing PI uptake and becoming PI positive and thus showed only red fluorescence (Wlodkowic et al., 2009).

Cell mortality of 36%, 53.4%, and 44.8% was observed for neat PTX drug, whereas cell mortality rates of more than 53%, 65%, and 57% were obtained for synthesized PTX-Au-Mag.MWCNTs complex at 10, 50, 100 µg/mL concentrations respectively. In both the cases, the percentage of necrotic cells was negligible (Q1 is mostly empty). This indicated that the complex induced the apoptosis of MCF-7 breast cancer cells with higher efficiency as compared with neat PTX, in vitro (P significant at 5%). This result confirmed that the efficacy of PTX has been ameliorated even at lower concentration of the drug in the complex, i.e., only 56% than that of neat PTX because there is only 56% w/w bound PTX PTX-Au-Mag.MWCNT complex (Fig. 13.9).

We can envisage that if we can load 100% PTX drug on the complex, three times better efficacy may be observed than that of the neat drug, which will consequently reduce dosing frequency and toxic side effects.

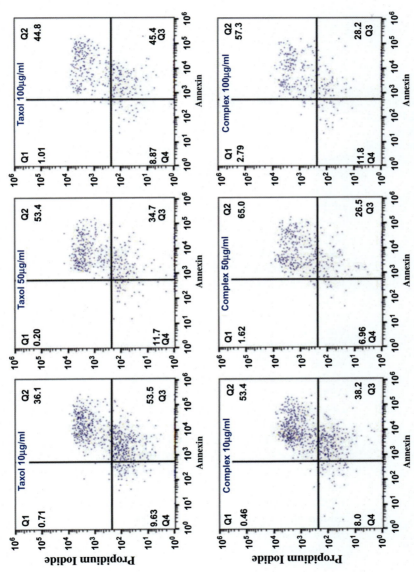

Fig. 13.8 FCM assay for apoptosis induction in MCF-7 cells by neat PTX and PTX-Au-Mag.MWCNTs complex containing 56% of PTX with three different concentrations, stained with FITC labeled annexin V and counter-stained with PI.

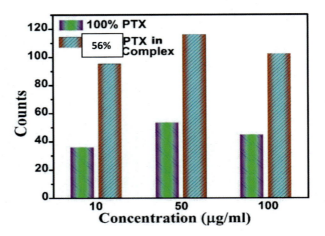

Fig. 13.9 Comparison between apoptosis induced in MCF-7 cells due to neat 100% PTX and bound 56% PTX in PTX-Au-Mag.MWCNT complex at 10, 50, 100 μg/mL.

This increased cell mortality of the PTX complex may be because of the improved drug delivery by nanomaterials due to improvement in solubility of hydrophobic drugs, such as PTX, and reduction in toxicity as well (Drbohlavova et al., 2013). Abraxane, an albumin-bound NP-PTX formulation with the particle size of ~130 nm, was approved by the FDA in 2005 for the treatment of metastatic breast cancer. This formulation showed better solubility and reduced toxicity compared to PTX-Cremophore market drug (Gradishar, 2006). But Abraxane is a high-cost formulation and showed a rapid elimination of PTX from the blood circulation and did not improve the pharmacokinetics of PTX (Taxol) (Sparreboom et al., 2005). Thus, the search for better alternatives is still active.

In an earlier report (Berlin et al., 2010) used two cancer cell lines: A549 (a lung cancer derived cell line, reported erroneously as breast cancer cell line in this paper) and H1975 (a nonsmall cell lung cancer derived cell line, reported erroneously as breast cancer cell line in this paper). These were treated for 3 days with commercial PTX, neat NPs, namely, PEG-HCCs (poly-ethylene glycol-hydrophilic-carbon-clusters) and PTX/PEG-HCC noncovalent complexes. The cell-killing efficacy of the PTX/PEG-HCCs complex treatment and that of neat PTX treatment was nearly equivalent. So, no better results were obtained after combining PTX with NPs. Same type of an equivalence response was reported for in vivo studies as well.

Bernabeu et al. (2014) used poly(ε-caprolactone)-tocopheryl polyethylene-glycol-succinate (PCL-TPGS) copolymer complexed with PTX for in vitro anticancer efficacy studies with two breast cancer cell lines. The PTX complexed with NPs showed at least 10 times more cytotoxicity than neat PTX. However, here the PCL-TPGS NPs themselves had very high cell toxicity for both the cell lines. Hence, authors postulated that the cytotoxicity of PTX complex is a combination of PTX toxicity plus PCL-TPGS NP toxicity.

Delivery of drug through tumor vessel walls and interstitial spaces is desired for high tumor treatment and target efficacy. CNTs are quite effective in these aspects, which could be due to the quasi 1-D shape of these materials. Interestingly, the length of nanotubes (20–300 nm currently) appears to be a factor that favors long duration of blood circulation (since the average length of the nanotubes exceeds the threshold for renal clearance). Thus drug delivery using CNTs is an effective cancer treatment with lower side effects (Liu et al., 2011; Bianco et al., 2005). The gold NPs used in our complex have demonstrated anticancer activity due to their potential ability to interrupt and slow down the activities of abnormally expressed signaling proteins, such as Akt and Ras (Rajeshkumar, 2016; Sriram et al., 2010). In addition, we have embellished PTX on the surface of magnetic CNTs; leading to increased cellular uptake when put under the influence of external magnetic field (data not shown here). Our complexes have shown better rates of apoptosis than any reports earlier. Thus, this magnetically driven, highly efficacious drug delivery system will have a lot of promise.

13.3.2.2 Assay for determination of apoptosis of cancer stem cells

Compelling body of evidence suggests that, small populations of tumor initiating cells within breast tumors called breast cancer stem cells (BCSCs) are responsible for tumor relapse and metastasis (Chen et al., 2013). Expression of surface biomarkers (CD44, CD133, EpCAM, and ALDH1), self-renewal pathways (Wnt, Notch, TGF-b, Hedgehog, Bmi-1, and STAT3), drug efflux transporters (ABCG2 and ABCB1), antiapoptotic proteins (Bcl2, survivins, etc.), autophagy, altered metabolism, and specialized microenvironment are the important biological features of BCSCs. It is, therefore, necessary to eradicate BCSCs to achieve radical cure in breast cancer (Chan, 2016; Chen et al., 2013; Pindiprolu et al., 2018). Many reports have described a number of anti-BCSC agents but most of them have failed in clinical setting due to their poor water solubility, short circulation time, inconsistent stability, and off-target effects (Baker et al., 1988; Buckley et al., 1995). Hence a suitable drug delivery system is required to improve bioavailability and carry anti-BCSC agents specifically to BCSCs without off-target effects. Nanocarriers are at the forefront in this regard due to site specific delivery and ability to improve bioavailability and stability of drugs.

It is a well-known strategy where overexpressed receptors in cancer stem cells can be used for selectively killing these cells. CD44 is a trans-membrane cell surface protein that is highly expressed on cell membranes of various types of cancer cells such as breast, pancreas, colon, prostate, and stomach etc. and it is also an important protein determinant for tumorigenic and metastatic properties of cancer cells (Bernabeu et al., 2014). As shown in Fig. 13.10, conventional PTX and complex showed antiproliferative effect on CSCs of MCF-7. The CSCs of MCF-7 cells were tagged with CD-44 antibody and then treated with two different formulations at two different concentrations, and effect was determined by flow cytometry. As seen in Fig. 13.10, in MCF-7 cells, there were 38.35% and 44.47% dead CSCs, after treatment with neat PTX at 10 and 50 μg/mL respectively. When treated with complex loaded with only 56% of PTX, the percentage of dead CSCs was found to be 23.62% and 28.58%. Thus, if we extrapolate this data for 100% PTX loading in the complex, we can say that the complex showed more efficacy in suppressing the growth of CSCs than pure PTX. Combined with water and DMSO solubility of the complex, this makes it an attractive proposition for future detailed experiments. Antiproliferative effect of neat PTX on CSCs was significantly lower than that on nonstem cancer cells, which was consistent with reported findings that CSCs were more resistant to chemotherapeutic drugs than nonstem cancer cells (Bosco et al., 2015; Liu et al., 2015). The antiproliferative effect was markedly increased after treatment

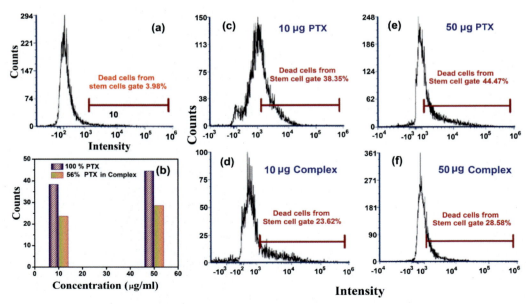

Fig. 13.10 Comparative studies on apoptosis induction in caner stem cells by using neat PTX and complex containing 56% PTX.

with complex. The results show that the delivery system of Au-Mag. MWCNTs was able to augment the antiproliferative effect, effectively reducing the number of CSCs of MCF-7.

The increased efficacy of the complex than the corresponding neat drug is probably due to slow release of PTX from the complex as well as its accumulation within the cancerous cells due to EPR effect. In addition, the complex may also prevent relapse and tumor metastasis as it could effectively suppress the proliferation of cancer stem cells. Hence, the complex as a therapy may provide a promising strategy for breast cancer treatment by targeting both cancer cells and cancer stem cells.

13.4 Conclusions

This research proposes a formulation of a theranostic agent by using combination of gold embellished magnetic MWCNTs, and PTX as an anticancer drug to achieve enriched antiproliferative effect against MCF-7 breast cancer cells and cancer stem cells in vitro. Flow cytometry measurements suggested that the complex showed enrichment in the antiproliferative and apoptosis-inducing effect of PTX, even at lower concentration of drug loading on MWCNTs. Au NPs have a higher affinity to cancer cells, and they selectively target cancer cells. Additionally, gold NPs have a higher affinity for cancer cells and they probably guided the complex toward cancer cells.

Additionally, the complex targeted cancer stem cells and this may lead to prevention of tumor regeneration and metastasis. Hence, the complex as a therapy may provide a promising strategy for breast cancer treatment by targeting both cancer cells and cancer stem cells.

Acknowledgments

P.G. acknowledges Dr. Babasaheb Ambedkar Research and Training Institute (BARTI), Pune, India for the financial assistance. The authors would also like to acknowledge the financial assistance from ERIP/ER/1003883/M01/908/2012/D (R&D)/1416 project. AAK would like to acknowledge SPPU-DRDP for financial assistance. The authors are grateful to Dr. Pankaj Poddar, NCL for his valuable help on the PPMS measurements of samples.

Conflict of interest statement

Authors have no conflict of interest with anyone or any organization.

References

Abu Samaan, T.M., Samec, M., Liskova, A., Kubatka, P., Büsselberg, D., 2019. Paclitaxel's mechanistic and clinical effects on breast cancer. Biomol. Ther. 9, 789. https://doi.org/10.3390/biom9120789.

Ahn, Y., Choi, E.J., Kim, E.H., 2003. Superparamagnetic relaxation in cobalt ferrite nanoparticles synthesized from hydroxide carbonate precursors. Rev. Adv. Mater. Sci. 5, 477–480.

Al-Hajj, M., Wicha, M.S., Benito-Hernandez, A., Morrison, S.J., Clarke, M.F., 2003. Prospective identification of tumorigenic breast cancer cells. Proc. Natl. Acad. Sci. 100 (7), 3983–3988.

Atyabi, F., Sobhani, R., Adeli, M., Dinarvand, R., Ghahremani, M., 2011. Increased paclitaxel cytotoxicity against cancer cell lines using a novel functionalized carbon nanotube. Int. J. Nanomedicine 6, 705–719.

Baker, P.B., Merigian, K.S., Roberts, J.R., Pesce, A.J., Kaplan, L.A., Rashkin, M.C., 1988. Hyperthermia, hypertension, hypertonia, and coma in a massive thioridazine overdose. Am. J. Emerg. Med. 6 (4), 346–349.

Berlin, J.M., Leonard, A.D., Pham, T.T., Sano, D., Marcano, D.C., Yan, S., Fiorentino, S., Milas, Z.L., Kosynkin, D.V., Price, B.K., et al., 2010. Effective drug delivery, in vitro and in vivo, by carbon-based nanovectors noncovalently loaded with unmodified paclitaxel. ACS Nano 4 (8), 4621–4636.

Bernabeu, E., Helguera, G., Legaspi, M.J., Gonzalez, L., Hocht, C., Taira, C., Chiappetta, D.A., 2014. Paclitaxel-loaded PCL-TPGS nanoparticles: in vitro and in vivo performance compared with abraxane®. Colloids Surf. B: Biointerfaces 113, 43–50.

Bianco, A., Kostarelos, K., Prato, M., 2005. Applications of carbon nanotubes in drug delivery. Curr. Opin. Chem. Biol. 9 (6), 674–679.

Bobo, D., Robinson, K.J., Islam, J., Thurecht, K.J., Corrie, S.R., 2016. Nanoparticle-based medicines: a review of FDA-approved materials and clinical trials to date. Pharm. Res. 33 (10), 2373–2387.

Bosco, D.B., Kenworthy, R., Zorio, D.A.R., Sang, Q.-X.A., 2015. Human mesenchymal stem cells are resistant to paclitaxel by adopting a non-proliferative fibroblastic state. PLoS One 10 (6), e0128511.

Bray, F., Ferlay, J., Soerjomataram, I., Siegel, R.L., Torre, L.A., Jemal, A., 2018. Global cancer statistics 2018: globocan estimates of incidence and mortality worldwide for 36 cancers in 185 countries. CA Cancer J. Clin. 68 (6), 394–424.

Buckley, N.A., Whyte, I.M., Dawson, A.H., 1995. Cardiotoxicity more common in thioridazine overdose than with other neuroleptics. J. Toxicol. Clin. Toxicol. 33 (3), 199–204.

Chabner, B.A., Roberts, T.G., 2005. Chemotherapy and the war on cancer. Nat. Rev. Cancer 5 (1), 65–72.

Chan, K.S., 2016. Molecular pathways: targeting cancer stem cells awakened by chemotherapy to abrogate tumor repopulation. Clin. Cancer Res. 22 (4), 802–806.

Chen, K., Huang, Y., Chen, J., 2013. Understanding and targeting cancer stem cells: therapeutic implications and challenges. Acta Pharmacol. Sin. 34 (6), 732–740.

Comşa, S., Cimpean, A.M., Raica, M., 2015. The story of MCF-7 breast cancer cell line: 40 years of experience in research. Anticancer Res. 35, 3147–3154.

Dean, M., Fojo, T., Bates, S., 2005. Tumour stem cells and drug resistance. Nat. Rev. Cancer 5 (4), 275–284.

Devi, T.S.R., Gayathri, S., 2010. FTIR and FT-Raman spectral analysis of paclitaxel drugs. Int. J. Pharm. Sci. Rev. Res. 2 (2). Article 019.

Diehn, M., Cho, R.W., Lobo, N.A., Kalisky, T., Dorie, M.J., Kulp, A.N., Qian, D., Lam, J.S., Ailles, L.E., Wong, M., et al., 2009. Association of reactive oxygen species levels and radioresistance in cancer stem cells. Nature 458 (7239), 780–783.

Drbohlavova, J., Chomoucka, J., Adam, V., Ryvolova, M., Eckschlager, T., Hubalek, J., Kizek, R., 2013. Nanocarriers for anticancer drugs—new trends in nanomedicine. Curr. Drug Metab. 14 (5), 547–564.

Dreaden, E.C., Austin, L.A., Mackey, M.A., El-Sayed, M.A., 2012. Size matters: gold nanoparticles in targeted cancer drug delivery. Ther. Deliv. 3 (4), 457–478.

Economopoulou, P., Kaklamani, V.G., Siziopikou, K., 2012. The role of cancer stem cells in breast cancer initiation and progression: potential cancer stem cell-directed therapies. Oncologist 17 (11), 1394–1401.

Elmore, S., 2007. Apoptosis: a review of programmed cell death. Toxicol. Pathol. 35 (4), 495–516.

El-Sayed, I.H., Xiaohua Huang, A., El-Sayed, M.A., 2005. Surface plasmon resonance scattering and absorption of anti-EGFR antibody conjugated gold nanoparticles in cancer diagnostics: applications in oral cancer. Nano Lett. 5 (5), 829–834.

Frosina, G., 2009. DNA repair in normal and cancer stem cells, with special reference to the central nervous system. Curr. Med. Chem. 16 (7), 854–866.

Gonzalez-Angulo, A.M., Morales-Vasquez, F., Hortobagyi, G.N., 2007. Overview of Resistance to Systemic Therapy in Patients With Breast Cancer. Springer, New York, NY, pp. 1–22.

Gradishar, W.J., 2006. Albumin-bound paclitaxel: a next-generation taxane. Expert. Opin. Pharmacother. 7 (8), 1041–1053.

Hao, R., Xing, R., Xu, Z., Hou, Y., Gao, S., Sun, S., 2010. Synthesis, functionalization, and biomedical applications of multifunctional magnetic nanoparticles. Adv. Mater. 22 (25), 2729–2742.

Heo, D.N., Yang, D.H., Moon, H.-J., Lee, J.B., Bae, M.S., Lee, S.C., Lee, W.J., Sun, I.-C., Kwon, I.K., 2012. Gold nanoparticles surface-functionalized with paclitaxel drug and biotin receptor as theranostic agents for cancer therapy. Biomaterials 33 (3), 856–866.

Holliday, D.L., Speirs, V., 2011. Choosing the right cell line for breast cancer research. Breast Cancer Res. 13, 2.15.

Hu, C., Niestroj, M., Yuan, D., Chang, S., Chen, J., 2015. Treating cancer stem cells and cancer metastasis using glucose-coated gold nanoparticles. Int. J. Nanomedicine 10, 2065–2077.

Huang, X., El-Sayed, M.A., 2010. Gold nanoparticles: optical properties and implementations in cancer diagnosis and photothermal therapy. J. Adv. Res. 1 (1), 13–28.

Huang, X., El-Sayed, I.H., Qian, W., El-Sayed, M.A., 2006. Cancer cell imaging and photothermal therapy in the near-infrared region by using gold nanorods. J. Am. Chem. Soc. 128 (6), 2115–2120.

Kwok, P.C.L., Chan, H.-K., 2014. Nanotechnology versus other techniques in improving drug dissolution. Curr. Pharm. Des. 20 (3), 474–482.

Lay, C.L., Liu, H.Q., Tan, H.R., Liu, Y., 2010. Delivery of paclitaxel by physically loading onto poly(ethylene glycol) (PEG)-graftcarbon nanotubes for potent cancer therapeutics. Nanotechnology 21 (6), 065101–065111.

Li, D., Müller, M.B., Gilje, S., Kaner, R.B., Wallace, G.G., 2008a. Processable aqueous dispersions of graphene nanosheets. Nat. Nanotechnol. 3 (2), 101–105.

Li, G.-Y., Jiang, Y.-R., Huang, K.-L., Ding, P., Yao, L.-L., 2008b. Kinetics of adsorption of *Saccharomyces cerevisiae* mandelated dehydrogenase on magnetic Fe_3O_4–chitosan nanoparticles. Colloids Surf. A Physicochem. Eng. Asp. 320 (1–3), 11–18.

Li, X., Lewis, M.T., Huang, J., Gutierrez, C., Osborne, C.K., Wu, M.-F., Hilsenbeck, S.G., Pavlick, A., Zhang, X., Chamness, G.C., et al., 2008c. Intrinsic resistance of tumorigenic breast cancer cells to chemotherapy. J. Natl. Cancer Inst. 100 (9), 672–679.

Li, R.-J., Ying, X., Zhang, Y., Ju, R.-J., Wang, X.-X., Yao, H.-J., Men, Y., Tian, W., Yu, Y., Zhang, L., et al., 2011. All-trans retinoic acid stealth liposomes prevent the relapse of breast cancer arising from the cancer stem cells. J. Control. Release 149 (3), 281–291.

Liu, Y., Lu, W.-L., Guo, J., Du, J., Li, T., Wu, J.-W., Wang, G.-L., Wang, J.-C., Zhang, X., Zhang, Q., 2008a. A potential target associated with both cancer and cancer stem cells: a combination therapy for eradication of breast cancer using vinorelbine stealthy liposomes plus parthenolide stealthy liposomes. J. Control. Release 129 (1), 18–25.

Liu, Z., Chen, K., Davis, C., Sherlock, S., Cao, Q., Chen, X., Dai, H., 2008b. Drug delivery with carbon nanotubes for in vivo cancer treatment. Cancer Res. 68 (16), 6652–6660.

Liu, Y., Chipot, C., Shao, X., Cai, W., 2010. Solubilizing carbon nanotubes through noncovalent functionalization. Insight from the reversible wrapping of alginic acid around a single-walled carbon nanotube. J. Phys. Chem. B 114 (17), 5783–5789.

Liu, Z., Robinson, J.T., Tabakman, S.M., Yang, K., Dai, H., 2011. Carbon materials for drug delivery & cancer therapy. Mater. Today 14 (7–8), 316–323.

Liu, H., Lv, L., Yang, K., 2015. Chemotherapy targeting cancer stem cells. Am. J. Cancer Res. 5, 880–893.

Mehdipoor, E., Adeli, M., Bavadi, M., Sasanpour, P., Rashidian, B., 2011. A possible anticancer drug delivery system based on carbon nanotube–dendrimer hybrid nanomaterials. J. Mater. Chem. 21 (39), 15456–15463.

Mukherjee, P., Bhattacharya, R., Wang, P., Wang, L., Basu, S., Nagy, J.A., Atala, A., Mukhopadhyay, D., Soker, S., 2005. Antiangiogenic properties of gold nanoparticles. Clin. Cancer Res. 11 (9), 3530–3534.

Ngo, V.K.T., Nguyen, H.P.U., Huynh, T.P., Tran, N.N.P., Lam, Q.V., Huynh, T.D., 2015. Preparation of gold nanoparticles by microwave heating and application of spectroscopy to study conjugate of gold nanoparticles with antibody *E. coli* O157:H7. Adv. Nat. Sci. Nanosci. Nanotechnol. 6 (3), 035015–035024.

Ojea-Jiménez, I., Campanera, J.M., 2012. Molecular modeling of the reduction mechanism in the citrate-mediated synthesis of gold nanoparticles. J. Phys. Chem. C 116 (44), 23682–23691.

Ojea-Jiménez, I., Romero, F.M., Bastús, N.G., Puntes, V., 2010. Small gold nanoparticles synthesized with sodium citrate and heavy water: insights into the reaction mechanism. J. Phys. Chem. C 114 (4), 1800–1804.

Patel, V.R., Agrawal, Y.K., 2011. Nanosuspension: an approach to enhance solubility of drugs. J. Adv. Pharm. Technol. Res. 2 (2), 81–87.

Patole, S.P., Alegaonkar, P.S., Lee, H.-C., Yoo, J.-B., 2008. Optimization of water assisted chemical vapor deposition parameters for super growth of carbon nanotubes. Carbon 46 (14), 1987–1993.

Pindiprolu, S.K.S.S., Krishnamurthy, P.T., Chintamaneni, P.K., Karri, V.V.S.R., 2018. Nanocarrier based approaches for targeting breast cancer stem cells. Artif. Cells Nanomed. Biotechnol. 46 (5), 885–898.

Qiao, R., Yang, C., Gao, M., 2009. Superparamagnetic iron oxide nanoparticles: from preparations to in vivo MRI applications. J. Mater. Chem. 19 (35), 6274–6293.

Quail, D.F., Joyce, J.A., 2013. Microenvironmental regulation of tumor progression and metastasis. Nat. Med. 19 (11), 1423–1437.

Rajeshkumar, S., 2016. Anticancer activity of eco-friendly gold nanoparticles against lung and liver cancer cells. J. Genet. Eng. Biotechnol. 14 (1), 195–202.

Romberg, B., Hennink, W.E., Storm, G., 2008. Sheddable coatings for long-circulating nanoparticles. Pharm. Res. 25 (1), 55–71.

Rowinsky, E.K., Donehower, R.C., 1995. Paclitaxel (taxol). N. Engl. J. Med. 332 (15), 1004–1014.

Ruoslahti, E., 2002. Antiangiogenics meet nanotechnology. Cancer Cell 2 (2), 97–98.

Sparreboom, A., van Tellingen, O., Nooijen, W.J., Beijnen, J.H., Cao, Q., Chen, X., Dai, H., 1996. Nonlinear pharmacokinetics of paclitaxel in mice results from the pharmaceutical vehicle cremophor EL. Cancer Res. 56 (9), 2112–2115.

Sparreboom, A., Scripture, C.D., Trieu, V., Williams, P.J., De, T., Yang, A., Beals, B., Figg, W.D., Hawkins, M., Desai, N., 2005. Comparative preclinical and clinical pharmacokinetics of a cremophor-free, nanoparticle albumin-bound paclitaxel (ABI-007) and paclitaxel formulated in cremophor (taxol). Clin. Cancer Res. 11 (11), 4136–4143.

Sriram, M.I., Kanth, S.B.M., Kalishwaralal, K., Gurunathan, S., 2010. Antitumor activity of silver nanoparticles in dalton's lymphoma ascites tumor model. Int. J. Nanomedicine 5, 753–762.

Stobinski, L., Lesiak, B., Kövér, L., Tóth, J., Biniak, S., Trykowski, G., Judek, J., 2010. Multiwall carbon nanotubes purification and oxidation by nitric acid studied by the FTIR and electron spectroscopy methods. J. Alloys Compd. 501 (1), 77–84.

Varanda, L.C., Júnior, M.J., Júnior, W.B., 2011. Magnetic and multifunctional magnetic nanoparticles in nanomedicine: challenges and trends in synthesis and surface engineering for diagnostic and therapy. In: Biomedical Engineering, Trends in Materials Science. Intechopen, pp. 397–427, https://doi.org/10.5772/13059.

Ventola, C.L., 2017. Progress in nanomedicine: approved and investigational nanodrugs. P T 42 (12), 742–755.

Vermeulen, L., de Sousa e Melo, F., Richel, D.J., Medema, J.P., 2012. The developing cancer stem-cell model: clinical challenges and opportunities. Lancet Oncol. 13 (2), e83–e89.

Wani, M.C., Taylor, H.L., Wall, M.E., Coggon, P., McPhail, A.T., 1971. Plant antitumor agents. VI. The isolation and structure of Taxol, a novel antileukemic and antitumor agent from *Taxus brevifolia*. J. Am. Chem. Soc. 93 (9), 2325–2327.

Wicha, M.S., Liu, S., Dontu, G., 2006. Cancer stem cells: an old idea—a paradigm shift. Cancer Res. 66 (4), 1883–1890.

Wlodkowic, D., Skommer, J., Darzynkiewicz, Z., 2009. Flow cytometry-based apoptosis detection. In: Erhardt, P., Toth, A. (Eds.), Methods in Molecular Biology. vol. 559. Humana Press, pp. 19–32. Clifton, NJ.

Wylie, A.G., Skinner, H.C.W., Marsh, J., Snyder, H., Garzione, C., Hodkinson, D., Winters, R., Mossman, B.T., 1997. Mineralogical features associated with cytotoxic and proliferative effects of fibrous talc and asbestos on rodent tracheal epithelial and pleural mesothelial cells. Toxicol. Appl. Pharmacol. 147 (1), 143–150.

Xu, C., Wang, B., Sun, S., 2009. Dumbbell-like Au–Fe_3O_4 nanoparticles for target-specific platin delivery. J. Am. Chem. Soc. 131 (12), 4216–4217.

Yang, Y., Gao, N., Hu, Y., Jia, C., Chou, T., Du, H., Wang, H., 2015. Gold nanoparticle-enhanced photodynamic therapy: effects of surface charge and mitochondrial targeting. Ther. Deliv. 6 (3), 307–321.

Yasun, E., Kang, H., Erdal, H., Cansiz, S., Ocsoy, I., Huang, Y.-F., Tan, W., 2013. Cancer cell sensing and therapy using affinity tag-conjugated gold nanorods. Interface Focus 3 (3), 20130006–20130015.

Zhao, L., Gao, L., 2004. Filling of multi-walled carbon nanotubes with tin(IV) oxide. Carbon 42 (15), 3269–3272.

Index

Note: Page numbers followed by *f* indicate figures and *t* indicate tables.

A

Abraxane, 279
Abscisic acid (ABA), 245
Acremonium species, 241–242
Activated charcoal (AC), 185
Active Pharmaceutical Ingredient (API), 207
Adaptive neuro-fuzzy inference system-genetic algorithm, 144–145
A disintegrin and metalloproteases (ADAM) protein, 258
Agrobacterium rhizogenes, 196–197
Alanine aminotransferase (ALT), 82–83
Alkylating agents, 14–15
Alternaria alternata, 237
Angiogenesis, 36–37
Angiosperms, 160, 167–168
Anthracyclines, 15–16
Antimetabolites, 15
Antioxidant, 219–220
Anti-tumor antibiotics, 15–16
Apoptosis, 4, 293–294, 299–300
Apoptosis signaling regulating kinase 1 (ASK1), 53
Apoptotic cell death, 107–108
Arabidopsis thaliana, 237–238
Artemisia annua, 236
Aspartate aminotransferase (AST), 82–83
Aspergillus aculeatinus, 244
Austrotaxus spicata, 160, 161*f*
5-Azacytidine, 143

B

Bacillus subtilis, 237
Bacteriobot, 90–91
Bcl-2 family, 51–52, 54*f*, 56–57, 63
Benzoic acid, 244–245
β-tubulins isotypes, 62–63

Bevacizumab, 277
Biopharmaceutical classification system (BCS), 73–74
Biosurfactants, 20
Biosynthesis, 206, 208, 214–216
Blood-brain barrier (BBB), 105–106
Botryodiplodia theobromae, 219–220
Bovine serum albumin (BSA), 108–109
Breast cancer
  adverse events, 287–288, 289–292*t*
  anti-cancer drugs, 287–288
  bevacizumab, 287–288
  chemotherapeutic drugs, 287–288
  efficiency, 300
  inflammatory breast cancer (IBC), 295
  locally advanced breast cancer (LABC), 295–296
  metastatic breast cancer (MBC), 294–295
  molecular mechanism, 293–294
  molecular subtypes, 296–297
  paclitaxel treatment
    resistance, 299–300
    side-effects, 298–299
  trastuzumab, 287–288
  treatments, 287–288
  types, 288–293
Breast cancer stem cells (BCSCs), 325
Bristol-Myers Squibb (BMS), 233

C

Ca^{2+} induced Ca^{2+} release (CICR), 55
Calcein, 106–107
Calcium-dependent apoptosis

Bcl-2 and calcium homeostasis, 56–57
chemotherapeutic agents, 53–54
endoplasmic reticulum (ER), 54–55
Calcium homeostasis, 56–57
Camarosporomyces flavigenus, 242
Camptothecin (CPT), 19
Cancer
  abnormal cells, 1
  biology, 3–6
  hallmarks, 1–2
  hormone secretion, 3
  infective agents, 1
  management, 2–3
  post-surgery, 2–3
  radiation treatment, 2
  therapeutic interventions
    chemotherapy (CT), 12–19
    drug delivery vehicles, 7
    immunotherapy, 11–12
    radiotherapy (RT), 9–11
    surgical excision, 7–8
  treatment, 19–22
  types, classification and grading, 6–7
Cancer stem cells (CSCs)
  antiproliferative effect, 312–313
  apoptosis, 325–327, 326*f*
  chemotherapy, 310
  Fe_3O_4 and Au NPs, 311–312
  GLOBOCAN 2018 database, 310
  Kaposi's sarcoma, 310–311
  magnetic nanoparticles, 312
  mitogens, 309–310
  multiwalled carbon nanotubes (MWCNTs), 312–313
  nanocarriers, 312
  nanotechnology, 311
  number of markers, 309–310
  parthenolide-loaded liposomes, 310

333

334 Index

Cancer stem cells (CSCs)
 (Continued)
 physiochemical properties,
  312–313
 proinflammatory cytokines,
  309–310
 radio and pharmacological
  interventions, 309–310
 tumor-suppressor genes and
  oncogenes, 309
 vinorelbine-loaded liposomes,
  310
Cancer therapy
 anti-mitotic mechanisms,
  270–271, 271*f*
 drug combination, 277–279
 innate immune system,
  274–275, 275*f*
 micronucleation, 273–274, 273*f*
 microtubule stabilizing drugs,
  270–271, 271*f*, 279
 mitotic catastrophe, 271–272
 non-mitotic mechanisms,
  272–273
 paclitaxel, 275–277
 taxol/paclitaxel, 269–270
Candidate trees (CTs), 175–176
Capxol™, 111
Carbohydrate source, 135–136
Carbon dots, 96
Carbon nanotubes (CNTs), 311
Carcinogenesis, 4, 40–41
Caspases, 296–297
Cationic liposomes (CL), 85–87
Cell culture studies, 270
Cell cycle arrest, 293–294
Cell cycle kinase drugs, 278–279
Chemoresistance
 ABC transporters, 252–253
 antioxidative systems, 255–256
 genetic factor, 253–254
 IRF9, 256
 microtubule dynamic, 252
 mitogen-activated protein
  kinase (MAPK), 255
 PI3K/AKT activation, 254–255
Chemotherapeutic drug, 251
Chemotherapy (CT)
 alkylating agents, 14–15

anthracyclines, 15–16
antimetabolites, 15
history, 12–13
mechanisms, 13–14, 13*f*
plant-derived anticancerous
 agents, 16–19
Circulating tumor cells (CTCs), 10
c-Jun N-terminal kinase (JNK)/
 stress-activated protein
 kinase (SAPK), 52–53
Cladosporium oxysporum, 219
Colletotrichum species, 241–242
Compact yellow (CY), 187–188
Coniothyrium palmarum, 243
Conjugated linoleic acid (CLA),
 106
Coronatine (Cor), 139, 244
Corylus
 C. avellana, 133, 167–168, 205,
  242, 244–245
 C. mandshurica, 205
CPT. *See* Camptothecin (CPT)
Cyclodextrin, 139–140, 243
Cyclooxygenase-2 (COX-2),
 219–220
Cytochrome C (Cyto C), 37
Cytotoxic T-lymphocyte associated
 protein-4 (CTLA-4), 11
Cytotoxic T lymphocytes (CTLs),
 78–80

D
Dendrimers, 91–92
Diolyl phosphatidylethanolamine
 (DOPE), 87
DNA methylation, 142–143
Docetaxel, 18*f*
Doxorubicin, 277
Drug encapsulation, 84–85
Drug resistance, 269–270
Dysplasia, 6

E
Ectomycorrhizal mats (ECM), 174
Elicitation strategy, 137–140, 138*f*
Embryo culture, 182
Encapsulation efficiency (EE),
 84–85, 94
Endophytes

anticancer drugs, 203–204
anticancer properties, 216–220,
 217–218*t*
antitumor drugs, 204
approaches, 205–207
cancer, 203–204
chemostructural classes, 204
host plant species, 204, 207–216,
 209–213*t*
industrial production
 approaches, 204
microbial sources, 204
nature, 204–205
pharmacological and
 commercial importance,
 204
tumors, 203–204
Endophytes-plant cells, 144
Endophytic fungi, 34
Endoplasmic reticulum (ER),
 54–55
Enhanced permeability and
 retention (EPR), 74–75,
 85–87, 311
Epidermal growth factor (EGF),
 296–297
Epidermal growth factor receptor
 (EGFR), 256, 295
Epithelial-mesenchymal
 transition, 36–37
Estrogen receptor (ER), 257,
 296–297
Ethosomes, 95
Ethylene inhibitors, 136–137
Ethylene responsive factor (ERF),
 236
Exosomes, 20–21
Extracellular signal-regulated
 kinase (ERK), 38–39
Extracellular vesicles (EVs),
 20–21

F
Fermentation, 214–216
5-Fluorouracil (5-FU), 13–14
Flow cytometry
 antiproliferative studies,
  315–316
 detection of apoptosis, 315

Index 335

Forkhead box molecule 1 (FOXM1), 253
Fourier transform infrared spectroscopy (FT-IR), 314
Fungal endophytes, 132–133, 137–139, 143–144
Fungal fermentation, 132–133
Fusarium
  F. mairei, 241–242
  F. oxysprum, 241–242
  F. solani, 219

G

General regression neural network-fruit fly optimization algorithm, 144–145
Genome-based methods, 21–22
Geranylgeranyl diphosphate synthase (GGPPS), 237
Geranylgeranyl pyrophosphate (GGPP), 234–235
Gigaspora gigantean, 178
Glioblastoma (GBM), 82–83
Glomus intraradices, 178
Green callus (CG), 187–188
Guanosine triphosphate (GTP), 35–36
Gymnosperms, 160

H

Hairy root culture, 230–231
HER2 receptor, 295
Heterologous expression systems, 132
High performance liquid chromatography (HPLC), 214–216
High performance liquid chromatography/liquid chromatography-mass spectrometry (HPLC/LC-MS), 214–216
Human serum albumin (HSA), 73–74
Hydrophobic-lipophilic balance (HLB), 94
3-Hydroxyl-3-methylglutaryl-CoA (HMG-CoA), 237–238

I

Image-guided radiotherapy (IMRT), 9
Immobilization, 140–141
Immune checkpoint inhibitors, 11
Immune therapy, 278
Immunogenic cell death (ICD), 78–80
Immunomodulation effects
  immune system, 57
  macrophages, 58–59
  macrophages polarization, 61–62
  mitotic inhibitor and microtubule stabilizer, 57
  regulatory T cells (Tregs), 57–58
  TLR4-dependent pathway, 59–61
Immunotherapy
  boosted by metronomic chemotherapy, 12
  immune checkpoint inhibitors, 11
Indocyanine green (ICG), 78–80
Inflammatory breast cancer (IBC), 295
In situ product removal (ISPR), 140, 141*f*
Intensity-modulated radiotherapy (IMRT), 9
Interferon (IFN), 256, 274–275
Interferon (IFN) regulatory factors (IRFs), 256
Interleukin 12 (IL-12), 39
Intestinal mucosa, 105–106
Isopentenol utilization pathway (IUP), 235
Isopentenyl diphosphate, 237
Isopentenyl pyrophosphate (IPP), 235

J

Jinxishan (JXS), 235–236

L

Lactose monohydrate (LMH), 89–90
Large unilamellar vesicles (LUVs), 84–85

Leukemia, 2, 6
Lewis lung carcinoma (LLC), 91
Liposomal taxane, 279
Liposomes
  advantages, 84
  hydrophilic/lyophobic drugs, 83–84
  pharmaceutical aspects, 84–85
  preclinical studies, 85–91
  size, 83–84
Liquid chromatography electrospray ionization tandem mass spectrometric (LC-ESI-MS), 214–216
Locally advanced breast cancer (LABC), 257, 295–296
Long-chain triglyceride (LCT), 89
Low-dose metronomic (LDM) chemotherapy, 38
Lymphoma, 2, 6

M

Macrophages, 58–59
  polarization, 61–65
Magnetic hyperthermia, 8, 22
Magnetic resonance imaging (MRI), 22
Malt extract agar (MEA), 214–216
Matrix metalloproteinase (MMP9), 219–220
Medium-chain triglyceride (MCT), 89
Metabolic engineering
  abiotic and biotic elicitors, 230–231
  abraxane, 230–231
  Alzheimer's and psoriasis, 230–231
  animal models, 233
  anticancer agents, 231
  arterial stents, 229–230
  bio-active compounds, 234
  biopharmaceutical products, 230–231
  biotechnological approaches, 230–231, 234
  cancer types, 229–230
  chemical structure, 233

Metabolic engineering (*Continued*)
- chemical synthesis, 234–235
- chemotherapeutic agent, 234–235
- commercial level, 230–231
- effectors, 235–236
- elicitation, 238–239, 240t
- endophytic fungi, 238–239
- heterologous systems, 237–238
- mechanisms of action, 233
- microtubules, 233
- MJ treatments, 245
- molecules, 238–239
- overexpression of genes, 230–231
- paclitaxel, 141–142, 229–230
- phyto-metabolites, 236
- plant cell cultures, 238–239
- plant cell suspension culture process, 230–231, 232f
- precursor access, 235
- taxol, 231, 233, 234f
- *T. baccata* cells, 239–241
- *T. chinensis*, 239–241
- *T. cuspidata*, 239
- transcriptional factors, 236
- vanadyl sulfate (VSO$_4$), 242–243

Metastasis, 3–4, 6
Metastatic breast cancer (MBC), 294–295
Methyl-β-cyclodextrin (MBCD), 243
Metronomic chemotherapy, 12
Mevastatin (MVS), 244
Micelles, 92–93
Microbe-associated molecular patterns (MAMPs), 241–242
Microbial fermentation approaches, 34
Microfluidics, 78
Microparticles, 95–96
Micropropagation
- acclimatization, 194–197, 197t
- callus proliferation, 192–194, 193f
- embryo and seed culture, 182–184
- explant sources, 188–191, 189f
- organogenesis, 192–194, 193f
- rooting, 194–197, 197t
- shoot topophysis, 194, 195f
- somatic embryogenesis, 184–188
- *Taxus* spp., 178, 179–181t, 182

MicroRNAs (miRNAs), 38
Microtubular cell system, 129
Microtubule-associated proteins (MAPs), 35–36
Microtubules, 17, 35–36, 48, 216, 251, 276
- stabilization, 48, 49f
Mitochondrial dysfunction, 298–299
Mitochondrial permeability transition pore (mPTP), 51
Mitogen-activated protein kinase (MAPK), 255
Mitosis, 270–272, 278–279
Mitotic centromere-associated kinesin (MCAK), 38
Mitotic slippage, 48–50
Mononuclear cell medium (MCM), 185
Multi-drug resistance (MDR), 62, 105, 252–253
Multi-lamellar large vesicles (MLVs), 84–85
Multilayer perceptron-genetic algorithm, 144–145
Multivesicular bodies (MVBs), 20–21
Multiwalled carbon nanotubes (MWCNTs)
- characterizations, 314
- detection of cytotoxicity, 314–315
- flow cytometry (*see* Flow cytometry)
- in vitro, 314
- paclitaxel loading, 313
- preparation, 313
- PTX-Au-Mag
  - abraxane, 324
  - absorbance curve, 318
  - antiproliferative activity, 319–321
  - cancer cell lines, 324
  - cell-killing efficacy, 324
  - cell mortality, 322, 324
  - cell viability, 319, 321f

- delivery of drug, 325
- embryogenesis, 319–321
- FCM assay, 319–321, 323f
- Fe-O and Au bond, 316–317
- FITC-annexin V binding, 321–322
- flow cytometer, 319–322, 322f
- FTIR spectroscopy, 316, 316f
- HR-TEM images, 319, 320f
- magnetic properties, 318, 318f
- magnetization (*M*s) values, 318
- MCF-7, 319
- necrotic cells, 319–321
- PCL-TPGS, 324–325
- phosphatidylserine (PS), 319–321
- superparamagnetic behavior, 318
- UV-Vis spectroscopy, 317, 317f

N

NADPH oxidase (NOX), 255–256
Nanocarriers, 74–75, 105–110
Nanomedicines, 20
Nanoparticle (NP), 311
- advantage, 75–76
- pharmaceutical aspects, 76–78
- polymers, 75
- preclinical studies, 78–83
Nanotubes, 93
National Cancer Institute, 233
Neuropathic symptoms, 298
Nicotiana spp., 238
Nicotinamide adenine dinucleotide phosphate (NADPH), 39
Nigella sativa L., 296–297
9-*cis*-epoxycarotenoid dioxygenase (NCED), 245
Niosomes, 94
Nonamplified human epidermal growth factor receptor 2 (HER2) expression, 296–297
Non-small cell lung cancer (NSCLC), 58
Nuclear envelope, 272–274
Nuclear magnetic resonance (NMR), 214–216
Nucleotide binding domains (NBDs), 252–253

Index **337**

O
Ovarian cancer, 269
Overall survival (OS), 294–295

P
Paclitaxel (PTX), 17–18, 18*f*, 32–33.
    See also Taxol
  anticancer drug, 171
  anti-cancer efficacy, 73–74
  bioprocessing strategies,
    146–147
  biosynthesis
    optimization, mathematical
      modeling, 144–145, 145*f*
    plant cell culture (*see* Plant
      cell culture)
  chemotherapeutic agent, 129
  clinical trials, 96–105
  discovery and evolution, 47
  dose-dependent effect, 49–50
  micropropagation (*see*
    Micropropagation)
  mitotic cell cycle arrest, 48–50
  nanocarriers, 74–75, 97–102*t*,
    105–110
  nanoparticle (NP), 73–83
  patents, 110–113
  plant growth media, 172
  requirement, 171
  resistance
    ABCB1 transporter, 259
    clinical markers, 256–258
    ovarian cancer patients,
      257–258
    signal transduction
      pathways, 259–260
    tubulin inhibitors, 258
  solubilization, 73–74
  sources
    C. avellana, 133
    chemical synthesis, 131
    fungal endophytes, 132–133
    heterologous expression
      systems, 132
    nursery cultivated *Taxus*, 131
    plant cell culture, 133–134
    semisynthesis, 131
  structures, 130*f*
  taxane nucleus, 73
  T. baccata, 171–172

T. brevifolia, 171
vegetative propagation
  auxins, 175
  cultivars and species, 174
  erect shoots, 173–174
  factors influence, 175–176
  Gymnosperms, 173
  phytohormones, 175
  rooting efficiency, 178
  seedling stocks, 173
  stem cutting experiments, 173
  T. baccata, 173, 175–176, 177*f*
  T. cuspidata, 174
  T. globosa, 176
  T. mairei, 176–178
  T. sumatrana, 176
  T. wallichiana, 174–176, 175*f*
Paraconiothyrium variabile, 244
Pathological complete response
  (pCR), 295–296
Permeability transition pore
  (PTP), 37
Pestalotiopsis
  P. microspora, 219
  P. pauciseta, 219
  P. terminaliae, 214–216
P-glycoprotein (P-gp), 74–75
Phenylalanine (Phe), 242–243
Phenylalanine ammonia-lyase
  (PAL), 244
Phosphatase and tensin homolog
  (PTEN), 296–297
Phosphate buffered saline (PBS),
  313
Phosphatidylcholine (PC), 88–89
Phosphatidyl glycerol (PG), 88–89
Phosphatidylinositol 3-kinase
  (PI3K), 254–255
Phosphodiester moiety, 113
Phospholipid vesicles, 83–84
Photodynamic therapy, 20
Phyllosticta tabernaemontanae,
  214–216
Phytohormones, 136
Plant cell culture
  co-culture, fungal endophytes,
    143–144
  culture conditions
    carbohydrate source, 135–136
    ethylene inhibitors, 136–137

    phytohormones, 136
    two-stage culture, 135
  elicitation strategy, 137–140,
    138*f*
  high-producing cell lines, 134
  immobilization, 140–141
  in situ product removal and
    two-phase culture, 140,
    141*f*
  metabolic engineering, 141–142
  paclitaxel biosynthesis, 135*f*
  paclitaxel production, 133–134
  precursor feeding, 137
  reactivation, paclitaxel
    biosynthesis pathway,
    142–143
Plant cell fermentation (PCF)
  technology, 207
Plant-derived anticancerous
  agents, 16–19
Plasma membrane calcium
  ATPase 2 (PMCA2), 256
Podocarpus gracilior, 205
Poly(ε-caprolactone)-tocopheryl
  polyethylene-glycol-
  succinate (PCL-TPGS),
  324–325
Polyethylene glycol (PEG), 81
Progesterone receptor (PgR),
  296–297
Programmed death receptor-1
  (PD-1), 11
Progression free survival (PFS),
  294–295
Proliposomes, 89–90
Proniosomes, 94
Proton beam therapy, 9
Pseudomonas syringae, 244
Pseudotaxus chienii, 161*f*, 162

R
Radiotherapy (RT)
  cancer, 9–11
  utilization rate, 9–10
Reactive oxygen species (ROS),
  54–55
Regulatory T cells (Tregs), 57–58
Reticuloendothelial system (RES),
  75, 311
Robotic-assisted surgery, 8

338 Index

S

SAC. *See* Spindle assemble checkpoint (SAC)
Sarcomas, 2, 6
Self-assembly approach, 109–110
Severe combined immunodeficiency (SCID), 21
Signal transduction pathways, 259–260
Single nucleotide polymorphisms (SNPs), 257, 298–299
Small interfering RNAs (siRNAs), 21–22
Solvent evaporation method, 77–78
Soya phosphatidylcholine (SPC), 89–90
Soy lecithin (SL), 85
Spindle assemble checkpoint (SAC), 48
Stem cell therapy, 3
Stereotactic radiotherapy (SRT), 9
Superoxide dismutase, 219–220
Superparamagnetic iron oxide (SPIOs), 82–83
Superparamagnetic iron oxide nanoparticles (SPIONs), 22
Surface plasmon resonance (SPR), 311–312

T

Taxadiene synthase (TS), 235–237, 245
Taxane-2α-*O*-benzoyl transferase (TBT), 244
Taxane 13α-hydroxylase (T13αH), 244
Taxanes, 16–17, 230–231, 309, 316–317
Taxol. *See also* Paclitaxel (PTX)
  albumin nanoparticles/pro-drugs, 30
  angiosperms, 167–168
  anticancer compound, 30
  calcium-dependent apoptosis, 53–57

chemistry, 34–35, 35*f*
discovery, 31–33
enumeration, 160–167
gene-directed apoptosis
  Bcl-2 family, 51–52, 54*f*
  JNK/SAPK, 52–53
history, 155–156
immunomodulation effects, 57–62
invention, 29
mechanism of actions, 35–39, 40*f*, 48, 49*f*
microtubules stabilization, 48, 49*f*
mode of action, 29
natural compounds, 29–30
natural resources, 33–34
production, 30
resistance mechanisms
  Bcl-2 family proteins, 63
  β-tubulins isotypes, 62–63
  multi-drug resistance (MDR-1), 62
semi-synthetic approaches, 30
signaling pathways and macrophage polarization, 60*f*
Taxomyces
  T. andreanae, 208, 214–216
  T. brevifolia, 205
Taxus
  classification, 157–159
  description, 156–157
  modern classification, 159–160
  paclitaxel sources, 131
  T. baccata, 161*f*, 162, 163*f*, 175
  T. brevifolia, 161*f*, 162
  T. canadensis, 161*f*, 162–164
  T. chinensis, 161*f*, 164
  T. cuspidata, 164, 165*f*, 236
  T. floridana, 165*f*, 166
  T. globosa, 165*f*, 166
  T. meirei, 165*f*, 166, 235–236
  T. sumatrana, 165*f*, 167
  T. wallichiana, 165*f*, 167
10-deacetyl baccatin III (10-DAB), 176–178

Thermal ablation, 8, 22
Thermosensitive liposomes (TSL), 90–91
Thidiazuran (TDZ), 191
Thin-layer chromatography (TLC), 214–216
Three-dimensional conformal radiotherapy (3D-CRT), 9
Thrombospondin-1 (TSP-1), 38
Time to progression (TTP), 294–295
TLR4-dependent pathway, 59–61
Transmembrane domains (TMDs), 252–253
Triple negative breast cancer (TNBC), 103–105, 296–297
Tubulins, 276–277
Tumor-associated macrophages (TAM), 58–59
Tumor microenvironment (TME), 309
Tumor necrosis factor alpha (TNF-α), 39
Tumor-suppressor genes (TSGs), 253–254
Two-phase culture, 140

U

Ultraviolet (UV) immunity analysis, 214–216

V

Vanadyl sulfate (VSO$_4$), 242–243
Vascular endothelial growth factor (VEGF), 38, 296–297
Vesicular arbuscular mycorrhiza (VAM), 174
Vibrating sample magnetometer (VSM) studies, 319
Vinca alkaloids (VAs), 16

W

Wisconsin Alumni Research Foundation, 231
Woody plants medium (WPM), 186–187, 190–191
World Health Organization, 1

Printed in the United States
by Baker & Taylor Publisher Services